# Solder Paste in Electronics Packaging

# Solder Paste in Electronics Packaging

Technology and Applications in Surface
Mount, Hybrid Circuits, and Component
Assembly

**Jennie S. Hwang, Ph.D.**
SCM Metal Products, Inc.

 **VAN NOSTRAND REINHOLD**
_____ **New York**

Van Nostrand Reinhold
115 Fifth Avenue
New York, NY 10003

Van Nostrand Reinhold International Company Limited
11 New Fetter Lane
London EC4P, 4EE, England

Van Nostrand Reinhold
480 La Trobe Street
Melbourne, Victoria 3000, Australia

Macmillan of Canada
Division of Canada Publishing Corporation
164 Commander Boulevard
Agincourt, Ontario M1S 3C7, Canada

16 15 14 13 12 11 10 9 8 7 6 5 4 3 2 1

**Library of Congress Cataloging-in-Publication Data**

Hwang, Jennie S.
    Solder paste in electronics packaging

    Bibliography: p.
    Includes index.
    1. Printed circuits—Design and construction.
2. Solder pastes.    3. Surface mount technology.    I. Title.
TK7868.P7H82    1989        621.381'74        88-28015
ISBN-13: 978-94-011-6052-0        e-ISBN-13: 978-94-011-6050-6
DOI: 10.1007/ 978-94-011-6050-6

To

My parents and grandfather
for their past encouragement and support


Leo

for his assistance, endurance, and encouragement


Raymond and Rosalind

for their future

# Acknowledgments

My grateful thanks are tendered to the following companies and individuals who have provided photographs and/or technical data:

AMI-Presco
AT&T—Federal Systems Division
Billiton Witmetaal B.V.
Bourns, Incorporated
Haake Buchler Instruments, Incorporated
Camelot Systems, Incorporated
Detrex Corporation
Dow Chemical Company
The Du Pont Company
Dynapert—HTC
EFD, Incorporated
Electrovert, USA Corporation
GEC Research
Heller Industries
Hollis Automation Corporation
IBM Corporation
IPC
Leybold Technologies, Incorporated
William E. Loeb & Associates
Micromeritics Instrument Corporation
Naval Weapons Center, China Lake
Philips Corporation—Signetics
Radiant Technology Corporation
SCM Metal Products, Incorporated
Texas Instruments, Incorporated

3M Company
Tin Research Institute
Donald Utz Engineering, Incorporated
Vanzetti Systems, Incorporated
Weltek International, Incorporated
Xetel Corporation
Dr. Wesley L. Archer
Dr. Ahmet N. Arslancan
Mr. George E. Blumb
Mr. Kenneth J. Cavallaro
Dr. Gareth D. Davies
Mr. William H. Down
Mr. David Gault
Mr. Charles A. Harper
Mr. Herold Hyman
Ms. Kathryn Johnson
Dr. W. G. Kenyon
Mr. Donald W. Kiser
Mr. Earl Lish
Dr. William E. Loeb
Mr. Michael F. Marchitto
Mr. Steven Meeks
Mr. Kenneth L. Miller
Mr. Wayne L. Mouser
Mr. Peter Naumchik
Mr. Marc Peo
Mr. Gary G. Petersen
Dr. Michael Romberg
Mr. Bryan B. Russell
Ms. Petrine Schipperijn Tan
Mr. Donald Utz
Dr. Riccardo Vanzetti
Mr. D. R. Wallis
Mr. Eugene R. Wolfe

And a collective thanks to all those individuals and publications cited
in the references of each chapter.

# Foreword

One of the strongest trends in the design and manufacture of modern electronics packages and assemblies is the utilization of surface mount technology as a replacement for through-hole technology. The mounting of electronic devices and components onto the surface of a printed wiring board or other substrate offers many advantages over inserting the leads of devices or components into holes. From the engineering viewpoint, much higher lead counts with shorter wire and interconnection lengths can be accommodated. This is critical in high performance modern electronics packaging. From the manufacturing viewpoint, the application of automated assembly and robotics is much more adaptable to high lead count surface mounted devices and components. Indeed, the insertion of high lead count parts into fine holes on a substrate might often be nearly impossible. Yet, in spite of these surface mounting advantages, the utilization of surface mount technology is often a problem, primarily due to soldering problems. The most practical soldering methods use solder pastes, whose intricacies are frequently not understood by most of those involved in the engineering and manufacture of electronics assemblies. This publication is the first book devoted exclusively to explanations of the broad combination of the chemical, metallurgical, and rheological principles that are critical to the successful use of solder pastes. The critical relationships between these characteristics are clearly explained and presented.

In this excellent presentation, Dr. Hwang highlights three important areas of solder paste technology. First, she clearly explains the fundamental and interdisciplinary aspects of solder pastes, including chemical and physical characteristics of solder pastes, metallurgical aspects of solder pastes, and the rheology or flow characteristics of paste materials. Secondly, she discusses the numerous application techniques and considerations of importance in applying solder pastes. Subjects here include application of solder pastes, solder reflow methods and comparisons, and last but not least, the critical topic of contamination and contamination control. The comparison made in these sections will be invaluable as guides for selection of solder operations for individual needs. The third critical area covered is that of reliability and testing, often considered the most important subject area in surface mount technology. Topics here include solder joint reliability and inspections, industry soldering problems, quality assurance and tests. This area is especially well presented, covering problems which are often unrecognized or misunderstood.

In summary, this is a most excellent presentation of a subject whose importance is critical in modern, high performance electronics systems. Readers of this book cannot help but benefit from it, and most likely improve their operations.

Charles A. Harper
Technology Seminars, Incorporated
P. O. Box 487
Lutherville, Maryland 21093

# Preface

In view of the increased importance of solder paste as an interconnecting material for electronics packaging, this book is intended to cover a full spectrum of technologies, techniques, and applications of solder paste by providing coherent and integrated knowledge of multiple disciplines in both theoretical and practical aspects. Separate parts of the book cover fundamental science/technologies related to the subject area (Part II), practical techniques and know-how in every phase of solder paste application (Part III), and reliability/quality (Part IV), as well as a brief market introduction (Part I) and future demands (Part V).

It is hoped that the book will fill the knowledge gap in the field and facilitate information flow in the industry. It is also hoped that the book is able to stimulate growing innovations in the material, design, and process of electronics interconnection technology. Furthermore, it is hoped that those who are electronics designers, manufacturing engineers, metallurgists, chemists, or mechanical engineers, business managers or production personnel, will find this book a genuine example to an interdisciplinary study that is increasingly needed in order to meet our technology and product challenges for years to come.

I must acknowledge a debt of gratitude to the considerable number of persons (the enumeration of whom would be impossible here) who have contributed in important ways to the knowledge, insights, and inspiration that have enhanced my career and nurtured me as a person. I gratefully acknowledge my indebtedness to Mrs. Martha Payne for her outstanding work in typing the entire manuscript in a skillful and patient manner. The author would also like to express her sincere thanks to the publisher's staff for services. Finally, I must express my appreciation to SCM Metal Products, Inc. for all the support given to me.

Cleveland, Ohio                                            JENNIE S. HWANG

# About the Author

Jennie S. Hwang is presently the technology manager and product manager at SCM Metal Products, Hanson PLC. Prior to joining SCM she has held several managerial and research positions with Martin Marietta Corporation and Sherwin-Williams Company.

She received her Ph.D. in materials science and metallurgical engineering from the Engineering School of Case Western Reserve University, an M.S. in liquid crystal chemistry from the Liquid Crystal Institute of Kent State University, and a second master's in inorganic chemistry from Columbia University. She also has a bachelor's in chemistry from Cheng-Kung University.

In conjunction with multidisciplinary trainings, her 12-year experience as technology/business manager covering various industries has brought her a unique breadth and depth of knowledge in materials and processes from both theoretical and practical aspects.

Dr. Hwang's major professional interests are global technological trends and market development of the electronics industry, and electronic materials in general. In specific, her interests include interconnection and packaging materials/process, surface mount technology, paste technology, and solder paste.

Dr. Hwang is a member of the International Society of Hybrid Microelectronics, the International Electronics Packaging Society, the American Ceramic Society, the American Chemical Society, the American Welding Society, the American Society of Metals, the American Society of Testing and Materials, and the Adhesion Society. She has served as the chair of ASM Electronic Materials–Cleveland, and is serving the membership committee of the American Society of Metals.

Dr. Hwang has been a speaker and a lecturer at major international and national conferences on the subject. She holds several U.S. and overseas patents in the area of paste technology.

# Contents

# List of Tables

# Solder Paste in Electronics Packaging

# PART I

# Overview

# Introduction

## 1.1 PURPOSE OF THE BOOK

In the immense electronic materials world, solder paste is considered relatively minor in volume, yet it plays an important role in electronics packaging. Solder paste as a joining material, provides electrical, thermal, and mechanical functions in an electronics assembly. The performance and quality of solder paste are crucial to the integrity of a solder joint, which in turn is vital to the overall function of the assembly.

As the realm of microelectronics/electronics continues to strive for top performance and quality at every level of materials, designs, and processing, and continues to grow and explore, fundamental knowledge and continued innovation in solder paste will be increasingly demanded. Until recently, solder paste has been considered an "art" or even "black magic" in the industry. In fact, solder paste is a complex product of high technology. It is a result of multiple integrated scientific disciplines.

With respect to global microelectronics/electronics markets, solder paste is serving three segments: hybrid circuits, printed circuits, and component manufacture. The demonstrated feasibility and benefits contributed by surface mount technology in printed circuit assembly make solder paste an increasingly important material. To fully take advantage of the benefits that surface mount technology can provide, all aspects of materials, equipment, designs, and processes must be optimized in performance and quality. Solder paste and soldering processes are essential parts of the surface mounting process, which interconnects components/devices directly onto the surface of printed circuit boards. In addition, solder paste is equally important in hybrid circuits and component segments, as the surface mount components become more available and the higher I/O integration circuits become more prevalent.

This book is intended to cover a full spectrum of technologies, techniques, and applications of solder paste, from fundamental illustrations to nitty-gritty, real world practice. The purposes of this book are: to provide an introduction to fundamental sciences/technologies involved in the solder paste area, to communicate the information in an integrated manner to those concerned, and to supply some solutions for current and potential problems in using solder paste. It is hoped that this publication will further the interest and innovation in the material aspects of electronics interconnecting and packaging technology in both industry and academia concerned with hybrid circuits, printed circuits, and components-making, and in those who have any general interest in electronic materials.

In order to facilitate reading, this introductory chapter includes a brief illustration of components, land patterns of current electronics packaging, and surface mount technology. The market and trends in the electronics industry are also briefly discussed. Detailed treatment is beyond the scope of this book, and the reader is referred to the references listed at the end of the chapter.

## 1.2   ELECTRONICS INDUSTRY IN GENERAL

Since the invention of the transistor in 1947, the electronics industry has continued to grow and change through revolutionary and/or evolutionary developments. According to Electronic Industries Association,[1] 1986 U.S. factory sales were estimated to be $209.7 billion, and that figure is expected to rise at the rate of 7–15% over the next five years. The worldwide electronics sales reportedly reached $400 billion in 1987. Some specific end-use products, materials, components, or processes are expected to grow at an even faster rate than overall. Those expected to have a growth of 20% or more include the supercomputer, surface mount capacitors, surface mount resistors, multilayer printed boards, and surface mount assembly processes.

When considering solder paste end-users, the market may be divided into three segments: hybrid circuit assembly, printed circuit board (PCB) assembly, and components assembly. In both hybrid and printed circuit assemblies, solder paste is used to interconnect active and passive components/devices, connectors, and other components onto the board; and in component assembly, solder paste serves as a joining material for attaching leads, lead frames, and the structural units to make end-use components. From the end-use point of view, the industry is generally divided into computer/data processing, communication, consumer, industrial, and military sectors.

While on the subject of electronics, it is inevitable to think of the semiconductor, which is the brain of electronics. The worldwide semiconductor market in 1988 is projected to be $34.9 billion, with a U.S. share of approximately 32%,[2] as shown in Figure 1.1. The new trends in this area include the evolution of a new generation of integrated circuit devices with submicron geometry and the accessory materials and equipment by which the intricate devices are produced. In addition, automation, improved quality and new substrate materials are also on the agenda.

The hybrid circuits—primarily demanded by military and aerospace applications, and other applications under harsh environment, and requiring high reliability—are reported to be $6.5 billion, based on the U.S. end-use market. Figure 1.2 indicates the growth rate among the military, commercial, and consumer sectors predicted for both before and after the stock market crash (October 19, 1987).[3] The new trends involve new materials and new processes as driven by the increased density and complexity of circuitry, and the surface mount technology which is considered to be both an opportunity and a threat to hybrid circuits.

The printed circuit industry has been experiencing tremendous impact from surface mount technology, as discussed in Section 1.4.

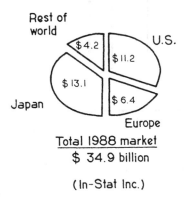

Rest of world
$4.2
U.S.
$11.2
$13.1
Japan
$6.4
Europe
Total 1988 market
$ 34.9 billion
( In-Stat Inc.)

**Figure 1.1** Worldwide 1988 semiconductor market.[2]

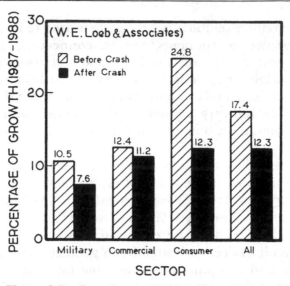

**Figure 1.2**   Growth rate of hybrid circuit sectors.[3]

The projected U.S. PCB assembly market between 1983 and 1991 is shown in Figure 1.3; the analysis indicates the growing trend of utilizing offshore and domestic contract assembly in place of in-house assembly.

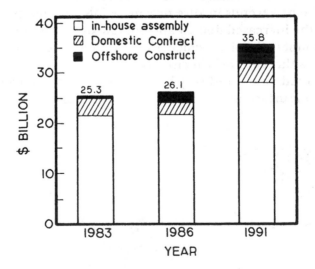

**Figure 1.3**   U.S. printed circuits assembly market.[4]

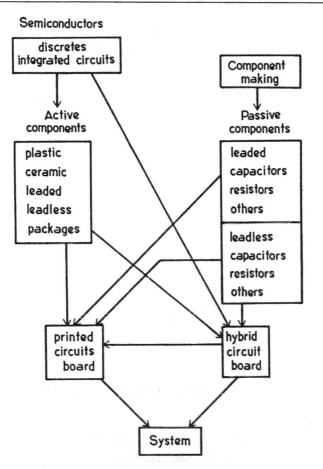

**Figure 1.4** Hierarchy of electronic packaging and interconnection.

## 1.3 ELECTRONICS PACKAGING AND MARKET

Electronics interconnection and packaging may be considered to be based on active and passive components. Figure 1.4 illustrates the hierarchy of electronics packaging and interconnection. The integrated circuits (IC), in terms of worldwide consumption, are shown in Figure 1.5. In this report, the plastic dual-in-line package (DIP), which is used for traditional through-hole PCB industry, shows drastic reduction in consumption in contrast to the significant growth in plastic chip carrier and quad chip carrier packages. The usage of ceramic chip carrier is shown to have a slight increase, and cerdip is expected to maintain relatively steady consumption.

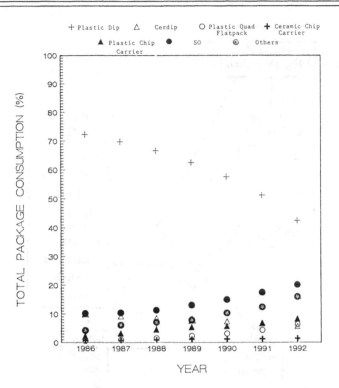

**Figure 1.5**   Worldwide IC package consumption.[5]

## TABLE 1.1

### Worldwide Demand for Capacitors and Resistors[6]

| | 1985 | | | 1990 | | |
|---|---|---|---|---|---|---|
| Component | Total Units* | Surface Mount Units* | Surface Mount % | Total Units* | Surface Mount Units* | Surface Mount % |
| Capacitors | 79.2 | 20.6 | 26 | 110.5 | 48.6 | 44 |
| Resistors | 78.6 | 17.3 | 22 | 101 | 50.5 | 50 |

*In billions.

For passive components, Philips Electronic Components and Materials provides information on worldwide demand for total capacitors and resistors and the corresponding surface mount packages, as shown in Table 1.1.[6] The surface mount chips are projected to have significant growth with 40–50% of total units in 1990.

## 1.4  SURFACE MOUNT TECHNOLOGY

As the name implies, surface mount technology (SMT) is the application of science and engineering principles to board-level assembly, by placing components and devices on the surface of circuit board instead of through-the-board. This concept has been utilized in hybrid assembly since the 1960s by interconnecting chip resistors, chip capacitors, and bare semiconductor dies on hybrid substrates. Since the flatpack package became available in the 1970s, interconnection to boards has ultilized the surface mounting concept. Nevertheless, the potential of surface mounting was not fully utilized and explored until the early 1980s. The impact of surface mounting on printed circuit assembly has been most dramatic since the inception of through-hole printed wiring boards in thc mid-1950s. It is also one of the most significant developments in the electronic era.

The merits of surface mount technology are quite straightforward. In simple terms, surface mount technology provides superior performance/cost ratio for printed circuit board manufacturing. Specifically, the benefits include

increased circuit density,

decreased component size,

decreased board size,

reduced weight,

shorter leads,

shorter interconnection,

improved electrical performance,

facilitating automation,

lower costs in volume production.

Figure 1.6 compares the relative size of two Texas Instruments one-megabyte memory expansion boards made with surface mount assembly and through-hole assembly process, respectively, and indicates a significant reduction in size with surface mount technology. In this assembly, the DIP board is eight layers, and the PLCC (plastic leaded chip carrier) board is four layers. The savings in board area with PLCC board is 60%, and the cost saving of PLCC board is approximately 55%, as compared with DIP board.[7]

Integrated Circuit Engineering has provided a comparison between DIP package and SMT package on PCB area as shown in Figure

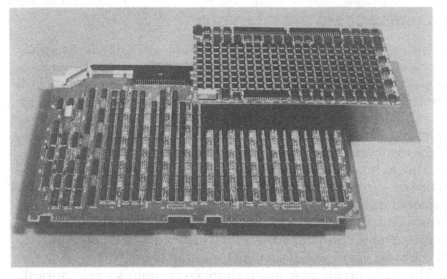

**Figure 1.6**  Comparison of DIP and PLCC one-megabyte memory expansion boards. *(Courtesy of Texas Instruments, Incorporated.)*

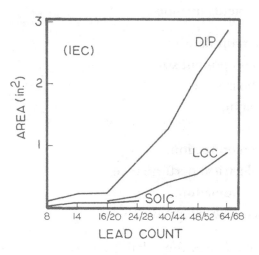

**Figure 1.7**  Comparison of PCB area
with DIP package and SMT packages.[8]

1.7. As can be seen, when lead count increases, the area differential becomes more pronounced. Figure 1.8 compares the weight difference between surface mount chip carriers and the equivalent DIPs for different lead counts, showing drastic weight reduction in surface mount package for all lead counts. Figure 1.9 illustrates the propagation delays among different device packages.

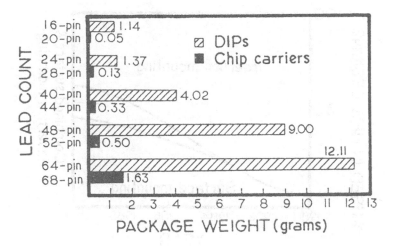

**Figure 1.8** Comparison of weight difference between chip carriers and DIPs.[8]

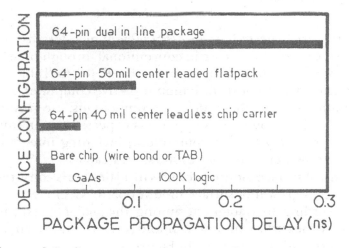

**Figure 1.9** Comparison of propagation delays among packages.[8]

## 1.5  SURFACE MOUNT TECHNOLOGY MARKET

After a period of learning and development in every aspect of surface mount assembly process through the efforts of pioneers and followers, the feasibility of surface mounting has been proved, and tangible successes are evidenced. The surface mount assemblies are expected to grow at an extraordinary rate. The analysis of Integrated Circuits Engineering indicates that the surface mount packaged semicon-

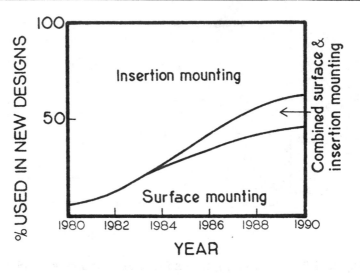

**Figure 1.10**  Growth of surface mount packaged semiconductors.[8]

ductors are expected to represent 50% of new designs by 1990, with
a corresponding decrease in conventional through-hole (insertion)
package, as indicated in Figure 1.10. Refer to Table 1.1 for the
expected growth of surface mount passive components.

With respect to major devices/components, the growth of worldwide surface mount consumption is expected to continue into the
1990s for every type of component, including integrated circuits
(IC), transistors, diodes, resistors, and capacitors. The significant
growth of surface mount packages in all devices is consistent with the
predicted growth of surface mount board assemblies.

According to Dataquest, among the major electronic end-use
sectors, the surface mount technology is most utilized in computer/
data processing,[5] as shown in Figure 1.11.

## 1.6  SURFACE MOUNT COMPONENTS

Due to the inherent difference between surface mount technology
and conventional through-hole assembling in soldering, cleaning,
and handling, components used for most surface mounting processes are subjected to more severe conditions with respect to temperature, heat exposure time, and chemical exposure. The Institute
for Interconnecting and Packaging Electronic Circuits (IPC) has
developed the guideline *ANSI/IPC-SM-782 on Surface Mount Land*

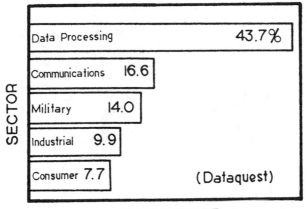

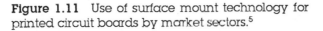

PERCENTAGE

**Figure 1.11** Use of surface mount technology for printed circuit boards by market sectors.[5]

*Patterns —Configurations and Design Rules. ANSI/IPC-SM-782* provides some requirements for such types of components in termination materials, resistance to cleaning, and resistance to soldering, as well as for handling and automation.[9]

The selected surface mount components under JEDEC (Joint Electronic Device Engineering Council) Committee or IPC standards are listed as follows, and their configurations illustrated in Figure 1.12.

**Resistors.** Types 0805, 1206, 1210

**Capacitors.** Types 0805, 1206, 1210, 1812, 1825

**Inductors.**

**Small outline transistors.** SOT-23, SOT-89

**Small outline integrated circuits (SOIC) with gull wing leads.** SOIC-8, SOIC-14, SOIC-16, SOIC-16L, SOIC-20L, SOIC-24L, SOIC-28L

**Small outline integrated circuits with J leads.** SOIC-14, SOIC-16, SOIC-18, SOIC-20, SOIC-22, SOIC-26, SOIC-28

**Dual and quad flatpacks.** 14-84 terminals

**PLCC (plastic leaded chip carriers).** PLCC-28, PLCC-44, PLCC-52, PLCC-68

**LCCC (leadless ceramic chip carriers).** 16-156 terminals

**Miniature PLCC.** 84-244 terminals

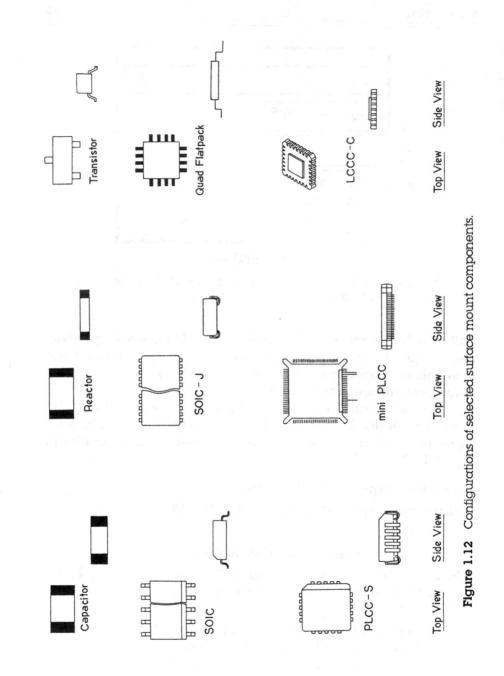

**Figure 1.12** Configurations of selected surface mount components.

14

Figure 1.13 shows cutaway view of an SO-14 package and a PLCC-44 package.

## 1.7  SURFACE MOUNT LAND PATTERNS

According to *ANSI/IPC-SM-782 on Surface Mount Land Pattern—Configurations and Design Rules,* the following outlines some selected information; for details consult *ANSI/IPC-SM-782.* [9]

A.  For chip resistors, the equations to determine the land patterns are

Land width = $W_{max} - K$
Land length = $H_{max} + T_{max} + K$
Gap between lands = $L_{max} - 2 T_{max} - K$

where

$W_{max}$ is the maximum width of component, $H_{max}$ is the maximum component height, $L_{max}$ is the maximum component length, $T_{max}$ is the maximum width of the solderable termination, and $K$ is a constant of 0.25 mm (0.010 in.).

B.  For chip capacitors, equations for land width and gap between lands are the same as for resistors, but

Land length = $H_{max} + T_{min} - K$

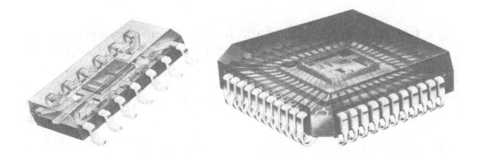

**Figure 1.13**  Cutaway view of SO-14 and PLCC-44. *(Courtesy of Philips Corporation–Signetics.)*

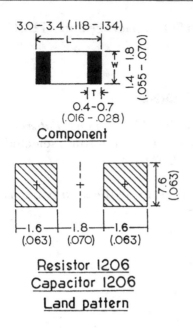

3.0 – 3.4 (.118 –.134)

Component

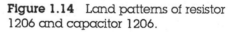

Resistor 1206
Capacitor 1206
Land pattern

**Figure 1.14** Land patterns of resistor 1206 and capacitor 1206.

As examples, Figure 1.14 illustrates the land pattern for resistor type 1206 and capacitor type 1206.

C. For SOIC, the recommended land pattern is

Land width = 0.63 mm (0.025 in.)
Land length = 2.0 mm (0.080 in.)
Gap between lands = $F - K$

where

$F$ is the body dimension and $K$ is a constant recommended to be 0.25 mm (0.010 in.). The land patterns for SOIC–gull wing lead are shown in Figure 1.15.

D. For flatpacks, Figure 1.16 shows a typical quad flatpack land pattern.

E. For PLCC, the IPC recommended land pattern is

Land width = 0.63 mm (0.025 in.)
Land length = 2.0 mm (0.080 in.)
$B = C + K$

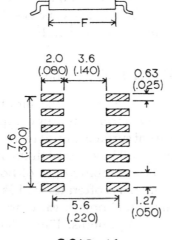

SOIC - 14

**Figure 1.15** Land patterns of SOIC-gull wing..

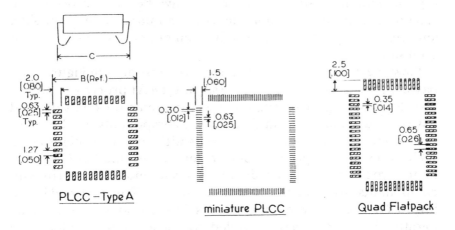

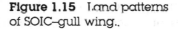

PLCC - Type A

miniature PLCC

Quad Flatpack

**Figure 1.16** Land patterns of typical quad flatpack, PLCC (JEDEC-Type A), and miniature PLCC.

where

$K$ is a constant recommended to be 0.75 mm (0.030 in.), and $B$ and $C$ are specified in Figure 1.16. The miniature PLCC land pattern can also be seen in Figure 1.16.

F. For LCCC, the land pattern is the same as that of PLCC except the value of $K = 1.75$ mm (0.070 in.) if on the ceramic substrate, and $K = 2.25$ mm (0.090 in.) if on glass epoxy substrate where thermal expansion coefficients do not match.

## 1.8  SURFACE MOUNT PROCESSES

Although the exclusive use of surface mount components (i.e., board is made with 100% surface mount components) is the way to fully utilize the benefits of surface mount technology, the majority of surface mount assembly still involves both through-hole and surface mount components. The combined use of both through-hole and surface mount components is often termed as "mixed technology." The extent of the mixed technology varies with the availability and cost of surface mount components. The existence of through-hole equipment also facilitates the continued use of through-hole components when appropriate.

For surface mounting processes, the industry has identified three classes and two types of processes. The three classes A, B, and C, represent an increasing level of complexity; the type 1 process indicates the surface mount components are mounted on only one side, and the type 2 process indicates that components are mounted on both sides.

Flow charts for the three processes are illustrated in Figure 1.17, based on IPC guidelines. As can be seen, Type A1 processes produce boards consisting of through-hole components on one side, and passive and small active surface mount components on the other side. They are limited to components with simple configurations such as resistors, capacitors, SOT components, and SOIC components. Type B1 processes produce fully surface mounted boards, with components either on two sides or on one side. The two-sided board involves two passes of reflow processes. In such cases, residue cleaning after the first pass may be needed to avoid a residue problem. The mixed-technology Type C2 processes produce boards consisting of both through-hole and surface mount components on one side and the surface mount components with simple configurations on the other side.

## 1.9  SURFACE MOUNT DEVICE RELIABILITY

With the growing adoption of surface mount devices, natural questions are: How reliable are the surface mount devices? How do they compare with DIP devices in terms of reliability?

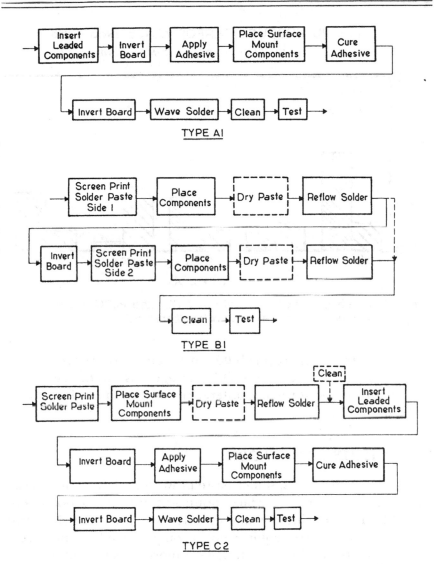

**Figure 1.17**  Process flow charts of Type A1, Type B1, Type C2 surface mount technology.[9]

Signetics Corporation has reported a direct comparison on the performance of these two types of packages under severe test conditions.[10] The tests covered include (1) dynamic or static high temperature life test (D/SHTL), (2) temperature-humidity bias stress (THBS), (3) pressure pot stress (PPOT), (4) thermal shock (TMSK), and (5) temperature cycling (TMCL). The test details can be found in Signetics' *Reliability Handbook*. Figure 1.18 summarizes the failure

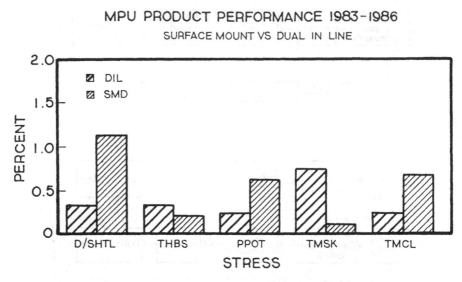

MPU PRODUCT PERFORMANCE 1983-1986
SURFACE MOUNT VS DUAL IN LINE

**Figure 1.18** Performance comparison of DIP and PLCC microprocessor products. *(Courtesy of Philips Corporation–Signetics.)*

percentage results (percent/1000 hours) on microprocessor product performance in PLCC and DIP packages. As can be seen, there is no significant performance difference between the surface mount and DIP packages.

## 1.10  UTILITY OF THE BOOK

In making solder joints, solder paste possesses merits which have a natural compatibility to surface mount technology for board-level assembly. The growing demand in usage and the increasing stringency in performance requirements prompt a need to understand solder paste technology and to have a more effective information flow. This book consists of 12 chapters intended to fulfill these needs. The interdisciplinary nature of solder paste technology is discussed in Chapter 2, followed by individual treatment of each disciplinary area, which is covered in Chapter 3 on chemical and physical characteristics, in Chapter 4 on metallurgical characteristics, and in Chapter 5 on rheology. Formulation technology, due to its proprietary nature, is limited to Section 3.11.

The remainder of the book is devoted to the application aspects of solder paste and to the techniques of using solder paste, from placing it to residue cleaning. Chapter 6 outlines the techniques of applying solder paste by pattern printing and dot dispensing. Chapter 7 discusses the reflow methodology covering state-of-the-art methods and practical issues. Chapter 8 discusses the residue cleaning processes and solvent systems. The emphasis of these chapters is on techniques and performance in the real world, as well as on the fundamentals upon which the processes are based.

Chapter 9 discusses the solder joint reliability, covering state-of-the-art studies and results, and potential solder failure phenomenon. Chapter 10 discusses the common concerns and problems encountered on the production floor, and provides some solutions to the problems.

One of the most prevalent questions about the use of solder paste in mass production is how to control and assure the quality of solder paste. Chapter 11 is therefore devoted to tests recommended for quality assurance from the user's point of view, rather than the vendor's. The tests are divided into five parts: paste, flux/vehicle, solder powder, reflow, and post-reflow characteristics.

The last chapter, Chapter 12, provides some thoughts on future efforts and studies warranted in this field, and some pending issues. The chapter also covers the complementary and emerging technologies which are directly or indirectly related to solder paste.

For readers' convenience, relevant specifications issued by military and federal agencies, and the professional committees are included in the Appendix.

## REFERENCES

1. Electronic Industry Association, *1987 Electronic Market Data Book.*
2. In-State, Incorporated, Scottsdale, Arizona.
3. W. E. Loeb & Associates, Soguel, California.
4. Electronic Trend Publications, Incorporated, Saratoga, California.
5. Dataquest, Incorporated, San Jose, California.
6. Alberto Socolovsky, *Electronic Business* (Jan. 1988): 72.
7. "Introduction to Surface Mount Technology" (Texas Instruments Incorporated, 1984).
8. Integrated Circuit Engineering, Scottsdale, Arizona.
9. *ANSI/IPC-SM-782*, Institute for Interconnecting and Packaging Electronic Circuits.
10. Peter Naumchik. Signetics Corporation, Sunnyvale, California.

# CHAPTER 2

# Interdisciplinary Approach

Solder paste is simple and plain in appearance, yet its fundamentals are broad and complex. It can be well understood only with an interdisciplinary approach.

This chapter stresses these interdisciplinary aspects of solder paste, covering the points of the basic role of solder paste in electronics assemblies, its constitution, and its technology, as well as the interplay of multiple technologies. It is hoped that this brief sketch will initiate positive concept about what solder paste is and the requirements to develop the technology, and therefore its products and processes.

## 2.1 BASIC FUNCTION

Solder paste is one of several interconnecting materials that can serve as a bonding agent between metallic surfaces of an assembly, under proper conditions. Solder paste, in the "deformable" viscoelastic form, can be applied in a selected shape and size and can be readily adapted to automation. Its "tacky" characteristic provides the capability of holding parts in position without additional adhesives before forming permanent bonds. The metallic nature of solder paste provides relatively high electrical and thermal conductivity. With these principal merits, solder paste is a viable interconnecting material, providing electrical, thermal, and mechanical properties applicable to electronics assemblies.

## 2.2 MATERIALS

The compositions of two metallic surfaces to be joined in the electronic end-use applications are numerous. The two sides of the material which are in direct contact with solder are the solder pads on board substrates and the interconnecting leads, pads, or terminations of components and devices. For board substrate side, the common compositions for solder pads on hybrid circuits, printed wiring boards, and component/module substrates are to be considered. For the other side, the common compositions of leads and pads of integrated circuit packages (e.g., PLCC, SOIC, and LCCC), and active and passive components (e.g., chip capacitor, chip resistor, diode, rectifier, LED, and transistor) are to be considered. The metallic surfaces that can be in direct contact with solder include the following compositions:

| Substrate | Component |
|-----------|-----------|
| gold | tin |
| silver | tin/lead |
| palladium/silver | nickel |
| platinum/silver | gold |
| palladium/gold | palladium/silver |
| platinum/gold | platinum/silver |
| copper | copper |
| tin/lead | |
| tin | |
| nickel | |

Electronic-grade solder paste is intended to make solder joints between the aforementioned surfaces in order to provide electrical, thermal, and mechanical linkage between components and substrate. It is thus apparent that the performance and quality of solder paste are vital to the integrity of a solder joint, which is in turn crucial to the overall function of the assembly.

## 2.3 COMPOSITIONAL CONSTITUTION

Solder paste, by one definition, is a homogeneous and kinetically stable mixture of solder alloy powder, flux, and vehicle, which is capable of forming metallurgical bonds at a given set of soldering conditions and can be readily adapted to automated production in making reliable and consistent solder joints. In terms of functionality, a solder paste has three major components: solder alloy powder, vehicle system, and flux system. The vehicle is primarily a carrier for solder powder and provides a desirable rheology. Rheology is used here broadly to represent the flow and deformation of solder paste under a given set of conditions (e.g., temperature, force), which is discussed in Chapter 5. The flux causes the alloy powder and the surfaces to be joined to maintain a clean and metallic state, so that good wetting and metallic continuity between solder alloy and the surfaces joined can be formed.

Both vehicle and flux are fugitive in nature at the completion of the soldering process, either partially escaping during the heating (reflow) stage through volatization, decomposition, and reaction, or being removed during the subsequent cleaning step. Nevertheless, they are crucial to the formation of a reliable, permanent bond. On a permanent basis, the alloy powder part is the only functional component in the final metallurgical bond.

Solder alloys and their metallurgical characteristics are discussed in Chapter 4; chemical and physical properties of the flux/vehicle system are discussed in Chapter 3.

## 2.4  TECHNOLOGY

From a technological point of view, pastes involve the interplay of several scientific disciplines. Sciences and technologies utilized include metallurgy/particle technology, chemistry/physics, rheology, and formulation technology. Figure 2.1 illustrates the spirit of paste technology. Based on this technology, a number of existing and/or potential application product lines can be derived, in addition to solder paste. Compositionally, these product lines consist of organo-polymeric vehicle and metallic/nonmetallic particulates, ranging from PM (powder metallurgy) injection molding and EMI (electromagnetic interference) shielding composite, to cermet thick film, polymer thick film, and brazing paste and adhesive. Each of these product lines has its uniqueness in terms of function, composition, and required processing parameters; however, one thing in common is paste technology.

## 2.5  INTERPLAY OF TECHNOLOGY

Figure 2.2 is an attempt to summarize solder paste technology in one chart. It is hoped that this chart reflects the essence of solder paste

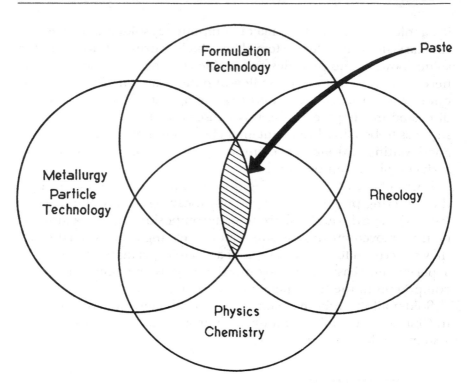

**Figure 2.1**  Paste technology.

technology by displaying the correlation of starting components and final performance of solder paste, and the incorporation of fundamental sciences and technologies.

In addition to metallurgy and particle technology, the physical, chemical, thermal, and rheological properties of vehicle/flux systems, are equally important to the performance and characteristics of the resulting paste. Furthermore, how the ingredients are composed (i.e., formulation) and how the formula is put together (i.e., processing) are crucial to the properties and parameters of vehicle/flux systems. With respect to performance, solder paste is categorized into four areas: applicability, solderability, residue characteristics, and joint integrity. Applicability refers to the ability of a paste to be adapted to a specific paste-applying technique such as dot-dispensing, screen printing, or stencil printing. Solderability is intended to be used in a broad sense—the ability of a paste to wet the surfaces to be joined with complete coalescence of solder powder particles, and to achieve a reliable metallurgical bond. Residue characteristics cover the physical and chemical properties of the

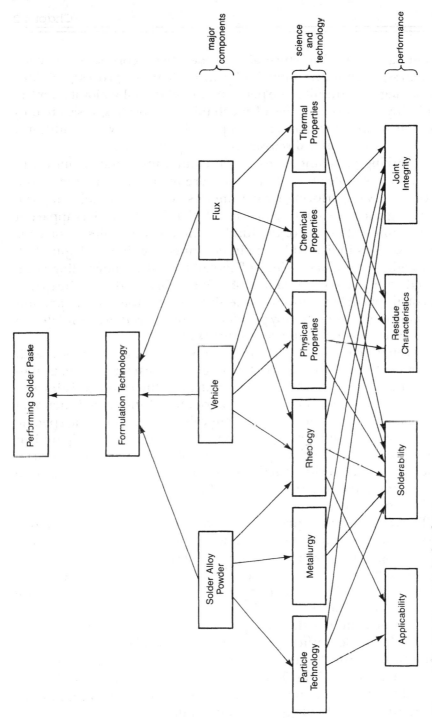

**Figure 2.2** Solder paste technology.

resulting chemical mixture after solder (e.g., corrosivity, activity, tackiness, hardness, compatibility with cleaning process). Solder joint integrity is the ultimate performance of the solder joint after the soldering process in terms of mechanical properties, resistance to adverse environment, and its compatibility with service conditions. Each of these areas is discussed in subsequent chapters.

A working paste under real-world conditions is quite complex in nature. Its complexity and variability are further augmented by the dependency of performance parameters upon the variables of paste handling and the soldering process. With this in mind, it is apparent that an understanding of the fundamental technologies involved is a necessity, as indicated by the interconnecting lines of Figure 2.2. Through the understanding of metallurgy, the solder alloy is selected with the consideration of solderability and joint integrity. Through particle technology, the size distribution, the shape, and the morphology of the alloy powder are considered to formulate the desired paste applicability and solderability.

Through chemistry, chemical properties (e.g., reactivity with solder powder and surfaces to be joined, and reactivity in relation to temperature) which affect solderability, residue characteristics and joint integrity can be better understood. In addition, the functional groups and the structure of chemicals in relation to a specific performance characteristic can be correlated and anticipated in principle. The physical properties, including surface and interfacial phenomena of individual ingredients and of a system as a whole, have significant effect on the paste performance in solderability and residue properties.

The rheology, not only as a result of a designed composition of a flux/vehicle system and solder powder, but also as a result of paste processing, controls the paste applicability, solderability, and even joint integrity. Thermal properties, such as stability versus temperature and reactivity versus temperature, contribute to the residue characteristics and solderability.

It is worth noting that the paste is considered mostly as being kinetically stable, rather than thermodynamically stable. This is in contrast to a true "solution" or to other multicomponent systems, such as microemulsions. Therefore, formulation and processing are crucial to the consistency and properties of the paste, and in most cases, even to its rheology and shelf stability.

A viable solder paste should be constituted by utilizing fundamental technologies in selecting starting raw materials, in anticipating the interaction among these raw materials, and in understanding interrelations between starting materials and end-use performance. Solder paste should be handled and used with an understanding of its characteristics, techniques, and technologies involved to assure the optimum performance.

# Basic Technologies

PART II

Basic
Technologies

# CHAPTER 3

# Chemical and Physical Characteristics

Regarding chemical and physical characteristics of solder paste, the coverage in certain areas is inevitably discussed in general terms in order to avoid proprietary information.

## 3.1 PERFORMANCE PARAMETERS

The performance parameters of a solder paste are many. For convenience, they are grouped into three states of the soldering process: paste state, reflowing state, and post-soldering state. The following outlines the parameters at each state, some of which apply to all types of applications, and some of which only specific applications.

### Paste State

- physical appearance
- stability and shelf life
- viscosity
- cold slump
- dispensability through fine needles
- screen printability
- stencil printability
- tack time
- adhesion
- exposure life
- quality and consistency

## Reflowing State

- compatibility with surfaces to be joined
- flow property before becoming molten
- flow property at and after becoming molten
- wettability
- dewetting phenomenon
- solder balling phenomenon
- bridging phenomenon
- wicking phenomenon
- leaching phenomenon

## Post-soldering State

- residue cleanability
- residue corrosivity
- electromigration
- joint appearance
- joint voids
- joint strength
- joint microstructure
- joint integrity versus mechanical fatigue
- joint integrity versus thermal fatigue
- joint integrity versus thermal expansion coefficient differential
- joint integrity versus intrinsic thermal expansion anisotropy
- joint integrity versus creep
- joint integrity versus corrosion-enhanced fatigue

In order to achieve a proper set of performance parameters, in addition to the solder powder part, which is covered in Chapter 4, the basic characteristics of the flux/vehicle system have to be designed accordingly. Factors to be considered are

- boiling point of liquid phase
- melting point of solid phase
- softening point of resins
- vapor pressure of component as well as the system as a whole
- chemical functional group and reactivity

- flux activation temperature
- flux activation time
- compatibility with solder alloy powder
- compatibility with the surface of substrates
- tackiness and adhesion
- environmental stability
- thermal decomposition and degradation
- rheology
- metal load acceptability
- compatibility with heat transfer mechanism
- compatibility with common cleaning solvents and equipment

A typical composition for the vehicle/flux system lies in the range of 5–20 ingredients. Each ingredient, or group of ingredients, provides a necessary function such as binding agent, fluxing agent, rheological controller, suspending agent, or specific function modifier. Every ingredient plays a role in the final performance of the paste. Most of the time, all ingredients are interrelated with respect to performance. For example, solder balling can be significantly aggravated by a slight difference in composition. Figure 3.1

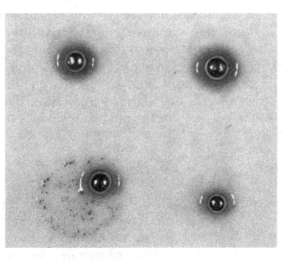

**Figure 3.1** Solder balling sensitivity to a slight composition difference.

shows the comparison of solder balling results of two pastes, A and B, tested in accordance with the procedure outlined in Section 11.28. I and II are the results of paste A with and without an additional ingredient, and III and IV are of paste B with and without the same ingredient. This added ingredient, even in small amount (less than 1 wt%) changed the paste B from solder ball–free to severe solder balling. The sensitivity to the slight compositional change depends on the type of paste as shown in the comparison of II and IV. Paste A, however, was only slightly altered.

## 3.2  PHYSICAL PROPERTIES

The basic physical properties to be considered include melting point, boiling point, softening point, glass transition temperature, vapor pressure, surface tension, viscosity, and miscibility. All these properties are controlled by intermolecular forces (or cohesive energy) that are important secondary bond forces, although not of great importance in the formation of stable chemical bonds.

For example, the tendency of a molecule to volatilize from its liquid is a function of a total transitional energy which in turn depends on temperature. The boiling point is dependent on the relation of such transitional energy and the cohesive energy as a result of intermolecular interactions. For a polymer, the relation is a function of molecular weight, which determines the occurrence of decomposition or volatilization as temperature increases. The melting point, in addition to cohesive energy, is influenced by the orderliness of molecules (entropy). Although melting point and boiling point in relation to structure is complicated, the higher boiling point in general is associated with higher melting point, and symmetrical molecules melt at higher temperature due to the low entropies of fusion, when other factors being equal. The thermodynamic equation

$$\Delta G = \Delta H - T\Delta S,$$

which defines free energy ($\Delta G$), enthalpy ($\Delta H$), and entropy ($\Delta S$), is applicable (where $T$ is absolute temperature). When dealing with polymers or oligomers, the melting point increases with increasing molecular weight. The viscosity of a polymer in solution depends on the molecular weight and molecular weight distribution at a given temperature, in addition to the characteristics of polymer and solvent. The viscosity is related to molecular weight by

$$\eta = KM^a$$

where $K$ and $a$ are empirical constants and $M$ is the molecular weight.[1]

Figure 3.2 illustrates the general relationship between the molecular weight of a polymer and the resulting viscosity in an otherwise identical system. The softening point also increases with molecular weight and increasing crystallinity. For amorphous polymers, the softening temperature is near the glass transition temperature, whereas for highly crystalline polymers it is close to the melting point. The glass transition temperature and the melting point are normally related as

$$\frac{T_g}{T_m} \cong \frac{2}{3} \text{ for unsymmetrical polymers, and}$$

$$\frac{T_g}{T_m} \cong \frac{1}{2} \text{ for symmetrical polymers}^2$$

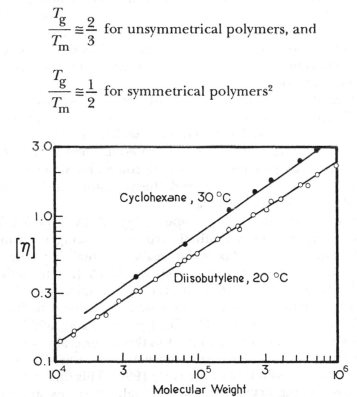

**Figure 3.2** Dependency of viscosity on molecular weight of polyisobutylene in two solvents.[1]

In a paste, the vapor pressure of every ingredient affects more than one aspect of paste performance, including the selection of reflow method and the void development in the solder joint. The boiling point of the liquid phase as a whole, or the boiling point of an individual liquid ingredient, can alter the flux activity and the residue characteristics, as well as the compatibility of reflow method. The melting point of the solid phase as a whole, or the melting point of individual solid ingredients, also can alter the flux activity.

## 3.3   FLUXES AND FLUXING

The function of fluxes in solder paste is to chemically clean the surfaces to be joined, to clean the surface of the solder powder, and to maintain the cleanliness of both substrate surface and solder powder surface during reflow so that a metallic continuity at the interface and a complete coalescence of the solder powder during reflow can be achieved. Therefore, broadly speaking, any chemicals capable of providing good reactivity with metal oxides and of cleaning the metals commonly used in soldering area, should be potential candidates as fluxes.

Customarily, two functional types of chemicals are used for the fluxing function. One is the rosin which serves as the primary flux, and the other is called the activator, which enhances the flux strength of the system. Rosins have been established as an effective and reliable fluxing chemical for readily solderable systems and are well accepted in the industry, and the activator can be a single chemical or a group of chemicals.

Several common flux types—Type R, Type RMA, Type RA, Type OA, Type SA, and other synthetic types—are being used in the industry. The R (rosin) flux is the weakest, containing only rosin without the presence of activator(s); the RMA (mildly activated rosin) flux is a system containing both rosin and activator(s). The RA flux is a fully activated rosin or resin system, having a higher flux strength than the RMA type. The OA type is organic acid flux and possesses high fluxing activity. Type OA flux is generally considered as corrosive. And the SA (synthetic activator) was disclosed by DuPont as a non-rosin flux in the early 1980s. This flux is designed to improve fluxing activity on hard-to-solder surfaces, and in the meantime it can be readily cleaned in common chlorofluorocarbon/alcohol solvent such as Freon TMS. The SA flux displays better wetting ability than RA flux and is equivalent to OA flux as measured by meniscograph.

It should be noted that the RMA type is normally recognized as the system which does not consist of halide-containing chemicals. However, the compositions may vary with vendors. Some commercial flux systems classified under Type RMA do contain halides such as hydrobromide or hydrochloride salts of amines.

According to Federal Specification QQ-S-571E, four types of fluxes are classified: R, RMA, RA, and AC. The AC type consists of highly active fluxes, including strong organic acids, organic halides, inorganic halides and other salts. The AC flux normally does not contain rosin and is highly corrosive; R flux contains only rosin without the presence of activator; RMA flux contains rosin and activator system. Both R and RMA fluxes do not consist of halide-containing chemicals. RA flux may be composed of halide-containing activator. The use of AC flux is not allowed for electronic applications.

In addition to specific fluxes in their chemical compositions, fluxes can be classified based on their activity. The activity can be measured by water extract resistivity (Section 11.16), copper mirror test (Section 11.17), halide test (Section 11.18) and surface insulation test (Section 11.33). The permitted level of flux activity should be specified in accordance with the requirements of the assembly in terms of cleanability, service conditions and reliability.

## 3.4 FLUX CHEMISTRY

From a chemical point of view, the inorganic chemicals such as strong acids, strong bases, and certain salts, are highly reactive, and thus are not suitable for electronic applications. Organics containing active functional groups such as carboxylic (-COOH) and amine (-NH$_2$, -NHR, -NR$_2$) are good fluxing agents. Commonly used organics include aliphatic acids, aromatic acids, aliphatic amines and their derivatives, hydrochloride salts of amines, and hydrobromide salts of amines. The fluxing strength depends on the molecular structure of the chemical and its physical properties, as well as on the surrounding medium. The effects of structure and medium on the strength of acids and bases are classified as inductive, resonance, hydrogen bonding, solvation, hybridization, and steric effects. For the commonly adopted inductive effect, the electron-withdrawing groups adjacent to the carboxylic group of a molecule enhance the acidity strength of the carboxylic group as a result of anion stabilization. Conversely, electron-releasing groups decrease the acidity. These effects are illustrated as follows:

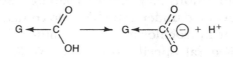

Electron-withdrawing group: increased strength.

$$G \longrightarrow C\begin{smallmatrix}O\\OH\end{smallmatrix} \longrightarrow G \longrightarrow C\begin{smallmatrix}O\\O\end{smallmatrix}^{\ominus} + H^+$$

Electron-releasing group: decreased strength.

Table 3.1 lists the acidity constants of some carboxylic acids, exemplifying this structural effect. As can be seen, by the substitution

## TABLE 3.1

### Acidity Constants of Organic Carboxylic Acids

| Chemical | $K^a$ |
|---|---|
| $CH_3COOH$ | $1.75 \times 10^{-5}$ |
| $ClCH_2COOH$ | $136 \times 10^{-5}$ |
| $Cl_2CHCOOH$ | $5530 \times 10^{-5}$ |
| $Cl_3CCOOH$ | $23{,}200 \times 10^{-5}$ |
| $CH_3CH_2CH_2COOH$ | $1.52 \times 10^{-5}$ |
| $FCH_2COOH$ | $260 \times 10^{-5}$ |
| $BrCH_2COOH$ | $125 \times 10^{-5}$ |
| ⬡—COOH | $6.3 \times 10^{-5}$ |
| $O_2N$—⬡—COOH | $36 \times 10^{-5}$ |
| $H_3C$—⬡—COOH | $4.1 \times 10^{-5}$ |

of α-hydrogen by electron-withdrawing groups such as -F, -Cl, -Br, the acidity is intensified. For aromatic acids, the substitution of hydrogen on the aromatic ring by an electron-withdrawing group such as $NO_2$ increases the acidity, and the electron-releasing group such as $-CH_3$ reduces the acidity.

Similarly, basicity is influenced by the electron inductive effect which renders more or less available the free pair of electrons for sharing with an acid. Therefore, the electron-releasing group increases the basicity of amine by providing more availability of electrons and by stabilizing the ion, and the electron-withdrawing group decreases its basicity. The mechanism is illustrated as follows:

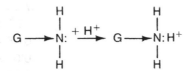

Electron-releasing group: increased strength.

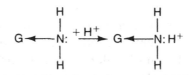

Electron-withdrawing group: decreased strength.

The basicity of some amines as listed in Table 3.2 demonstrates the effect of electron induction. In addition to electronic effect, solvation and steric effects also play a role, as shown in the comparison of primary, secondary, and tertiary aliphatic amines. The basicity of the tertiary amine in most cases is hindered by the solvation and steric effect. Temperature is another factor affecting the strength of acids and bases, which can alter the expected strength, as predicted from the structural effects. It is evident that organic chemistry can be well utilized in this area. Because in-depth discussion is beyond the scope of this book, further reading is recommended.[3]

In addition to the strength of flux agent, fluxing kinetics is another important aspect of fluxing performance. Regarding kinetics, temperature and time are two universal variables. Without getting into formulation specifics, the flux activation temperature and activation time during heating/reflow have to be designed to fit the process settings.

**TABLE 3.2**

**Basicity Constants of Amines**

| Chemical | $K_b$ |
|---|---|
| Methylamine | $4.5 \times 10^{-4}$ |
| Dimethylamine | $5.4 \times 10^{-4}$ |
| Trimethylamine | $0.6 \times 10^{-4}$ |
| N-Propylamine | $4.1 \times 10^{-4}$ |
| Di-N-Propylamine | $10 \times 10^{-4}$ |
| Tri-N-Propylamine | $4.5 \times 10^{-4}$ |
| Aniline | $4.2 \times 10^{-10}$ |
| Methylaniline | $7.1 \times 10^{-10}$ |
| Dimethylanilane | $11.7 \times 10^{-10}$ |
| O-Chloroaniline | $0.05 \times 10^{-10}$ |
| O-Bromoaniline | $0.03 \times 10^{-10}$ |
| O-Nitroaniline | $0.00006 \times 10^{-10}$ |

As previously stated, the function of the flux in solder paste is to clean the solder powder and to clean the surfaces of substrates to be joined during heating/reflow, as well as to maintain the cleanliness until the joint is completely formed (i.e., to prevent reoxidation during reflow cycle). The molecular structure and physical/chemical properties of the flux system respond to heating. In this sense, flux is also related to heat transfer.

The flux obviously has to be chemically compatible with the substrates to be joined and with the solder powder, and to be capable of removing oxides, debris, and other nonmetallic compounds that could hinder the wetting and the formation of metallic continuity. For example, the flux strength and flux design for copper or tin substrate may differ from that required for substrates such as nickel or silver/palladium and other thick film conductor compositions.

For applications in electronics, the Type R and Type RMA fluxes are considered to be appropriate due to their less "corrosive" nature. These two types are constituted to be non-halogen-containing. The Federal Specification QQ-S-571E calls for specific tests for a flux to be qualified as the R or RMA type. Tests include specific resistivity value for a water extract, a corrosion test on a copper mirror, and a halogen test. (The test method and specification are included in Appendix I.) The specification also requires that RA flux residue be completely removed after soldering and excludes the use of Type AC flux in electrical and electronic circuits.

## 3.5  ROSIN CHEMISTRY

Rosin is a natural resin obtained from pine trees. It is classified as gum rosin, wood rosin, and tall oil rosin. Gum rosin is the residue obtained from the oleoresin collected from living trees after separation from terpentine oil; wood rosin is the residue of the distillate extracted from stumpwood; and tall oil rosin is the product separated from fatty acid by fractional distillation of tall oil in pulping processes.

Rosin consists of several rosin acids, rosin acid esters, rosin anhydrides, and fatty acids. The major components of an unmodified rosin are abietic acid, isopimaric acid, neoabietic acid, pimaric acid, dihydroabietic acid and dehydroabietic acid with the following chemical structures, and general formula $C_{19}H_{29}COOH$ or $C_{19}H_{27}COOH$ or $C_{19}H_{31}COOH$

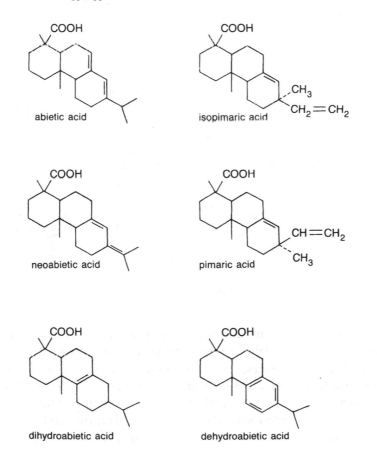

| | |
|---|---|
| abietic acid | isopimaric acid |
| neoabietic acid | pimaric acid |
| dihydroabietic acid | dehydroabietic acid |

All of these components are characterized by one carboxyl group, a condensed three-ring structure, and a double bond or conjugated double bond, except the fully hydrogenated acid. The reactive sites are the carboxyl group and conjugated double bond or the double bond. Therefore, the rosin can be readily modified through organic disproportionation, hydrogenation, polymerization, saponification, esterification, and the Diels-Alder reaction, as represented by the following reactions of abietic acid, which is most prevalent among the different compounds in natural rosins.

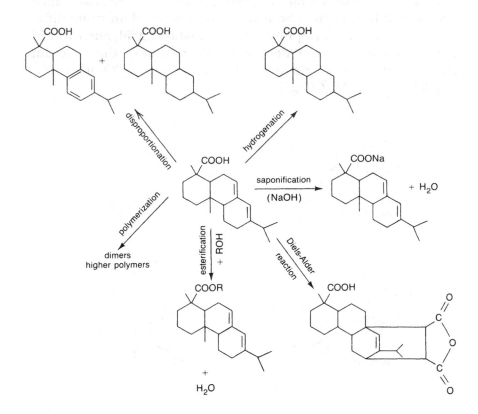

In view of these readily occurring reactions, rosins and rosin derivatives are available in many structures such as disproportionated rosin, rosin esters, hydrogenated rosin esters, polymerized rosins, and polymerized rosin esters, in addition to physically modified rosins. Crystallization and oxidation are two phenomena that often occur in the unmodified rosins. Crystallization can increase the melting point of the rosin, and oxidation normally degrades the properties of rosin.

Tall oil rosin is more susceptible to crystallization. Avoiding excessively heating the rosin at low temperature (below 140°C) minimizes the phenomenon. The resistance to crystallization and oxidation can be built into the rosin during manufacturing, to assure its quality.

In addition, at elevated temperature (above 300°C) the rosin may undergo the decarboxylation. Each rosin possesses different responses to temperature. Figure 3.3 illustrates the changes in two different rosins, as represented by weight loss with time at two different temperatures under conduction heating. At a temperature of 210°C, both rosins change with the time of heat exposure, and rosin A is more vulnerable to the temperature change as reflected in the weight loss than is rosin B, showing that rosin B has a better heat resistance. At a relatively high temperature, the two rosins degrade at a much faster rate and in a similar manner. Figure 3.4 is the response of rosins A and B to infrared heating. Figures 3.5 and 3.6 are the respective differential scanning calorimetry (DSC) thermograms of rosin A and B when exposed to nitrogen and air atmosphere.

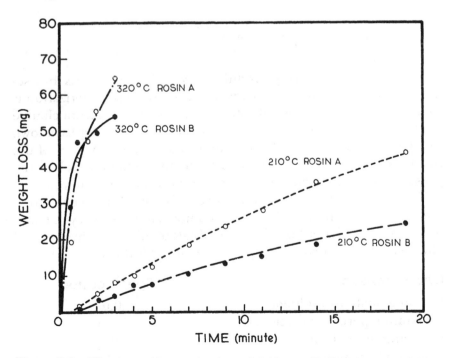

**Figure 3.3**  Changes of two rosins in weight loss with time in response to conduction heating.

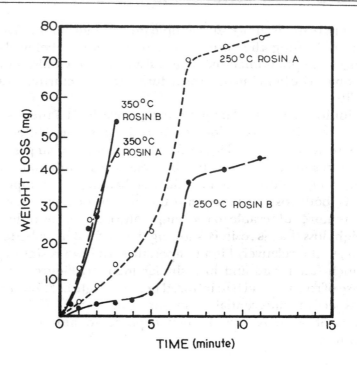

**Figure 3.4** Changes of two rosins in weight loss in response to infrared heating.

Under normal soldering conditions, W. L. Archer reports that iso-merization among the rosin acids occurs, and no polymerization is found in the rosin acid as indicated by the lack of significant change in molecular weight distribution before and after heating,[4] as shown in Figure 3.7. The physical and chemical behavior, however, of the rosin acids are expected to vary with the environment, such as the presence of strong acid, strong base, and the solvent system.

To meet Federal Specification QQ-S-571E, the rosin to be used to constitute solder paste is defined to conform to Class A, Type I, and grade WW or WG of Specification LLL-R-626D. The physical and chemical characteristics of Type I rosin are as follows:

| Characteristics | Specification |
| --- | --- |
| Insoluble matter in toluene | 0.05 |
| (Maximum percent by weight) | |
| Softening point, C, minimum | 70 |
| Acid number, minimum | 160 |
| Saponification number, minimum | 166 |

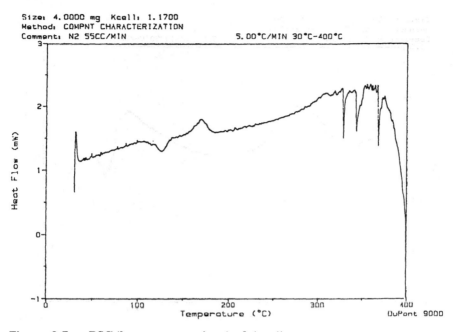

**Figure 3.5a**   DSC thermogram of rosin A in nitrogen.

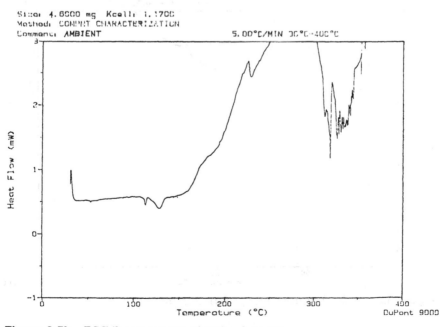

**Figure 3.5b**   DSC thermogram of rosin A in air.

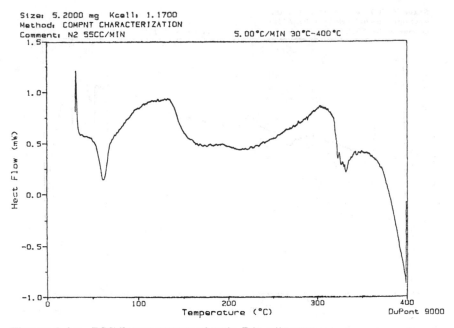

Size: 5.2000 mg   Kcell: 1.1700
Method: COMPNT CHARACTERIZATION
Comment: N2 55CC/MIN                        5.00°C/MIN 30°C-400°C

**Figure 3.6a**   DSC thermogram of rosin B in nitrogen.

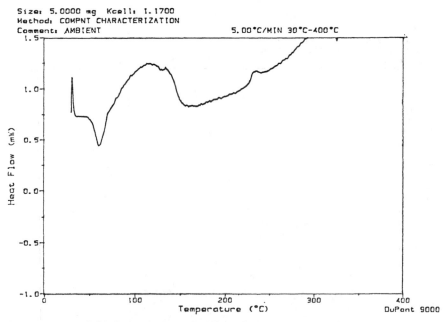

Size: 5.0000 mg   Kcell: 1.1700
Method: COMPNT CHARACTERIZATION
Comment: AMBIENT                            5.00°C/MIN 30°C-400°C

**Figure 3.6b**   DSC thermogram of rosin B in air.

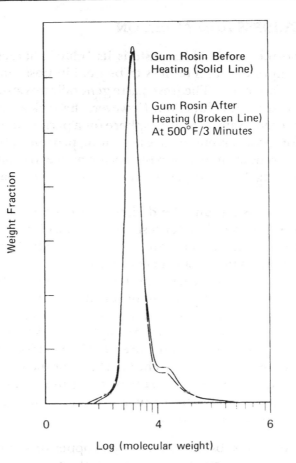

**Figure 3.7**  Molecular distribution of a gum rosin before and after heating.[4]

The carboxylic group of rosin acid is the primary source of acidity and fluxing activity. Although a higher acid number is expected to have higher fluxing strength, the fluxing activity in solder paste is usually complicated by other ingredients in the flux/vehicle system.

With respect to solubility, most rosins and their derivatives are soluble or partially soluble in aromatic hydrocarbons, esters, ketones, and some alcohols. The extent of solubility in a specific solvent system affects the physical properties of the resulting solution, as well as chemical reactivity in some cases.

## 3.6  TACKINESS AND ADHESION

One of the merits of solder paste is its "glue" characteristic that enables components and devices to be held in position before the solder joints are made. The fresh paste generally possesses adequate tackiness to serve this function. However, the tack ability after the paste is exposed to ambient atmosphere for a period of time tends to decline. For most current surface mounting processes, the tackiness is one of the requirements for paste performance in order to accommodate the lag between paste deposition on the board and the reflow.

The solder paste should be designed to have a proper holding force over a required time for a specific application. The criterion can be expressed in tack force and tack time. The IPC Standard IPC-SP-819 proposes a tack test procedure. Essentially, the tack force is measured by a stainless steel test probe, by bringing the test probe in contact with the printed paste sample of 0.64 cm (0.25 in.) diameter and 0.25 mm (0.010 in.) thickness at a rate of 0.1 ± 0.02 in./min and applying a force of 300 ± 30 g to the sample. After 5 sec following the application of this force, the measurement in peak force required to break contact is the tack force. The tack time for the measured force is the time elapsed between the paste printing and the measurement. Alternatively, a qualitative measuring technique can be conveniently set up. The procedure follows:

1. Prepare clean substrate plates (e.g., copper or stainless steel or alumina), the number depending on the test.
2. Print a given pattern (e.g., PLCC-28) on the substrate.
3. Expose the printed pattern at a series of time intervals (e.g., 4 h, 8 h, 12 h, 24 h, 48 h, 72 h) under specified conditions (e.g., 50 ± 10% relative humidity, 25 ± 2°C).
4. Place the PLCC-28 component on the exposed paste pattern and assure intimate contact between the component and paste by applying a given load (e.g., 5 g) on the component for a given time (e.g., 2 sec).
5. Turn the component/plate unit upside down and record the time when the component drops off by gravity.

This test is a qualitative measurement at best, yet the technique is simple and flexible, providing a good comparative performance.

The test parameters such as print pattern, print thickness, time span, exposure condition, contact load, and contact time can be selected to suit the specific assembling environment. The technique is particularly useful in determining the relative performance among pastes.

The tacky test equipment dedicated to the tack test of solder paste is available.[5] By using this equipment, the test is essentially to measure the torque force as expressed by the holding angle of the plate on the paste pattern, which has been subjected to an accelerated exposure condition. The key in this test is to correlate the test results to the actual performance of the paste in a real environment. For comparative ranking purpose, the results obtained from the tacky tester are expected to be in agreement with the qualitative measure as described in the preceding.

Different pastes impart different performance characteristics in the extent of tackiness and tack time. For a paste having limited tackiness and tack time, performance can be improved by the way that paste is handled and stored. Instead of storing the printed board or assembly under ambient atmosphere, cold storage (refrigeration), inert atmosphere storage (nitrogen), and solvent vapor storage are the options. Since most pastes are not highly hydroscopic, refrigeration would not cause any other problem due to moisture condensation. A solvent chamber saturated with volatile solvent vapor is found to be effective in prolonging the tack property of the paste. The solvent used needs to be compatible with the paste; ideally, use of the same solvent system as in the paste is recommended. It is apparent that selecting the paste which possesses the tack property suitable for the process is always most desirable.

## 3.7  THERMAL PROPERTIES

Most ingredients in vehicle/flux systems are organic in nature. Whether the ingredients are in the liquid or solid phase, they are sensitive to thermal effect at the reflow temperature. The phase transition (solid → liquid → vapor), chemical decomposition (covalent bond breaking), and chemical reaction (covalent bond breaking and formation) are common phenomena occurring in chemicals as temperature rises. The molecular structure, functional group, chemical reactivity, and molecular weight are factors to be considered.

The thermal effect on individual chemicals can be measured by several techniques. The instrumentation includes differential scanning calorimetry (DSC), thermomechanical analysis (TMA), thermogravimetric analysis (TGA), and differential thermal analysis (DTA). As a function of temperature, DSC measures the heat associated with the transition of materials and thus is very useful in determining boiling point, melting point, softening point, glass transition, phase transition, and heat of reaction. It is a convenient tool to monitor thermal stability of materials. Providing information about thermal stability, DTA measures temperature and semi-quantitative calorimetric properties. It is capable of measuring relatively high temperature properties. Measuring dimensional change as a function of temperature or time, TMA generates information about mechanical behavior of the material under external load, in response to temperature. Monitoring weight change as a function of temperature or time, TGA provides data on material degradation and decomposition. It is especially useful in providing compositional information. Although very limited information on solder paste in this regard is available, thermal analysis instrumentation is useful to paste users and paste researchers as a quality control and material characterization tool.

It should be noted that the thermal effect on flux/vehicle composition becomes pronounced when high lead-content solder alloys are used, which requires higher than 300°C peak temperature to reflow. At this temperature, most organics undergo some physical and chemical changes as indicated in Figures 3.3–3.6. These temperature-induced changes in organics in turn affect the residue cleanability and the properties of the resulting solder joint.

## 3.8 DRYABILITY

The flux/vehicle system can be constituted with dryable or non-dryable compositions as incorporated in paste under reflow conditions. Its relationship with reflow performance is discussed in Section 7.19.

## 3.9 RESIDUE

The properties of residue after reflow should be considered in its ionic activity and chemical reactivity such as corrosivity, its physical property such as soft or hard form at ambient temperature, and cleanability. The cleanability is discussed in Chapter 8. The ionic activity and chemical reactivity can be formulated in a required level. The residue can be made in a soft or hard state through formulation based on the intended reflow temperature and its correlation with ingredients selected. The soft residue is normally easier to be removed; however, hard residue is desired for noncleaning applications.

## 3.10  SOLDERABILITY

Solderability, in a broad sense, is the ability to achieve a clean, metallic surface on solder powder and on substrates during the dynamic heating process so that a complete coalescence of solder powder particles and good wetting of molten solder on the surface of the substrates can be formed. It is apparent that the solderability relies on both fluxing efficiency provided by solder paste and the quality of the surface of the substrates. Both areas should be evaluated to obtain optimum solderability. Because Section 3.3 has already discussed the fluxing of solder paste, this section focuses on substrates.

As molten solder comes into intimate contact with the substrate, the wetting ability in terms of wetting angle ($\theta$) may be represented in two extremes, as shown in Figure 3.8: for wetted phenomenon where wetting angle is smaller than 90°, and for unwetted phenomenon where wetting angle is larger than 90°.

Among the common substrates to be soldered, the demand on the flux for good wettability depends on

intrinsic wettability by molten solder (Sn, Sn/Pb > Cu > Ag/Pd, Ag/Pt, Ag/Pd/Pt > Ni),

ease of reaction between oxides of the metal substrate and the flux, and

other surface condition of substrate.

Therefore, even using the same flux system, the solderability may change due to the variation in the quality of the bare metal substrate.

With respect to solderability measurement, meniscograph tests have successfully provided the information on the relative solderability of the substrates by using a wetting balance to monitor the wetting

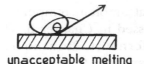
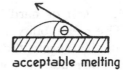

unacceptable melting          acceptable melting

## A) molten solder on substrate

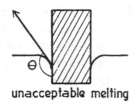
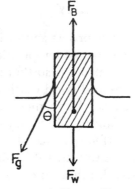

unacceptable melting

acceptable melting

## B) substrate dipping in molten solder

**Figure 3.8**  Unacceptable and acceptable wetting phenomena.

force of a specimen in a molten solder bath which normally has the composition of 60-Sn/40-Pb, at a given temperature as a function of time. As the specimen is in contact with or dipped into the molten solder, major parameters involved are the interfacial force between the molten solder and the perimeter of the specimen, the force due to Archimedean buoyancy, and the geometry and volume of the specimen. The forces involved, as shown in Figure 3.8, are expressed as

$$F_w \propto F_g \cos \theta - F_b = F_g \cos \theta - \rho \, gV$$

where $F_g$ is the interfacial force, $F_b$ is the buoyancy force, $\theta$ is the contact angle, $\rho$ is the density of molten solder, $g$ is the gravitational

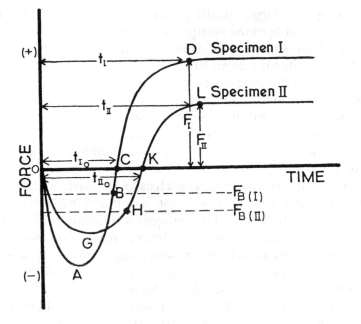

**Figure 3.9** Examples of wetting curve of meniscograph.

acceleration, $V$ is the immersed volume of specimen, and $F_w$ is the resultant wetting force. Figure 3.9 represents the principal features of wetting curves measured by the wetting balance technique. For the case of Specimen I, when the specimen is immersed in the molten solder bath at the preset depth, but prior to the commencement of wetting as indicated from point O to A, an upward force is exerted on the specimen. Wetting starts at A and proceeds through C, then reaches equilibrium at D. At B, the only force is that of upward buoyancy, and at C the upward buoyancy and downward wetting forces are balanced, resulting in a net zero force. The magnitude of force, $F_I$ represents the wetting force. In some cases, the specimen may undergo more induction period as shown in the curve of Specimen II between points O and G. During this period, the specimen is heated up to reach a wetting condition. The extent of the period depends on the size, conductivity, and surface condition of the specimen.

Useful information can be derived from the curve by comparing the initial wetting time $(t_{I_0}, t_{II_0})$, the equilibrium wetting time $(t_I, t_{II})$, and wetting force $(F_I, F_{II})$. In Figure 3.9, Specimen I demonstrates a faster wetting and larger wetting force than Specimen II. This principle is adaptable to the comparative evaluation of solderability of solder pads on substrate by using a given flux, as well as of the relative flux activity on a given substrate.

The criteria for the magnitude of wetting time and wetting force are to be determined for a specific application and/or components. There is no universal number to be followed. Under a given set of conditions in temperature, specimen dimension, and the speed and depth of immersion, the criteria for solderability can be readily established.

The solderability of solder pads such as on printed circuit boards can be readily tested by an edge-dip test, as outlined in Section 11.29. Solderability is then evaluated by the quality of wetting. The formation of a smooth, uniform and continuous solder coating on the surface of the solder pads without dewetting, nonwetting, and pin holes is ideal. Dewetting is the phenomenon of molten solder receding after it has coated the surface, leaving a rough and irregular surface with thick mounds of solder connected by thin solder film. In dewetting, the substrate surface is not exposed. Nonwetting is defined as the phenomenon of molten solder not adhering to, or only partially adhering to, the substrate surface, thereby leaving the substrate surface exposed. The molten solder in such case tends to form spotty adherence or form a single sphere with high wetting angle (> 90°).

For printed circuit boards, the copper solder pads as fabricated are commonly coated with tin-lead solder to prevent copper oxidation during storage and therefore facilitate solderability. However, the bare copper PCBs after being subjected to a proper chemical treatment can also provide a prolonged shelf life. A copper benzotriazole process has been recently developed for PCBs and found to be effective.[6] Although the chemical, benzotriazole, is a well-known corrosion inhibitor for copper, a properly set process is the key to its effectiveness. This process basically consists of cleaning, benzotriazole application, water rinsing, and air drying. The cleaning is an important step. It is accomplished by using an alkaline solution to remove oil and grease and by using sulfuric acid and sodium persulfate to remove oxide and to provide a slight etching action. The effectiveness of benzotriazole on copper is attributed to the formation of copper benzotriazole complex through the following equations:[7,8]

$$C_6H_4N_2NH \longrightarrow C_6H_4N_2N^- + H^+$$

$$C_6H_4N_2N^- + Cu^+ \longrightarrow C_6H_4N_2N-Cu$$

The copper benzotinazole complex is identified as having three layers composed of cuprous oxide, chemisorbed cuprous benzotriazole monolayer, and anodic corrosion layer of cuprous benzotriazole.[7]

Another practical aspect related to solderability is the soldering rate, which refers to the time required to complete the reflow of a solder paste on a given substrate. Assuming that the quality of the substrate is consistent, the solder rate is often compared among different grades of paste from the same supplier and/or different suppliers. Although faster rate is usually desirable, it should be noted that the comparison is to be made among similar types of paste. For example, dryable paste which contains low boiling volatiles normally solders faster than nondryable paste which does not contain low boiling volatiles. In such cases, the soldering rate has no bearing on the superiority or inferiority of solderability.

## 3.11  SOLDERABILITY: COMPONENTS

This section focuses specifically on the issue of plating and tinning of leads of components in relation to solderability. Component leads are commonly made of copper, copper alloys, alloy-42 (41-42.5 nickel, balance iron) and Kovar (29% nickel, 17% cobalt, 53% iron, 1% others).

Two types of "coated" leads are common: those having aqueous plating (electroplated coating) and molten solder dip. Each of these two processes has merits and limitations. Plating process normally provides more uniform thickness which is often porous, and molten solder dip produces a thicker and denser fused coating. The compositions used in tinning are in a range of tin-lead alloys.

The solderability of tinned leads depends on the following factors:

- preparation of base lead materials
- composition of virgin alloys
- surface finish and condition of coating
- age of coating
- storage of coating
- thickness of coating

A coating thickness of 0.0003 in. (7.6 microns) is most adopted, and thin coating is often associated with poor solderability. Nonetheless, the ideal coating thickness depends on many practical factors. The desirable compositions of coating alloy in a wide range of tin-lead ratio are discussed in the literature.[9,10]

Solderability of coated leads after a variety of shelf time and conditions is a concern. Assuming the coating is intact, this concern can be viewed from two aspects: first, how the solderability is affected by the surface degradation due to oxidation and/or contamination during shelfing period, and second, how the solderability is affected by the interaction between lead material and tin-lead coating during shelf time or treatment such as a burn-in test. The surface oxidation of tin-lead alloy normally is not an insolvable problem since the fluxing of some solder pastes is able to take care of it. The formation of copper-tin intermetallic compounds at copper-based leads and tin-based coating interface ($Cu_3Sn$, $Cu_6Sn_5$) is always a potential, although it would be extremely sluggish at room temperature as indicated in the Cu/Sn phase diagram. With this interaction, the tin content at coating/lead interface will be gradually consumed, resulting in solderability degradation. The consumption rate depends on temperature and time. In this regard, alloy-42 and Kovar leads are expected to do better than copper leads and high-tin coating. However, high-tin coating having higher surface tension normally provides better wettability.

It should be noted that, for leads coated by molten solder dip, copper-tin intermetallic compounds (namely, $Cu_6Sn_5$) can be formed rapidly during coating. However, due to its intactness, the molten solder dip coating is expected to be relatively more stable during storage than electroplated coating. When coating is porous, as made from electroplating, leads made of Alloy-42 and Kovar may experience deterioration with age due to moisture permeation through porous crystalline structure.

In either case, the degradation of the coating is obviously driven by a kinetically controlled process and depends on other practical and environmental factors. Therefore, to assure good quality of coated surface of leads, the shelf time and storage temperature must be minimized. Using the freshly coated leads is ideal.

The solderability test for components is outlined in Section 11.29. The components are subjected to accelerated aging in a live steam environment to determine the shelf life effect on the solderability of components.

For testing a very small specimen such as an individual lead of a chip carrier package, or a termination of discrete chips, the use of a small globule of molten solder instead of a solder bath, in conjunction with a sensitive surface contact detector, is found successful.[11] Figures 3.10 and 3.11 illustrate the testing of a chip capacitor using 200 mg solder and a single J lead of PLCC package using 25 mg solder, respectively.

## 3.12  METAL LOAD ACCEPTABILITY

Metal load is one of the factors affecting the shelf stability and the viscosity/rheology of paste, and therefore the applicability of paste deposition. It is also related to void development in the solder joint, as well as to residue cleanability.

Most solder pastes, for practical reasons, contain 85–92 wt % of metal content which corresponds to approximately 35–55 vol % depending on the solder alloy and flux/vehicle employed. Table 3.3 lists the conversion of weight percent to volume percent for 63-Sn/ 37-Pb alloy based on the alloy density being 8.4 g/cm$^3$, and on flux/ vehicle systems being 0.9 g/cm$^3$ and 1.1 g/cm$^3$, respectively.

The acceptability of metal load in the flux/vehicle system depends on the interactions between the metal particle and the organic vehicle/flux system in physical and/or chemical nature. Therefore, in addition to metal load in weight or volume, the particle size distribution and particle shape influence the acceptability of metal load in a given flux/vehicle system as well as in the properties of the resulting paste. Table 3.4 compares the viscosity of pastes composed of different powder sizes, size distribution, and particle shape in a given vehicle/flux system over a range of metal load. Paste viscosity

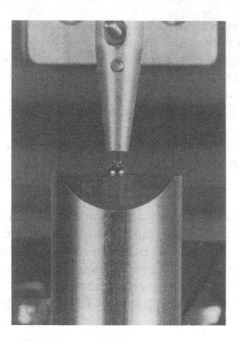

**Figure 3.10** Fixture of meniscograph solderability tester for 1206 chip capacitor. *(Courtesy of GEC Research–Billiton Witmetaal B.V.)*

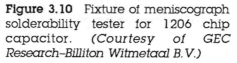

**Figure 3.11** Fixture of meniscograph solderability tester for J leads of PLCC. *(Courtesy of GEC Research–Billiton Witmetaal B. V.)*

## TABLE 3.3

### Metal Load in Weight and Volume Percents for 63-Sn/37-Pb Solder Pastes

| Vehicle/Flux Density 0.9 g/cm³ | | Vehicle/Flux Density 1.1 g/cm³ | |
|---|---|---|---|
| Weight % | Volume % | Weight % | Volume % |
| 85 | 37.7 | 85 | 42.6 |
| 87 | 41.8 | 87 | 46.8 |
| 88 | 44.0 | 88 | 48.9 |
| 89 | 46.4 | 89 | 51.4 |
| 90 | 51.4 | 90 | 54.3 |
| 92 | 55.4 | 92 | 60.0 |
| 95 | 67.1 | 95 | 71.5 |

## TABLE 3.4

### Dependency of Paste Viscosity on Physical Characteristics of Powder

| Metal load (weight %) | -200/+325 Mesh Spherical | -325 Mesh Spherical | -200 Mesh Irregular |
|---|---|---|---|
| 85 | 240,000 | 280,000 | 400,000 |
| 87 | 300,000 | 350,000 | — |
| 89 | 390,000 | 510,000 | — |
| 90 | 460,000 | 660,000 | — |
| 91 | 690,000 | 880,000 | — |
| 92 | 1,020,000 | 1,450,000 | — |

increases with increasing metal load. The finer powder provides higher viscosity of a paste, and irregular powder also delivers high viscosity when other conditions are equal. Since the solder powder has a high intrinsic density and high apparent density, the vehicle/flux system must be constituted to be able to provide good suspension power, specifically, for the applications which demand less viscous paste.

The increased metal load in paste imparts decreased volume of organic portion and therefore increased solder volume. The reduced organic portion normally coincides with an increase in viscosity, which in turn reduces the spread of paste, minimizing bridging problem. The increase in solder volume is also beneficial to void reduction, and in some cases, to residue cleaning. To achieve a high level of metal load is not difficult to do, but in the meantime, maintaining all other attributes needed for a viable paste could be a hurdle. For example, in most cases, the flux/vehicle system has its limit to accommodate the metal powder load to maintain a workable rheology. Too high a viscosity impairs applicability and other characteristics. Research efforts in this area to achieve a metal load exceeding 60% in volume is always appealing.

## 3.13  FLOW PROPERTIES

Chapter 5 discusses the rheological properties of paste in relation to paste application methods. This section deals with the flow properties of paste after it is applied. When heat is applied to the paste through any means, the paste tends to spread or slump due to gravity and to thermal energy generated,[12] as shown in Figure 3.12, Sample 2.

Spreading may cause solder bridging between the pads and inadequate solder joint standoff height. Depending on the characteristics of paste and the substrate, total spreading is usually the result of spreading in both the paste state and the molten solder state. In order to minimize or overcome this problem, addition of sufficient nonmetallic spheres to the paste as spacers that provide a uniform and adequate distance between the component and the substrate has been performed.[13] High-melting-point solder particulates which do not melt at the temperature required for reflow are incorporated into the lower-melting-point solder paste.[14]

In forming solder bumps for flip-chip devices, the technique using thick film glass dams or stop-offs to limit solder flow to the tip of the substrate metallization, is found to be successful in restraining the solder bump from collapsing or wetting out on the electrode land.[15, 16] Increasing the metal load, increasing the viscosity of the paste, and modifying the ingredients including the solvent system, rosin/resin selection and rheological additives are used as common means to improve the restriction on flow. Although increasing metal load generally reflects increasing slump resistance, each flux/vehicle system has a limit for metal powder tolerance (as discussed in

## FLOW PROPERTY COMPARISON
### high restrictivity vs. low restrictivity

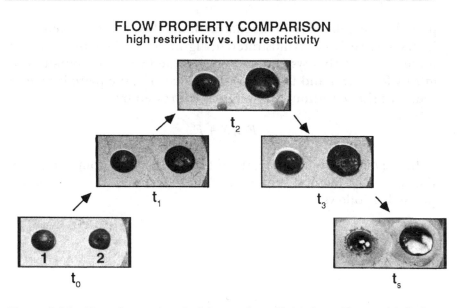

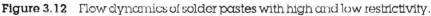

**Figure 3.12** Flow dynamics of solder pastes with high and low restrictivity.

Section 3.12) to deliver a proper rheology for paste application and adequate solderability.

The increase in viscosity, either as a result of higher metal load or a result of vehicle/flux formulation, may also create difficulty in paste application. This leads to a development of a flux/vehicle system with the intrinsic thermodynamic property of resisting the slump of paste, as the temperature is elevated toward the solder liquidus temperature, and of reducing further spreading of molten solder on the wettable substrates.[12] Figure 3.12 illustrates the flow dynamics of two types of paste under a typical reflow condition. Paste #1, incorporated with the flow control property through thermodynamics, exhibits shape retention. Paste 2, without such a property, shows spread in area as temperature rises.

Spreading in the paste state and in the molten solder state is dominated by different driving forces. In the paste state, the interparticle forces are dominating, and in the molten solder state, the relative interfacial tension (force per unit area) between flux/vehicle liquid, substrate, and molten solder, governs the spreading. Therefore, the objective is to design a flux/vehicle system capable of intimately wetting the surface of solder particles while providing adequate surface tension in the liquid phase so that the solder particles are imbedded in the matrix of the flux/vehicle system with high cohesive force, as illustrated in Figure 3.13. When the cohesive force $(F_c)$ in such a system is not overcome by the combination of gravity force $(F_g)$ and thermal disturbance $(F_t)$, the paste is able to retain its shape without slumping, as expressed by

$$F_c > F_g + F_t.$$

In the molten solder state, the interfacial tension dominates and, the spreading factor $S$, in terms of interfacial tensions may be expressed as follows:

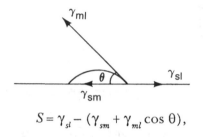

$$S = \gamma_{sl} - (\gamma_{sm} + \gamma_{ml} \cos \theta),$$

where $\gamma_{sl}$ is the interfacial tension of flux/vehicle and substrate, $\gamma_{sm}$ is the interfacial tension of molten solder and substrate, and $\gamma_{ml}$ is the interfacial tension of molten solder and flux/vehicle.

To illustrate the effect, the difference in the spreading factor between System A and System B, using Antonow's approximation is

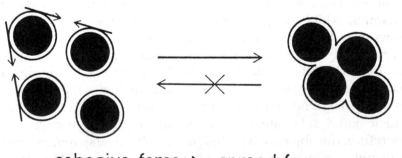

**cohesive force $>$ spread force**

**Figure 3.13.** Illustration of interparticle cohesive force for slump resistance.

Thus, when

$$\gamma_{lB} > \gamma_{lA}$$
$$\therefore S_A > S_B,$$

and when

$$\gamma_{lB} < \gamma_{lA}$$
$$\therefore S_A < S_B.$$

The spreading factor in a system with relatively higher surface tension is smaller than in that with lower surface tension.

Since most solder pads, as on printed circuit boards or on hybrid circuits, are individual pads surrounded by an unwettable surface (solder mask or bare alumina surface), it is perceived that the molten solder should have retracted back to its own pad after reflow due to its high surface tension. This perception does not always hold as in the cases where the solder volume exceeds the tolerance of the pad area, and a component exerts the weight on solder. Again, the balance of the forces is the criterion. Furthermore, when we examine the spread of phenomena which cause pad bridging or binding between two solder joints, the primary factor is a result of hot slump in the paste state before the solder melts. Hot slump is defined as the shape change when the paste sinks in height and expands in area in response to temperature rise before the solder melts. As the paste deposit between adjacent pads is in physical contact due to hot slump, it is likely to cause solder bridging between pads. The likelihood of this depends on the volume of solder applied in relation to pad dimension and spacing between pads, as well as on the weight of component exerted on the paste deposit per unit area. Thus, as the circuitry becomes more intricate and shrinks in size, it is more prone to bridging problem.

By monitoring the thermodynamic parameters of the flux/vehicle system, the flow property can be controlled.[12] It is indicated that the flow property is directly related to the surface tension of the liquid phase. Figure 3.14 indicates a correlation between the surface tension and the flow restrictivity of paste while other parameters are held constant.

The flow restrictivity as expressed in percentage is conveniently chosen as the ratio of the initial area of the paste deposit before heat is applied to the final area at the completion of heating in the paste state and in the molten state, respectively.

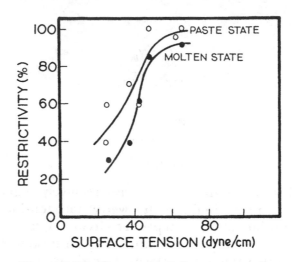

**Figure 3.14** Flow restrictivity versus surface tension.

$$\text{Flow restrictivity} = \frac{\text{initial area}}{\text{final area}} \times 100\%$$

To compare the systems at extreme conditions, Table 3.5 lists the data of the effect of surface tension on the flow restrictivity. The table also demonstrates the results of different reflow methods. The two systems distinguished by the thermodynamic parameter display significant difference in flow restrictivity in all three reflow methods.

Pad bridging can be essentially eliminated by using paste with built-in flow restrictivity. The occurrence of pad bridging in relation

**TABLE 3.5**

**Effect of Reflow Methods on Restrictivity**

| Surface Tension (Dyne/cm) | Restrictivity (%) | | | | | |
|---|---|---|---|---|---|---|
| | Conduction | | IR | | Vapor Phase | |
| | Paste | Molten | Paste | Molten | Paste | Molten |
| 66.5 | 96 | 93 | 95 | 90 | 100 | 100 |
| 26.3 | 59 | 30 | 54 | 41 | 57 | 47 |

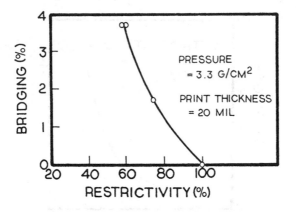

**Figure 3.15** Dependency of pad bridging reduction on restrictivity.

to flow restrictivity is shown in Figure 3.15, indicating that high flow restrictivity reduces pad bridging.

For pastes possessing low flow restrictivity, the extent of bridging is highly sensitive to the pressure exerted onto the paste deposit and onto the print thickness. Figures 3.16 and 3.17 indicate the effects of pressure and print thickness, respectively. Therefore, in order to minimize pad bridging, the selection of the flow property in paste should be compatible with the specific devices/components to be used. Increasing print thickness further aggravates the bridging probability of pastes that do not possess the flow restrictivity. With the growing adoption of fine pitch pattern in surface mount technology, it is critical to use paste having high flow restrictivity.

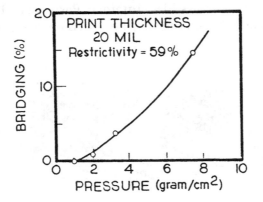

**Figure 3.16** Dependency of pad bridging on pressure.

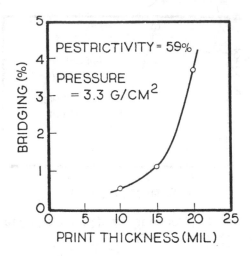

Figure 3.17   Dependency of pad bridging on print thickness.

## 3.14   EXPOSURE LIFE

Section 3.6 covers the life time for tackiness. Another closely related property is exposure life, which refers to solderability performance after a certain exposure time under a specified condition (e.g., temperature, humidity). The pastes which deliver long tack time in holding parts may not necessarily have good solderability performance after such a time lapse. Figure 3.18 shows solderability degradation of a paste having exposed under atmosphere (65% ± 5% humidity and 75° ± 5°F) for 24 hours in comparison with the fresh paste, while its tack life was still maintained. Paste users should obtain product information from the supplier to assure paste compatibility

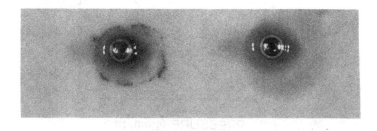

Figure 3.18   Solderability degradation of a paste after improper environmental exposure.

with the environment that paste will be exposed to, and the performance that is required for the application.

## 3.15  FORMULATION

The first step in formulating a paste product is to define the performance objective to be achieved. With a clear objective in mind, a paste can be designed to meet the performance parameters by utilizing the fundamental technologies, by understanding how the selected raw materials contribute to the characteristics of a paste, and by anticipating any synergistic or antagonistic interactions among raw materials. A product involves many performance parameters, and some of them may be trade-offs. For example, high metal content is beneficial to solder joint in volume and void and in residue, yet may make the paste more prone to crusting and difficult to apply. High viscosity paste may improve flow control against temperature, yet causes the paste to be difficult to apply. Using highly active fluxing chemicals may improve solderability in some cases, yet its use may leave more corrosive residue. In such a case, improving the solderability by selecting proper ingredients without the use of highly active fluxing chemicals is a skill. Increasing flux content does not always improve solderability in terms of wetting and/or solder balling elimination. (Figure 3.1 exemplifies the sensitivity of solder balling phenomenon.)

The ability to achieve an ultimate balance among the performance parameters and to prioritize them for a specific group of applications is the key. Once a prototype formula capable of providing all functions is constituted, the next step is to fine-tune the formula to coincide with specifications or any specific values designated.

With the accomplishment of the development of specific composition formula, it is equally important to develop a reproducible process for making a paste with consistent performance. It is not an exaggeration, but an indication of the importance of the role of process, to state that the identical composition formula can produce different products when the process is allowed to vary. The required skill is to be able to utilize the technologies to make the best balance of all desired performance parameters, and to meet the performance requirements in the real world, as well as to develop a controlled process in making end-use products.

## REFERENCES

1. Paul J. F. Flory, "Molecular Weights and Intrinsic Viscosities of Polyisobutylenes," *J. Am. Chem. Soc.* 65 (1943): 372–382.

2. R. F. Boyer, *J. Appl. Phys.* 25 (1954): 825.

3. Jerry March, *Advanced Organic Chemistry, Reactions, Mechanisms, and Structure,* (New York: McGraw-Hill, 1968).

4. Wesley L. Archer, T. D. Cabelks, and J. J. Nalazek. The Dow Chemical Company.

5. Austin American Technology Corporation, Leander, Texas.

6. Man Kei Ho, "Copper Surface Finish Promotes Solderability," *Electronic Packaging and Production* (Oct. 1987).

7. D. Chadwick and T. Hashemi, "Adsorbed Corrosion Inhibitors Studied by Electron Spectroscopy: Benzotriazole on Copper and Copper Alloys," *Corrosion Science* 18 (1978): 39.

8. P. G. Fox et al., "Some Chemical Aspects of the Corrosion Inhibition of Copper by Benzotriazole," *Corrosion Science* 19, (1979): 457.

9. Eric Fincham, "Finish with Solderability," *Circuits Manufacturing* (Feb. 1988).

10. R. Bowlby, "Finish First," *Circuits Manufacturing* (Nov. 1987).

11. Billiton Witmetaal and G.E.C. Research Limited, The Meniscograph Solderability Tester, Nederland.

12. Jennie S. Hwang and N. C. Lee, "A New Development in Solder Paste with Unique Rheology for Surface Mounting," proceedings, 1985 International Symposium on Microelectronics.

13. Wallace Rubin, "Smoothing the Way for Solder Creams," *Circuit Manufacturing* (Oct. 1984).

14. Tom Dixon, "SMT Forces Solder Paste Improvements," *Electronic Packaging and Production* (Aug. 1984).

15. L. S. Goldman and P. A. Totta, "Area Array Solder Interconnections for VLSI," *Solid State Technology* (Jun. 1983): 91.

16. P. A. Totta, "Flip Chip Solder Terminals," proceedings, 1971 Electronics Components Conference: 275.

# Metallurgical Aspects

From a metallurgical point of view, making a sound solder joint involves a series of steps, starting from selecting a proper solder alloy, to wetting ability on substrates, control of solidification from melt, and to external exposure after solidification. Metallurgical aspects of solder paste deal with not only the areas for conventional solder alloys, but also with the discrete solder powders which melt, coalesce, and solidify into a resulting solder joint. This chapter provides an overview of solder alloys with respect to their phase diagrams, wetting/spreading, solidification, heat treatment effects, and mechanical properties, as well as solder powder characteristics and microstructure.

## 4.1 ALLOYS IN GENERAL

Broadly, soldering is a process of joining two metallic surfaces of either bulk metals or of metallized or metallic surfaces of nonmetallic materials by means of a heating source, using solder alloy as the fillet material. One general definition for solder is fusible alloys with liquidus temperature below 400°C (750°F). The liquidus temperature is the temperature above which the alloy is in molten state. An alloy is a metallic material formed from a combination of two or more elements by means of dissolving and mixing in the liquid state. The liquid alloy thus freezes into the solid state through solidification. The resulting alloy may have a different crystal structure from any of the starting elements. In a binary alloy, the element in larger concentration is called the solvent, and the element in smaller concentration is solute. If the added solute can achieve a change in the property of the pure solvent element, this addition produces an alloy no matter how small the concentration is. Therefore, to form an alloy, no minimum concentration of solute is specified. Where solid solution can be formed, the solute atoms may directly substitute for the solvent atoms in the lattice structure, forming substitutional solution. Or solute atoms may enter the interstitial sites in the crystal structure instead of replacing solvent atoms, resulting in interstitial solution. Figure 4.1 is the schematic of substitutional and interstitial solutions.

## 4.2  SOLDER ALLOY

The elements commonly used in solder alloys are tin (Sn), lead (Pb), silver (Ag), bismuth (Bi), indium (In), antimony (Sb), and cadmium (Cd).  Their melting points are listed in Table 4.1.

In addition to tin-lead alloys, binary solder alloys include tin-silver, tin-antimony, tin-indium, tin-bismuth, lead-indium, lead-bismuth. Ternary alloys include tin-lead-silver, tin-lead-bismuth, and tin-lead-indium.  The solidus and liquidus temperatures of some commonly used compositions are listed in Table 4.2.

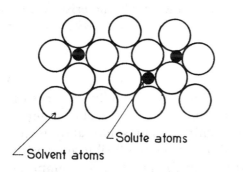
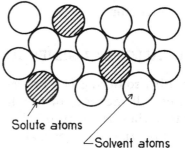

INTERSTITIAL SOLID SOLUTION   SUBSTITUTIONAL SOLID SOLUTION

**Figure 4.1**  Interstitial and substitutional structures.

**TABLE 4.1**

**Melting Points of Common Solder Elements.**

|      | Sn  | Pb  | Ag   | Bi    | In    | Sb    | Cd    |
|------|-----|-----|------|-------|-------|-------|-------|
| °C   | 232 | 328 | 961  | 271.5 | 156.6 | 630.5 | 321.2 |
| °F   | 450 | 620 | 1762 | 520   | 313   | 1167  | 610   |

**TABLE 4.2**

## Melting Range of Common Solder Alloys

| Alloy Composition | Melting Range | | | | Mushy Range | |
|---|---|---|---|---|---|---|
| | Solidus | | Liquidus | | | |
| | °C | °F | °C | °F | °C | °F |
| 70-Sn/30-Pb | 183 | 361 | 193 | 380 | 10 | 19 |
| 63-Sn/37-Pb | 183 | 361 | 183 | 361 | 0 | 0 |
| 60-Sn/40-Pb | 183 | 361 | 190 | 375 | 7 | 14 |
| 50-Sn/50-Pb | 183 | 361 | 216 | 420 | 33 | 59 |
| 40-Sn/60-Pb | 183 | 361 | 238 | 460 | 55 | 99 |
| 30-Sn/70-Pb | 185 | 365 | 255 | 491 | 70 | 126 |
| 25-Sn/75-Pb | 183 | 361 | 266 | 511 | 83 | 150 |
| 10-Sn/90-Pb | 268 | 514 | 302 | 575 | 34 | 61 |
| 5-Sn/95-Pb | 308 | 586 | 312 | 594 | 4 | 8 |
| 62-Sn/36-Pb/2-Ag | 179 | 355 | 179 | 355 | 0 | 0 |
| 10-Sn/88-Pb/2-Ag | 268 | 514 | 290 | 554 | 22 | 40 |
| 5-Sn/90-Pb/5-Ag | 292 | 558 | 292 | 558 | 0 | 0 |
| 5-Sn/92.5-Pb/2.5-Ag | 287 | 549 | 296 | 564 | 9 | 15 |
| 5-Sn/93.5-Pb/1.5-Ag | 296 | 564 | 301 | 574 | 5 | 10 |
| 2-Sn/95.5-Pb/2.5-Ag | 299 | 570 | 304 | 579 | 5 | 9 |
| 1-Sn/97.5-Pb/1.5-Ag | 309 | 588 | 309 | 588 | 0 | 0 |
| 96.5-Sn/3.5-Ag | 221 | 430 | 221 | 430 | 0 | 0 |
| 95-Sn/5-Sb | 235 | 455 | 240 | 464 | 5 | 9 |
| 42-Sn/58-Bi | 138 | 281 | 138 | 281 | 0 | 0 |
| 43-Sn/43-Pb/14-Bi | 144 | 291 | 163 | 325 | 19 | 34 |
| 52-Sn/48-In | 118 | 244 | 131 | 268 | 13 | 24 |
| 70-In/30-Pb | 160 | 320 | 174 | 345 | 14 | 25 |
| 60-In/40-Pb | 174 | 345 | 185 | 365 | 11 | 20 |
| 70-Sn/18-Pb/12-In | 162 | 324 | 162 | 324 | 0 | 0 |
| 90-Pb/5-In/5-Ag | 290 | 554 | 310 | 590 | 20 | 36 |
| 92.5-Pb/5-In/2.5-Ag | 300 | 572 | 310 | 590 | 10 | 18 |
| 97.5-Pb/2.5-Ag | 303 | 578 | 303 | 578 | 0 | 0 |

The alloy selection is based on

- metallurgical compatibility, consideration of leaching and potential formation of intermetallic compounds

- environment or service compatibility, consideration of silver migration
- temperature capability, consideration of service temperature, process
- wettability on substrate

## 4.3 PHASE DIAGRAMS

The addition of solute(s) into a solvent results in several basic systems:

- complete miscibility in solid and liquid states
- complete miscibility in liquid state and partial miscibility in solid state
- complete miscibility in liquid state and no miscibility in solid state
- systems containing intermediate phases

As a prelude to discussing these basic systems, some terminologies are defined as follows:

**Solid solution.** The crystalline, homogeneous mixture of two or more elements in the solid state over a range of compositions, in either substitutional or interstitial type.

**Intermetallic compound.** One type of intermediate phase; the mixture of two or more elements forming a new crystal structure at stochiometric ratio.

**Intermediate phase.** Solid solution formed in the intermediate ranges of composition that do not include the pure components.

The atomic size of solute and solvent in an alloy and the crystal structure of elements are important factors in the characteristics of the resulting alloy. Table 4.3 provides the atomic size and crystal structure of elements in solder alloys.

For a system with complete solid and liquid miscibility, the solvent can dissolve any amount of solute element in either the solid or liquid state. The alloys made of solute and solvent elements having similar size and lattice structure (e.g., copper-nickel alloys) normally fall in this category. In such an alloy, copper has an atomic radius of 1.57 Å and a crystal structure fcc (face-centered cubic), and nickel has an atomic radius of 1.62 Å and a crystal structure fcc. Figure 4.2 shows

**TABLE 4.3**

**Atomic Size and Crystal Structure of Solder Elements**

|                  | Sn    | Pb   | Ag   | Bi    | In    | Sb    | Cd    |
| ---------------- | ----- | ---- | ---- | ----- | ----- | ----- | ----- |
| Atomic Size (Å)  | 1.72  | 1.81 | 1.75 | 1.63  | 2.00  | 1.53  | 1.71  |
| Crystal Structure| tetra | fcc  | fcc  | rhom  | tetra | rhom  | hexa  |

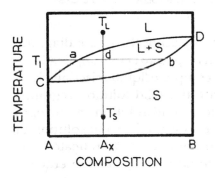

**Figure 4.2**  Phase diagram of complete miscibility in solid and liquid states.

the phase diagram of this system. It consists of a liquid region ($L$), a coexisting liquid-solid region ($L + S$), and a solid region ($S$). The curve $CaD$ above which the alloy is liquid is called the liquidus line, and the curve $CbD$ below which the alloy is solid is called the solidus line. For a given composition $A_x$, the alloy is a liquid at temperature $T_L$, and as temperature drops to $T_S$ the alloy turns into a solid one-phase region. At $T_1$, the two phases coexist; the intersection point ($a$) on the liquidus line represents the composition of the liquid, and the intersection point ($b$) on the solidus line is the composition of the solid crystal. The concentration of the two phases at temperature $T_1$ and composition $A_x$ is estimated by the Lever Law: *The ratio* ad/ab *is the proportion of the solid phase, and* bd/ab *is the proportion of the liquid phase.*

When elements have complete liquid miscibility and no solid miscibility, as in Figure 4.3, the phase diagram is characterized by two triangular two-phase regions having solid $A$ or $B$ plus liquid, a horizontal line $CED$ at the freezing temperature, and an eutectic point $E$ at which three phases coexist. The eutectic point reflects the lowest temperature at which any liquid can exist. The compositions on the left side of the eutectic point (e.g., $A_x$) are called hypoeutectic, and those on the right side (e.g., $A_y$) are hypereutectic. Since the elements do not have solid miscibility, the properties of these alloys

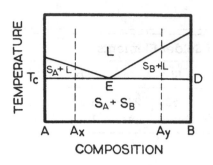

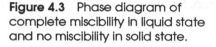

**Figure 4.3** Phase diagram of complete miscibility in liquid state and no miscibility in solid state.

cannot be altered by heat treatment. By using the phase diagram, useful information can be derived with respect to actual applications.

An alloy made from elements with complete liquid miscibility and partial solid miscibility is featured with the solid solution region, a liquid and solid solution region, and a solid solution mixture, as shown in Figure 4.4. The area *FCA* represents the solid solution of *B* in *A*, and area *GDB* is a solid solution of *A* in *B*. As indicated, the solubility of *B* increases as the temperature drops to the eutectic temperature, and then decreases as temperature continues to drop. The same applies to the solubility of *A* in *B*. The presence of a solid solution makes the properties of alloy dependent on heat treatment.

If the tendency to produce intermediate phases of different structures is absent, the solid solution of metals can be readily formed when the atomic volumes of the two metals are similar and two pure metals have the same or similar crystal structure.

More complex phase diagrams contain intermediate phases, including intermetallic compounds, allotropy (different crystalline states), and peritectoid and eutectoid (solid state analogs of eutectic and peritectic).

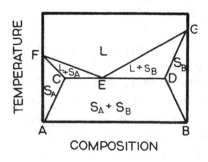

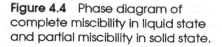

**Figure 4.4** Phase diagram of complete miscibility in liquid state and partial miscibility in solid state.

## 4.4  PHASE DIAGRAMS: SOLDER ALLOYS

Among solder alloys, indium and lead form a continuous series of solid solutions with the existence of a miscibility gap as indicated in Figure 4.5. There are two peritectic horizontals at 171.9°C and 159.2°C, and an intermediate phase $\alpha_1$.

Figure 4.6 is the phase diagram of bismuth-tin. A composition of 58 wt % Bi at temperature 139°C is the eutectic point.  Insignificant

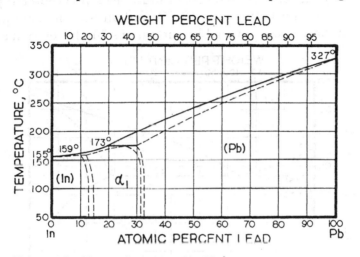

**Figure 4.5**  Phase diagram of In/Pb.[1]

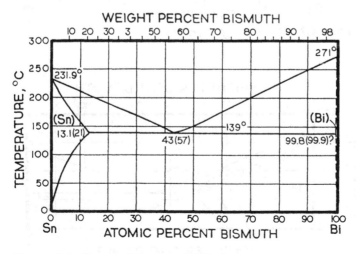

**Figure 4.6**  Phase diagram of Bi/Sn.[1]

solid solubility of Sn in Bi is found at the eutectic temperature, and solid solution of Bi in Sn is 21 wt %.

For tin-lead alloys, the eutectic point is 63 wt % Sn at a temperature of 183°C. The maximum solubility of Pb in Sn is 2.5 wt %, and Sn in Pb is 19 wt %. The phase diagram is shown in Figure 4.7.

For further reference, phase diagrams for bismuth-lead, indium-tin and antimony-tin are included in Figure 4.8, 4.9, and 4.10, respectively. The phase diagram of bismuth-lead shows eutectic point at 125°C, a peritectic point at 184°C, and an intermediate phase. Although there is an appreciable amount of solubility of Bi in Pb, Pb

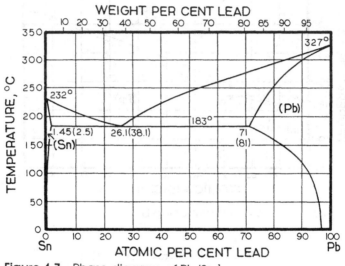

**Figure 4.7**   Phase diagram of Pb/Sn.[1]

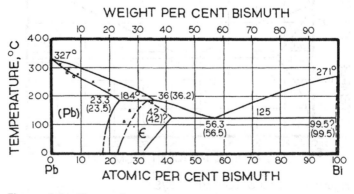

**Figure 4.8**   Phase diagram of Bi/Pb.[1]

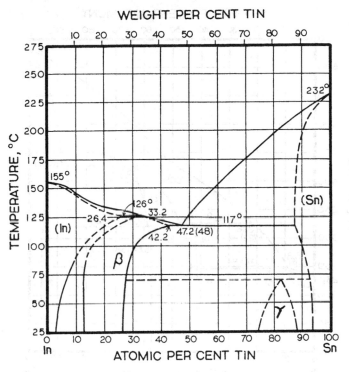

**Figure 4.9** Phase diagram of In/Sn.[1]

in Bi is negligible at the eutectic temperature. For indium-tin, a eutectic point is found at 48 wt % of Sn at 117°C, and there is mutual solid solubility. The phase diagram of antimony-tin has three per-titectic reactions around 246°C, 325°C, and 425°C at 10.5 wt % Sb, 22 wt % Sb, and 50 wt % Sb, respectively.

For silver-tin alloys, as shown in Figure 4.11, the eutectic point is found at 221°C with a composition of 96.5 wt % Sn and 3.5 wt % Ag. There is a complete miscibility of Ag in Sn in the liquid state. At the eutectic temperature, the solid solubility of Ag in Sn is insignificant (< 1%). Solubility of Sn in Ag, however, is around 10 wt %. Intermediate phases are present as silver concentration approaches 76 wt %.

More details on phase diagrams of solder alloys can be found in Reference 1 at the end of this chapter. The ternary phase diagrams are in Appendix II.

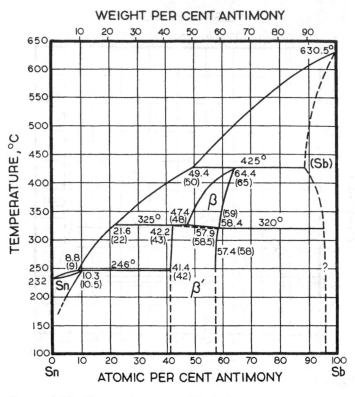

**Figure 4.10**  Phase diagram of Sb/Sn.[1]

## 4.5  WETTING AND SPREADING

Wetting is an important surface phenomenon.  The free surface of
a solid, as is well known, has higher energy than the interior of the
solid due to broken bonds at the surface.  In soldering, we deal with
metallic surfaces which possess high surface energy (assuming a
clean and pure state) as compared with oxides or organic materials.
The following shows the magnitude of energy of three basic categories for comparison.

| | |
|---|---|
| Metals | >200 cal/cm³ |
| Organics, nonmetals | 50–200 cal/cm³ |
| Fluorocarbons, low hydrocarbons | <50  cal/cm³ |

As shown in the following equation, free energy is expressed as
ergs per square centimeter, where an erg is equivalent to a dyne-
centimeter in physical magnitude.  That is, free energy has the units

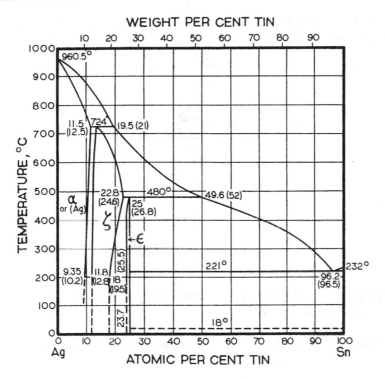

**Figure 4.11**   Phase diagram of Ag/Sn.[1]

of work, which are force times distance through which it acts—in this case, dyne times centimeters.  Here the surface energy and surface tension are used interchangeably.

$$\underset{\text{Surface Energy}}{\frac{\text{ergs}}{\text{cm}^2}} = \underset{}{\frac{\text{dynes} \times \text{cm}}{\text{cm}^2}} = \underset{\text{Surface Tension}}{\frac{\text{dynes}}{\text{cm}}}$$

For a system at constant temperature $(T)$, and pressure $(P)$,

$$\left(\frac{\delta G}{\delta A}\right)_{P,T} = \gamma,$$

where $G$ is the free energy, $A$ is the area, and $\gamma$ is the surface tension. The thermodynamic condition for spreading to occur is

$$\Delta G \;<\; 0.$$

The spreading of a liquid $l$ with negligible vapor pressure on a solid surface $s$, is pictured below.

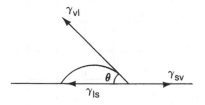

Thus,

$$-\left(\frac{\delta G}{\delta A}\right)_{P,T} = \gamma_{sv} - (\gamma_{ls} + \gamma_{vl} \cos \theta),$$

where $\gamma_{ls}$, $\gamma_{vl}$, $\gamma_{sv}$ are liquid-solid, liquid-vapor, and solid-vapor interfacial tension, respectively.

Letting $-\left(\frac{\delta G}{\delta A}\right)_{P,T}$ be the spreading coefficient, $S$, we have

$$S = -\left(\frac{\delta G}{\delta A}\right)_{P,T} = \gamma_{sv} - (\gamma_{ls} + \gamma_{vl} \cos \theta).$$

Therefore, for spreading to occur

$$\gamma_{sv} - (\gamma_{ls} + \gamma_{vl} \cos \theta) > 0$$

or

$$\gamma_{sv} > \gamma_{ls} + \gamma_{vl} \cos \theta.$$

This spreading condition holds true when the system does not involve significant chemical or metallurgical reactions at the interface.

The surface energy of the liquid can be readily obtained, but the surface energy of the solid and the interfacial energy between solid and liquid are difficult to measure. The surface free energy of solids can be calculated from heat of sublimation and heat of fusion data, and interfacial energy is calculated by measuring the dihedral angle between phases. The creep rate technique has been used to determine the surface free energy of solid metals directly.[2,3,4] In general, the surface energy of solids (metals) is slightly higher than that of the corresponding liquids (molten metals), which accounts for the cohesion energy in solids.

As a general guideline, for a system with liquid to wet the solid substrate, the spreading occurs only if the surface energy of the substrate to be wetted is higher than that of the liquid to be spread. Therefore, in order for wetting to occur, the surface energy of the liquid and the surface energy of the solid substrate are key factors in determining the spreading and wetting.

The surface tension of liquid metal was found to correlate well with the inverse of the atomic volume.[5] For liquid alloys, it is indicated that the surface energy follows approximately the linear mixture law when the components form solid solution, and exhibits negative deviation from the linear law when the components form intermetallic compounds.[6] Although it is well recognized that the surface energy of pure liquid alloy decreases with increasing temperature, the temperature dependence of the surface energy of liquid alloys can go either direction (decreasing or increasing), depending on the relative distribution of alloy components between surface and bulk as the temperature increases.

The spreading of liquid tin-lead solder on clean copper surfaces with zinc ammonium chloride flux has been studied.[7] The area of spread for tin-lead solders with varying compositions is shown in Figure 4.12. The spread increases with increasing tin content and

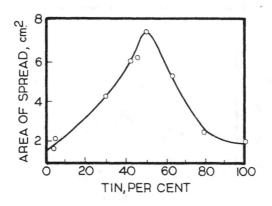

**Figure 4.12**   Spreading of tin-lead solders on copper.[7]

reaches a maximum at 50% lead–50% tin, and then decreases with
further increase in tin content. The study also showed that the spread
may be affected by a third additive. The addition of silver up to 3%
of the tin content did not influence the amount of spread, and antimony addition up to 6% of the tin content gave a marked reduction
in area of spread. When using a weaker flux, the areas of spread
remained in the same order as when using zinc ammonium chloride.

This phenomenon can be explained by the competition between
surface tension and the relative affinity of liquid metal to the copper
substrate. Beyond a certain concentration of lead, the inferior
affinity of lead for the copper substrate predominates and reduces
the spreading. This is further evidenced by the shift of maxima to
higher tin content as temperature increases, as shown in Figure 4.13.

It is known that liquid metal may cause embrittlement of solid
substrate through wetting action.[8] As a polycrystalline metal is wetted
by a liquid metal it forms a solid-liquid interface of low energy. If the
interfacial energy of solid-liquid is smaller than the grain boundary
energy, the liquid may penetrate along the grain boundaries. Since
the conditions for wetting the solid surface are

$$\gamma_{sv} > (\gamma_{ls} + \gamma_{vl} \cos \theta)$$

and

$$\gamma_{ls} < \tfrac{1}{2}\gamma_{gb}$$

where $\gamma_{gb}$ is the grain boundary energy.

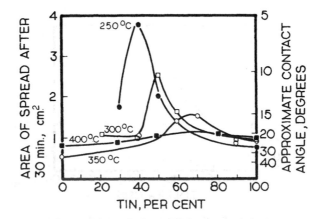

**Figure 4.13** Effect of tin concentration on
spreading at different temperatures.[7]

Thus, the condition for wetting the surface without wetting the grain boundaries is

$$\tfrac{1}{2}\gamma_{gb} < \gamma_{ls} < (\gamma_{sv} - \gamma_{vl}\cos\theta).$$

The surface energy is also related to other cohesive properties of solids. For example, under ideal conditions of Hooke's Law, the fracture stress, $\sigma_f$, is related to the surface energy by

$$\sigma_f = \sqrt{\frac{E\gamma}{a}} \, ,$$

where $E\gamma$ is Young's modulus, and $a$ is the lattice parameter. With the known surface energy and material parameters, the strength can be estimated. However, actual solids have much lower strength than the equation estimates.

## 4.6 SOLIDIFICATION

As a stage of the soldering process, the molten solder has to solidify to form the solder joint. This section outlines some fundamentals in solidification and the importance of this stage to the solder joint.

In general, the solidification of liquid metal occurs by nucleation and crystal growth. The free energy of nucleation, $\Delta G$, is generally represented by

$$\Delta G = \frac{4}{3}\pi r^3 \Delta G_v + 4\pi r^2 \gamma \, ,$$

where $r$ is linear dimension of the nucleus, $G_v$ is the free energy change per unit volume, and $\gamma$ is surface free energy per unit area. This equation clearly shows that when $r$ is small, the surface free energy dominates, and when $r$ is large, the volume free energy prevails.[9] Therefore, there is a critical size which corresponds to the unstable equilibrium between nucleus and the liquid matrix. The nucleation rate, $R_n$, follows the Arrhenius-like relation

$$R_n = CN_{\text{exp}}(-E/kT)$$

where $C$ is the frequency factor, $N$ is the number of nucleation sites, $E$ is the energy of nucleation at the critical size, $k$ is the Boltzmann constant, and $T$ is the absolute temperature.

For crystal to form from the melt, the plane interface between the nuclei and the liquid must have a lower temperature than the equilibrium temperature (supercooling). The growth rate is proportional to the magnitude of the temperature gradient between the equilibrium temperature and the actual temperature of the interface, as well as depending on the crystallographic orientation of the interface. Furthermore, the temperature gradient is directly related to the heat flow along the interface—that is, the relative rate of generation of latent heat and the rate of heat removal. Obviously, the thermal conductivity of the solid and liquid phase is a factor in the solidification process. When the temperature gradient in front of the interface is a rising one, assuming the temperature gradient is linear and perpendicular to the interface, the growing interface is stable and any localized instability formed on the interface would remelt. This is well illustrated by G. J. Davies as shown in Figure 4.14. The corresponding temperature distribution in solid and liquid phase is represented in Figure 4.15.

However, when the temperature falls in front of the interface mov-

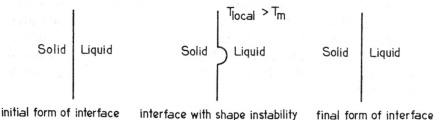

initial form of interface    interface with shape instability    final form of interface

**Figure 4.14**  Stable interface formation during solidification.[9]

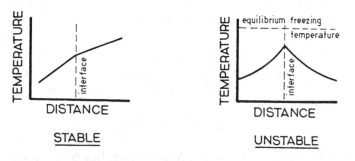

STABLE                          UNSTABLE

**Figure 4.15**  Temperature gradient in solid and liquid for stable and unstable interface.

ing into the liquid, a temperature inversion occurs as indicated in Figure 4.15.[10, 11, 12] In this case, the temperature of the liquid is well below the equilibrium freezing temperature due to supercooling, and the heat will flow from the interface into the supercooled liquid, resulting in a negative temperature gradient in the liquid. Because the latent heat is released at the interface, the temperature of the interface rises above that of both liquid and solid and drops as moving from the interface to the solid equilibrium freezing temperature.

As a result of the inverted temperature gradient, the interface becomes unstable and produces spikes into the liquid, and their secondary and tertiary branches are formed. This results in a dendritic structure of crystals. Figure 4.16 illustrates the dendritic generation during the solidification.

Dendritic growth during solidification is a common phenomenon in pure metals and alloys. In alloys, in addition to thermal supercooling—that is, a temperature gradient at the interface—impurities and alloying elements can cause a so-called constitutional supercooling.[13] The constitutional supercooling occurs when a solid freezes with a composition different from that of the liquid from which it forms. This composition difference lowers the liquidus temperature of the liquid in contact with the interface. For the phase diagram of a binary alloy as shown in Figure 4.17, the variation of concentration of solute and its corresponding change in liquidus temperature are represented in Figure 4.18, (a) and (b).

Figure 4.19 is the temperature distribution of constitutional supercooling in which the liquid is at a temperature above that of the interface but below its liquidus temperature, in contrast to the system

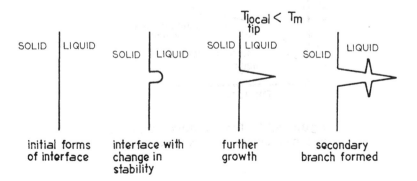

**Figure 4. 16**   Dendrite growth due to inverse temperature during solidification.[9]

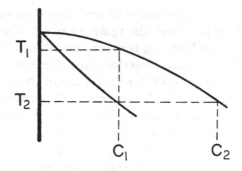

**Figure 4.17** Phase diagram of a binary alloy.

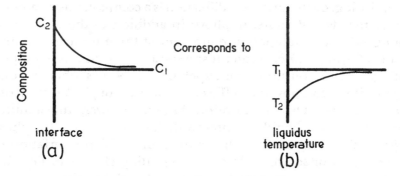

Corresponds to

**Figure 4.18** Composition and liquidus temperature change in constitutional supercooling.

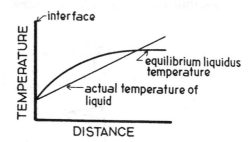
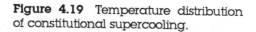

**Figure 4.19** Temperature distribution of constitutional supercooling.

in which the liquid is at a lower temperature than the interface, as occurs in thermal supercooling. The supercooling arises from a change in composition (not temperature) due to the fact that the solute element tends to remain in the liquid rather than join the solvent atoms in freezing, causing the liquid melt to be enriched with solute atoms. This phenomenon is called microsegregation. This makes the interface plane unstable since any protuberance formed on the interface would find itself in the supercooled liquid, which provides the driving force for further growth. Consequently, the freezing process is the formation of a cellular or a cellular dendritic structure. Such constitutional supercooling is more readily present in the alloys with significant size difference between solute and solvent atoms.

Walton, Tiller, Rutter and Winegard have investigated the instability of a smooth solid-liquid interface during solidification for a lead in tin system, and found that the cellular structure observed in binary alloys can be eliminated by the proper choice of growth conditions.[14] For a given solute concentration, there exists a critical ratio of temperature gradient ($G$) in the melt to the rate of solidification ($R$) of the crystal—that is, $G/R$. The cellular structure can be eliminated by exceeding a critical ratio as expressed by

$$\frac{G}{R} = - \, m \, \frac{C_o}{D} \left( \frac{1-k}{k} \right),$$

where $m$ is the slope of the liquidus line, $C_o$ is the initial concentration of solute in the melt, $D$ is the diffusion coefficient of the solute in the liquid, and $k$ is the ratio of concentration of solute in the solid being formed to the concentration in the liquid from which it forms ($k$ is assumed to be constant). This section gives only a brief overview in attempting to reflect the basic factors in the solidification process. The reader is referred to References 9, 14, 15, and 16 at the end of this chapter for detailed treatment on solidification of alloys.

As discussed, the heat flow during the solidification and the crystal structure of alloy are crucial factors to the properties and structure of the solidified alloy. The design of the assembly, such as the thermal conductivity of the system and the process of soldering such as cooling rate, is expected to affect the solder. Figure 4.20 demonstrates the microstructure of 63-Sn/37-Pb solder paste reflowed

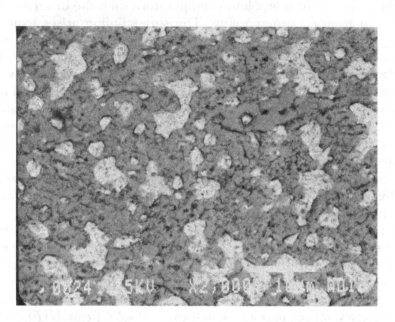

**Figure 4.20(a)**   SEM microstructures of 63-Sn/37-Pb melt under
(a) slow cooling.

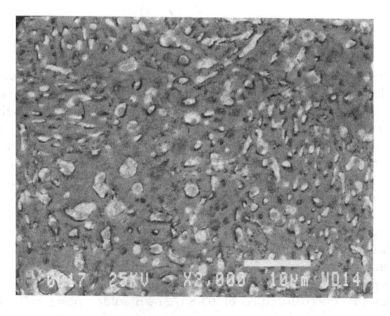

**Figure 4.20(b)**   SEM microstructures of 63-Sn/37-Pb melt under
(b) fast cooling.

under the same condition but cooled at a different rate. The difference in microstructure of the resulting solder is evident.

## 4.7   HEAT TREATMENT AND THERMAL EXCURSION

Heat treatment following solidification is often used to alter the physical and mechanical properties of alloys. Categorically, alloys with phase diagram containing a solvus line and multiple solid phases, as discussed in Section 4.3, are heat-treatable.

The general functions of heat treatment include removing segregation if the equilibrium structure of the alloy is a single phase, achieving hardening, changing porosity, and altering the grain size. As illustrated in Figure 4.21, homogenization can be achieved by heating the alloy with composition $C$ to temperature $T_1$ for a sufficient time to allow diffusion for homogenizing the structure. The solution treatment is accomplished by heating to high temperature $T_1$ for a long enough time to take all solute into solid solution and to disperse it uniformly, then rapidly quenching to temperature $T_2$ (usually room temperature) to retain all the solutes in the solid solution. The resulting supersaturated solid solution often achieves hardening with a fine dispersion of solute, by aging the alloy at $T_2$ or by tempering the alloy at slightly higher temperature. The technique is applicable to tin-lead alloys which are heat treatable and solution hardenable.

## 4.8   METAL POWDER

Metal powder can be made by several techniques:   (1) chemical reaction and decomposition, including solid-state reaction, reduc-

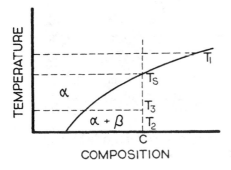

**Figure 4.21**   Solid solubility line of phase diagram.

tion, and precipitation from aqueous solution or organic solvent; (2) electrolytic deposition; (3) mechanical processing of solid metal; and (4) atomization of liquid metals.

Metal powders made from chemical reduction under high temperature are generally spongy and porous, such as copper powder (Cu) shown in Figure 4.22. The fine-particle noble metal powders shown in Figure 4.23 are frequently precipitated by reduction of the salts in aqueous solution with proper pH. The precipitate slurry is then filtered, washed, and dried under highly controlled conditions. A mechanical method is generally used to produce flake-like particles. The metals possessing high malleability, such as gold (Au), silver (Ag), copper (Cu), and aluminum (Al) are most suitable for making flakes; Figure 4.24 is a typical morphology of silver flakes. The electrolytic deposition process is characterized by dendritic particles and produces high purity powders. Its particle morphology is shown in Figure 4.25, and particle sizes are affected by selecting the type of reducing agent, the strength of reducing agent, the addition rate of reducing agent, and other reaction conditions. The charac-

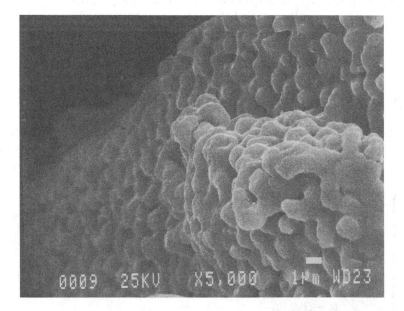

**Figure 4.22** SEM micrograph of copper powder made by high temperature chemical reduction process.

**Figure 4.23**  SEM micrograph of silver powder made by chemical precipitation process.

**Figure 4.24**  SEM micrograph of silver flake.

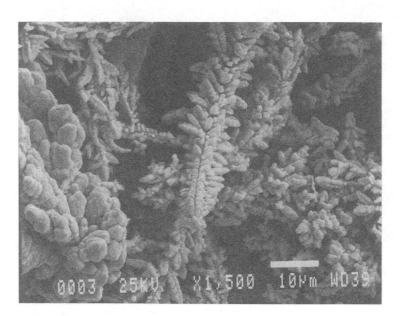

**Figure 4.25** SEM micrograph of copper powder made by electrolytic deposition process.

teristics of particles are also affected by the current density, electrolyte, additives, and temperature.[17]  The principle of atomization method is to disintegrate the molten metal under high pressure through an orifice into water or into a gaseous or vacuum chamber. The powders produced by this method have relatively high apparent density, good flow rate, and are spherical in shape as shown in Figure 4.26. Table 4.4 summarizes the typical particle size and distribution under each method employed.

Among these powder-making techniques, powders to be used in solder paste are mostly produced by atomization because of desirable inherent morphology and shape of resulting particles.  Therefore, the following extends the discussion on atomization technique only.

Figure 4.27 is the schematic of an inert gas atomization system with options of bottom pouring system and tilting crucible system.  The system consists of control cabinet, vacuum induction furnace, tundish, argon supply line, ring nozzle, atomization tower, cyclone, and powder collection container.[17]  Alloy is melted under inert gas at atmospheric pressure to avoid the evaporation of component ingre-

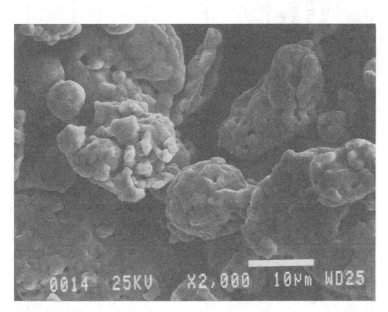

**Figure 4.26(a)**  SEM micrographs of (a) water atomized copper powder and (b) gas atomized solder powder.

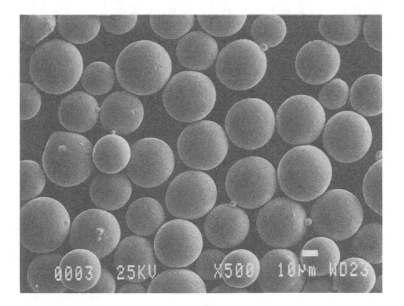

Figure 4.26(b)

**TABLE 4.4**

**Examples of Metal Powders Produced by Different Techniques**

| Technique | Powder (Shape) | Surface Area (m²/g) | Tap Density (g/cc) | Particle Size Distribution (microns) | | |
|---|---|---|---|---|---|---|
| | | | | 90% | 50% | 10% |
| Chemical Precipitation | Ag | 0.1–2.0 | 0.5–4.5 | 0.3–3.0 (avg.) | | |
| | 30-Ag/70-Pd | 6.0–8.0 | 0.6–1.0 | 4.0 | 0.8 | 0.3 |
| | 80-Ag/20-Pd | 2.0–3.5 | 1.0–1.7 | 5.0 | 2.5 | 0.5 |
| Solid-State High Temperature Reduction | Cu | 0.1 | 1.5–3.0 | 1–200 (avg.) | | |
| Mechanical | Ag (flake) | 1.0–2.0 | 2.0–4.5 | 0.5–15 (avg.) | | |
| | Cu (flake) | 1.3 | 1.4 | 18.3 | 8.3 | 3.2 |
| Electrolytic | Cu (dendritic) | – | 0.8–2.4 | 63–200 (avg.) | | |
| Atomization | Sn/Pb (spherical) | 0.1–1.0 | 4.0–6.0 | 5–75 (avg.) | | |

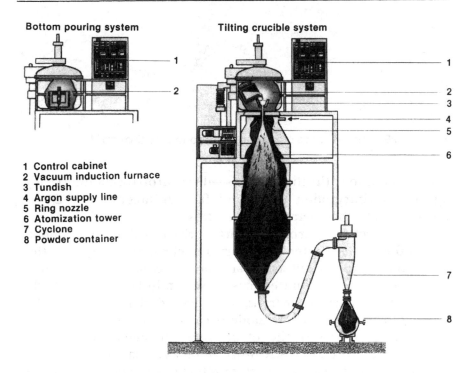

**Bottom pouring system**        **Tilting crucible system**

1 Control cabinet
2 Vacuum induction furnace
3 Tundish
4 Argon supply line
5 Ring nozzle
6 Atomization tower
7 Cyclone
8 Powder container

**Figure 4.27** Schematic of inert gas atomization system. *(Courtesy of Leybold Technologies, Incorporated.)*

dients. A high melt rate can be achieved. The molten material is then dosed into the atomization tower. The melt is disintegrated into powder under inert gas at atmospheric pressure by an energy-rich stream of inert gas. The process conducted in a closed system is able to produce high-quality powder.

In addition to inert gas and nitrogen atomization, centrifugal and rotating electrode processes have been extensively studied.[18, 19, 20, 21, 22, 23] The atomization mechanisms and the mean particle diameter are related to operating parameters—diameter of rotating electrode ($D$), melting rate of rotating electrode ($Q$), and angular velocity of the rotating electrode ($\omega$)—and to material parameters—surface tension of atomized liquid at melting point ($\gamma$), dynamic viscosity of atomized liquid ($\eta$), and density of atomized liquid at melting point ($\rho$). Relationships among these parameters are presented subsequently.

Three modes of liquid disintegration by rotating electrode process were identified by Champagne and Angers during the atomization of metallic powders as shown in Figure 4.28.[21, 22, 23] The relative

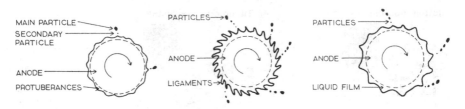

**Figure 4.28**  Three modes of rotating electrode atomization.[21]

prevalence among the three modes—direct drop formation (DDF), ligament disintegration (LD), and film disintegration (FD)—is determined by the operating parameters.

In the direct drop formation, particle size distribution is bimodal in that the number of the secondary particles is about equal to that of the main particles, but secondary particles constitute only a small portion of the total atomized mass as shown in Figure 4.29. As the melting rate increases, the transition from the DDF mode to the LD mode and finally to the FD mode occurs.  An increase in angular velocity and a decrease in the rotating mode diameter also favor such a transition.  The particle size distribution also changes accordingly as shown in Figure 4.29 for the atomization by the ligament mode. In addition, DDF mode leads to more spherical particles, the LD leads to many different sizes of particles with ellipsoidal shape, and the FD mode produces even broader size distribution. Large tank size and low melt flow rate generally favor the smooth particles, with few satellites (small particles attached on big particles).

Another two informative relations obtained by Champagne and Angers[23] are the mean volume-surface diameter of atomized particles and the particle size distribution in relation to operating and mate-

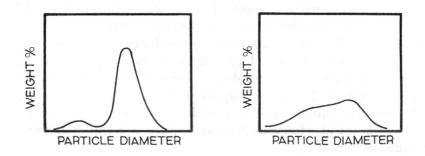

**Figure 4.29**  Particle size distribution of a powder atomized in (a) DDF mode and (b) LD mode.[20]

rial parameters. It has been found that the mean volume-surface diameter ($d$) is proportional to the surface tension of atomized liquid and melting rate, but inversely proportional to the angular velocity of the rotating electrode, the diameter of electrode, and the density of the atomized liquid, expressed by the following relation wherein the symbols were earlier defined:

$$d \propto \frac{\gamma^{0.50} Q^{0.02}}{\omega^{1.03} \rho^{0.50} D^{0.58}}$$

The mass proportion of secondary particles ($P_s$) is directly related to the angular velocity of the rotating electrode, the density of the atomized liquid, and the melting rate, but inversely proportional to the diameter of the electrode, and the surface tension of the atomized liquid, expressed by the following relation:

$$P_s \propto \frac{\omega^{0.33} \rho^{0.56} Q^{1.24}}{D^{0.15} \gamma^{1.05}}$$

Metal powder can also be produced by vacuum atomization, which is believed to yield clean and finer particles. For superfine alloy powder, it is reported that a new atomizing technique is available using pulverizing energy produced by a 50 MPa water pump concentrated at the apex of a conical jet by which the thin stream of molten metal is disintegrated into superfine droplets.[24]

Ultrasonic gas atomization is another technique which successfully produced metal powders.[25] A process of two-stage spinning cup atomization with a liquid quenchant is in development to produce fine particles with greater latitude in particle size control.[26]

## 4.9  SOLDER POWDER CHARACTERISTICS

Metal powder can be produced from different techniques as outlined in Section 4.8. Some characteristics imparted by the techniques are indicated in Table 4.4. This section discusses the physical properties and characteristics of solder powder suitable for solder paste.

## Melting Range

The melting range of commonly used solder alloys is listed in Table 4.2. Fisher-Johns melting point apparatus is a convenient tool to measure liquidus temperature. For the distinction between liquidus and solidus temperature of noneutectic alloys and other phase transitions, a thermal analyzer is a viable apparatus. The thermograms of alloys are referenced to the corresponding phase diagrams for possible phase transitions with consideration that phase diagram is derived from equilibrium condition and most processes occur under nonequilibrium condition. Figure 4.30 shows the thermograms obtained by using Differential Scanning Calorimeter DuPont Model 9000 under nitrogen for alloys 63-Sn/37-Pb (curve *a*), 96.5-Sn/3.5-Ag (curve *b*), and 10-Sn/88-Pb/2-Ag (curve *c*). Curve *c* clearly indicates the phase transitions of 10-Sn/88-Pb/2-Ag alloy, revealing a thermal fingerprint for an alloy. This technique can be used as a quality control tool for the melting range of the solder powder, which also reflects its composition.

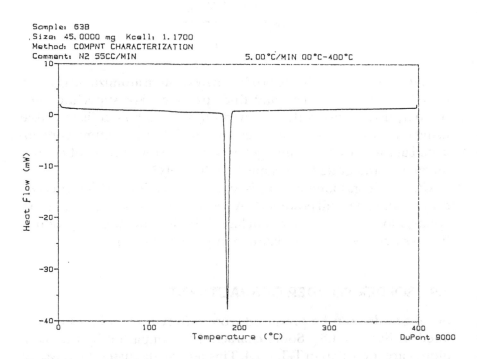

**Figure 4.30(a)**  Thermogram identification of solder powders: (a) 63-Sn/37-Pb.

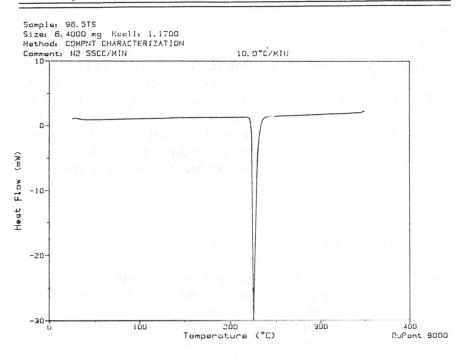

Figure 4.30(b)   Thermogram identification of solder powders:
(b) 96.5-Sn/3.5-Ag.

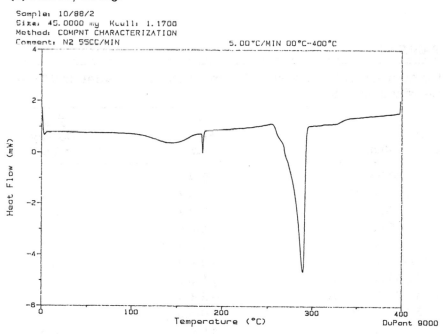

Figure 4.30(c)   Thermogram identification of solder powders:
(c) 10-Sn/88-Pb/2-Ag.

## Flow Rate

A standard method to measure flow rate is *ASTM B-213* ( *Test for Flow Rate of Metal Powders*). A given weight of powder is run through the orifice of a Hall flowmeter. The length of time elapsed is a measure of the powder flowability. Very fine powder may tend to agglomerate, and therefore hinder flow through the orifice. Agglomeration could also occur when powder is contaminated. For the range of particle size generally used in solder paste, agglomeration is not expected.

## Apparent Density

To measure apparent density, a density cup and Hall flowmeter are specified in *ASTM B-212* ( *Test for Apparent Density of Metal Powders*). Due to the inherent existence of interstices in particle packing, the tap density is lower than the true density. Table 4.5 compares the tap density of solder powders produced by inert gas atomization with the true density of bulk solder alloys.

**TABLE 4.5**

**Comparison of Tap Density of Common Solder Powder and True Density of Bulk Solder**

| Alloy Composition | Tap Density (g/cm³) | True Density (g/cm³) |
|---|---|---|
| 60-Sn/40-Pb | 4.5–4.7 | 8.5 |
| 62-Sn/36/-Pb/2-Ag | 4.5–4.7 | 8.4 |
| 63-Sn/37-Pb | 4.5–4.7 | 8.3 |
| 25-Sn/75-Pb | 5.4–5.6 | 10.0 |
| 10-Sn/75-Pb | 5.6–5.9 | 10.5 |
| 5-Sn/95-Pb | 6.0–6.2 | 10.8 |
| 10-Sn/88-Pb/2-Ag | 5.6–5.9 | 10.4 |
| 1-Sn/97.5-Pb/1.5-Ag | 6.2–6.4 | 11.3 |
| 96.5-Sn/3.5-Ag | 4.0–4.3 | 7.4 |
| 42-Sn/58-Bi | 4.8–5.0 | 8.6 |
| 43-Sn/43-Pb/14-Bi | 4.8–5.1 | 9.0 |

## Particle Morphology

As discussed in Section 4.8, solder powder produced by an atomization technique is suitable for use in solder paste. The spherical shape and smooth surface without satellites are desirable features to be used for solder paste. Figure 4.31 depicts the SEM micrographs of the spherical, smooth-surfaced particles with narrow size distribution (– 200/+325 mesh), spherical particles with small particle size (– 325 mesh), and irregular particle (– 200 mesh) of 63-Sn/37-Pb alloy.

Under close examination by a scanning electron microscope of the surface of the particles, alloy phases are indicated. Figure 4.32 shows the surfaces of particle for (a) 62-Sn/36-Pb/2-Ag, (b) 96.5-Sn/3.5-Ag, and (c) 43-Sn/43-Pb/14-Bi. The contrast phases in Figure 4.32 (a) are identified by x-ray energy dispersion technique as a dark tin-rich phase and a light lead-rich phase with semi-quantitative analysis, as shown in Figure 4.33 (a) for dark phase and (b) for light phase. Similarly, the x-ray energy dispersion analysis for the surface of 43-Sn/43-Pb/14-Bi is shown in Figure 4.34 (a) for dark phase and (b) for light phase.

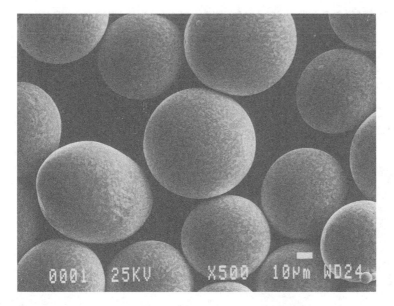

**Figure 4.31(a)**   SEM micrographs of spherical solder powders in: (a) – 200/+325 mesh.

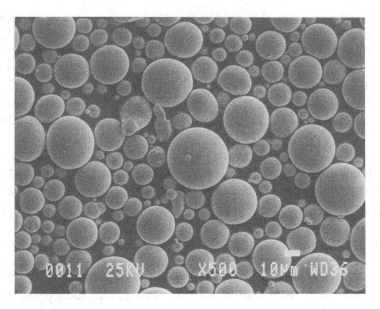

**Figure 4.31(b)**   SEM micrographs of spherical solder powders in:
(b) –325 mesh.

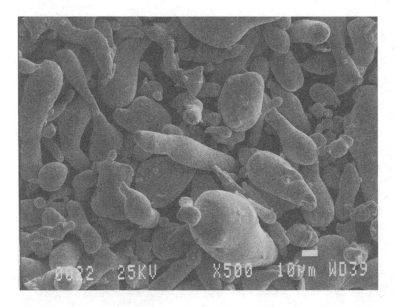

**Figure 4.31(c)**   SEM micrographs of spherical solder powders in:
(c) irregular –200 mesh.

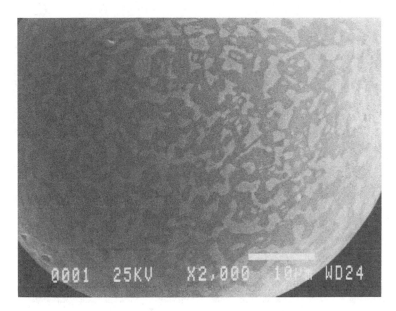

**Figure 4.32(a)**  SEM micrographs of particle surface of:
(a) 62-Sn/36-Pb/2-Ag.

**Figure 4.32(b)**  SEM micrographs of particle surface of:
(b) 96.5-Sn/3.5-Ag.

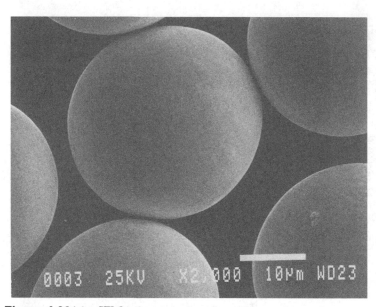

**Figure 4.32(c)** SEM micrographs of particle surface of:
(c) 43-Sn/43-Pb/14-Bi.

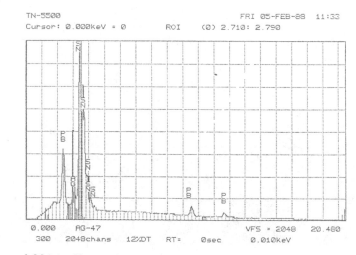

ZAF CORRECTION        25.00 KV        40.00 Degs

```
No. of Iterations  4
----   K      [Z]    [A]    [F]   [ZAF]  ATOM.%  WT.%
PB-M  0.194  1.070  1.090  0.999  1.166  12.60   20.11
AG-L  0.031  0.949  1.194  0.970  1.101   3.68    3.06
SN-L  0.774  0.986  1.131  1.000  1.116  83.72   76.82
 * - High Absorbance

SSQ:
```

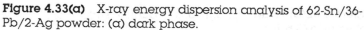

**Figure 4.33(a)** X-ray energy dispersion analysis of 62-Sn/36-Pb/2-Ag powder: (a) dark phase.

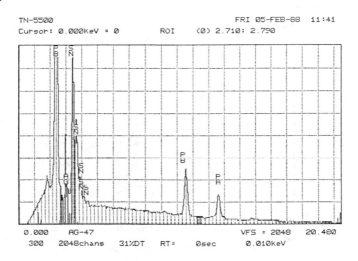

```
ZAF CORRECTION        25.00 KV        40.00 Degs

No. of Iterations  2
----      K      [Z]      [A]      [F]    [ZAF]   ATOM.%   WT.%
PB-M    0.661   1.035   1.042   0.999   1.079   48.14   61.85
AG-L    0.012   0.918   1.468   0.990   1.335    2.21    1.48  *
SN-L    0.325   0.953   1.365   1.000   1.302   49.64   36.67
 * - High Absorbance

SSQ:

     TN-5500                              FRI 05-FEB-88  11:41
     Cursor: 0.000keV = 0         ROI    (0) 2.710: 2.790
```

```
    0.000     AG-47                         VFS = 2048    20.480
    300    2048chans    31%DT    RT=     0sec     0.010keV
```

**Figure 4.33(b)** X-ray energy dispersion analysis of 62-Sn/36-Pb/2-Ag powder: (b) light phase.

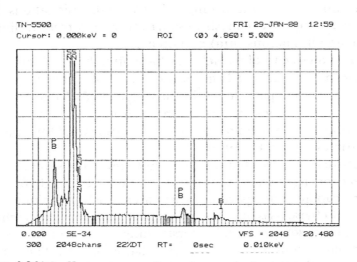

```
ZAF CORRECTION        25.00 KV        40.00 Degs

No. of Iterations  3
----      K      [Z]      [A]      [F]    [ZAF]   ATOM.%   WT.%
SN-L    0.648   0.975   1.196   0.999   1.167   77.79   66.76
PB-L    0.253   1.072   1.001   1.000   1.073   16.07   23.99
BI-L    0.097   1.071   0.998   1.000   1.070    6.14    9.25
 * - High Absorbance

SSQ:

     TN-5500                              FRI 29-JAN-88  12:59
     Cursor: 0.000keV = 0         ROI    (0) 4.860: 5.000
```

```
    0.000     SE-34                         VFS = 2048    20.480
    300    2048chans    22%DT    RT=     0sec     0.010keV
```

**Figure 4.34(a)** X-ray energy dispersion analysis of 43-Sn/43-Pb/14-Bi powder: (a) dark phase.

```
ZAF CORRECTION          25.00 KV          40.00 Degs

No. of Iterations  1
----      K       [Z]      [A]      [F]    [ZAF]   ATOM.%  WT.%
SN-L    0.021   0.923    1.569    0.998   1.448   5.26    3.09   *
PB-L    0.716   1.003    1.000    1.000   1.004   69.61   71.03
BI-L    0.261   1.002    0.998    1.000   1.000   25.12   25.88
  * - High Absorbance

SSQ:
```

```
     TN-5500                                    FRI  29-JAN-88   13:06
     Cursor: 0.000keV = 0          ROI     (0) 4.860: 5.000
```

```
     0.000     SE-34                           VFS = 2048    20.480
      300     2048chans      5%DT    RT=    0sec     0.010keV
```

**Figure 4.34(b)**  X-ray energy dispersion analysis of 43-Sn/43-Pb/14-Bi powder: (b) light phase.

## Particle Size and Size Distribution

As seen in *Test for Sieve Analysis of Granular Metal Powders, ASTM B-214* specifies the sieve analysis of granular metal powders. The weight fraction is classified according to ASTM mesh designations; Table 4.6 provides the ASTM mesh designation and corresponding size of opening in microns ($\mu$m) and inches.

Two mesh cuts of particle size have often been adopted in solder paste, namely -200/+325 mesh and -325 mesh. The -200/+325 mesh corresponds to particle size in the range of 45-75 microns, and -325 mesh corresponds to smaller than 45 microns. The particle size distribution is not limited to these two mesh cuts. Other particle size distributions can be specified.

Particle size distribution can be determined by several techniques based on light scattering, sedimentation, electrical zone sensing, permeability, and specific surface area. Each technique has its inherent characteristics. The result may vary from one technique to another to some extent. Understanding of the principle behind the techniques is important to the correlation of data. The comparison between the average particle size identified from a surface area

**TABLE 4.6**

### Selected Mesh Designation and Size of Opening

| Mesh Designations | | | Size of Opening | | |
|---|---|---|---|---|---|
| ASTM[1] | Tyler | England[2] | Microns | Inches | ISO[3] |
| 10 | 9 | 8 | 2000 | 0.0787 | 2 mm |
| 12 | 10 | 10 | 1680 | 0.06661 | 1.7 |
| 14 | 12 | 12 | 1410 | 0.0555 | 1.18 |
| 16 | 14 | 14 | 1190 | 0.0469 | 1 |
| 18 | 16 | 16 | 1000 | 0.0394 | 850 μm |
| 20 | 20 | 18 | 841 | 0.0331 | 710 |
| 25 | 24 | 22 | 707 | 0.0278 | 600 |
| 30 | 28 | 25 | 595 | 0.0234 | 500 |
| 35 | 32 | 30 | 500 | 0.0197 | 425 |
| 40 | 35 | 35 | 420 | 0.0165 | 355 |
| 45 | 42 | 44 | 354 | 0.0139 | 300 |
| 50 | 48 | 52 | 297 | 0.0117 | 250 |
| 60 | 60 | 60 | 250 | 0.0098 | 212 |
| 70 | 65 | 72 | 210 | 0.0083 | 180 |
| 80 | 80 | 85 | 177 | 0.0070 | 150 |
| 100 | 100 | 100 | 149 | 0.0059 | 125 |
| 120 | 115 | 120 | 125 | 0.00049 | 106 |
| 140 | 150 | 150 | 105 | 0.0041 | 90 |
| 170 | 170 | 170 | 88 | 0.0029 | 75 |
| 200 | 200 | 200 | 74 | 0.0029 | 75 |
| 230 | 250 | 240 | 63 | 0.0025 | 63 |
| 270 | 270 | 300 | 53 | 0.0021 | 53 |
| 325 | 325 | 350 | 44 | 0.0017 | 45 |
| 400 | 400 | 400 | 37 | 0.0015 | 38 |

[1]ASTM E11 (1970)
[2]British Standard Sieves BS 410 (1969)
[3]International Standard Organization (1970)

measurement with that found by sedimentation, light scattering, or electrical zone sensing provides the indication of the degree of particle agglomeration.

A typical particle size distribution of both mesh cuts are obtained by using Micromeritics SediGraph 5000 ET particle size analyzer.[27] The analyzer as shown in Figure 4.35 measures the sedimentation

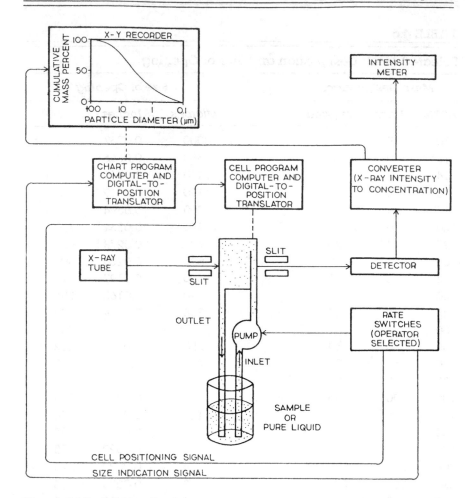

**Figure 4.35**  Schematic of particle size analyzer. *(Courtesy of Micromeritics Instrument Corporation.)*

rates of particles dispersed in a suitable liquid according to Stokes' Law, which shows that the viscous resistance of a sphere moving with uniform velocity is

$$F = 6\pi\, a\, \eta\, v$$

or

$$F = \frac{4\pi}{3}\, a^3\, (\sigma - \rho)\, g,$$

where $F$ is the force exerted on the sphere, $a$ is the radius of the sphere, $\eta$ is the viscosity of the liquid, $v$ is the velocity, $\sigma$ is the density of the sphere, $\rho$ is the density of the liquid, and $g$ the gravity acceleration constant.

The data are plotted as cumulative mass percent, cumulative number percent, and cumulative area percent versus equivalent spherical diameter. The concentration of particle is measured by low energy x-ray. Figure 4.36 (a) and (b) are the representative cumulative mass distribution for -200/+325 mesh cut and -325 mesh cut, respectively. Table 4.7 lists typical particle diameter analysis in cumulative mass percentage, for both mesh cuts.

## Oxide Content

Oxide content of solder powder has direct impact on solderability of solder paste, particularly on solder balling. To achieve an oxide-free or a minimal oxide formation during solder powder producing process and subsequent handling and storage has always been the objective.

To determine oxide content, a wet chemical method has been used to measure the weight difference between before-fusion and after-fusion of a specified weight of solder powder in a rosin solution.

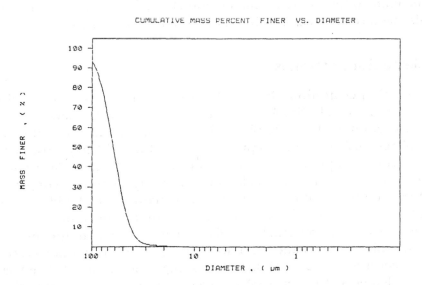

Figure 4.36(a)  Typical particle size distributions: (a) –200/+325 mesh.

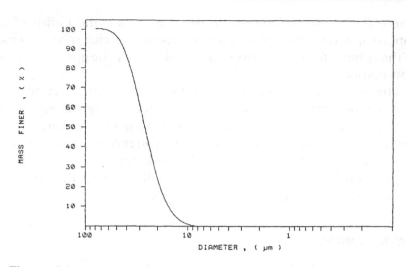

**Figure 4.36(b)** Typical particle size distributions: (b) –325 mesh solder powder.

Because of the nature of this method, the result obtained is usually an overestimation. However, with proper technique it reflects the cleanliness or purity of the powder, and is a viable indication of the oxide and other foreign contamination level. The oxide content can also be directly measured by an oxygen detection instrument.

## Elemental Impurities

Federal Specification QQ-S-571E specifies the limits of trace elements for Sn/Pb, Sn/Ag and Sn/Sb alloys. The elements under specification include Sb, Bi, Cu, Fe, Zn, Al, As, and Cd, and their limits are in Appendix I. Impurities of Zn, Al, and Cd may cause poor adhesion and joint grittiness; the presence of Fe and Cu beyond specified limits may also make the joint gritty. However, copper at 0.1% is found to improve the creep resistance of tin-lead solders. The addition of antimony up to 6% of tin content in tin-lead solder increases creep strength. Its small concentration retards tin pest (an allotropic transformation to amorphous gray tin). Phosphorous (P) and cerium (Ce) at 0.05-0.1% levels impart an antioxiding effect. The elements of arsenic (As) and bismuth (Bi) at the concentration larger than 0.3% cause dewetting.

**TABLE 4.7**

**Typical Particle Size Analysis of –200/+325 and –325 Mesh Solder Powder***

| Particle Diameter Interval (μm) | Cumulative Mass (finer %) | |
| --- | --- | --- |
| | –200/+325 Mesh | –325 Mesh |
| 100.00–95.00 | 91.1 | — |
| 95.00–90.00 | 88.2 | — |
| 90.00–85.00 | 83.9 | — |
| 85.00–80.00 | 79.2 | — |
| 80.00–75.00 | 72.2 | 100.3 |
| 75.00–70.00 | 63.9 | 100.1 |
| 70.00–65.00 | 54.6 | 99.9 |
| 65.00–60.00 | 44.0 | 99.3 |
| 60.00–55.00 | 33.0 | 98.2 |
| 55.00–50.00 | 21.8 | 96.0 |
| 50.00–45.00 | 13.3 | 92.4 |
| 45.00–40.00 | 6.7 | 85.5 |
| 40.00–35.00 | 2.8 | 75.7 |
| 35.00–30.00 | 1.1 | 61.0 |
| 30.00–25.00 | 0.5 | 42.7 |
| 25.00–20.00 | 0.1 | 23.6 |
| 20.00–15.00 | –0.2 | 8.4 |
| 15.00–10.00 | –0.1 | 0.7 |
| 10.00– 5.00 | — | –0.2 |

*Obtained with Micromeritics SediGraph 5000 ET

## Solder Balling Test

For a known flux/vehicle composition, solder balling is a complementary test of the performance characteristics of solder powder. The test may be conducted by applying a specific volume of solder paste containing the to-be-tested powder on a clean alumina substrate. The paste is then heated at normal soldering temperature. The molten solder alloy will coalesce into a clean sphere. Any discrete particles left outside the sphere are considered as potential cause of solder balling during the actual soldering. Under microscopic examination, a clean sphere without any debris or particles is an optimum. (Further discussion is in Section 11.28.)

## 4.10 PHYSICAL PROPERTIES OF SOLDER ALLOYS

The density, thermal conductivity, electrical conductivity, thermal expansion coefficient (TEC), hardness, surface tension of melt, and viscosity of melt are all important physical properties for solder joint. This section summarizes the available data on these physical properties of some common solder alloys as seen in Table 4.8.[28, 29 30]

In general, for tin-lead alloys, the surface tension decreases with increasing lead content, and viscosity increases with increasing lead content. It is expected that the hardness decreases with increasing lead content. The thermal and electrical conductivity decreases with increasing lead content, and the thermal expansion coefficient increases with increasing lead content. Surface tension of liquid metals is found to be inversely proportional to atomic volume,[31] yet how the surface tension of alloys relates to the volume remains to be determined.

The viscosity of metals in the molten state is one of the parameters controlling the flow property during soldering. Andrade has derived a formula relating the viscosity to the molecular weight of metals ($A$), and molecular volume at the melting point ($V$).[32, 33]

$$\eta_m = 5.1 \times 10^{-4} \frac{(AT_m)^{1/2}}{V^{2/3}}$$

$$\cong 5.1 \times 10^{-4} \left( \frac{T_m^{1/2} \rho^{2/3}}{A^{1/6}} \right),$$

where $\eta_m$ is the melting point viscosity, $T_m$ is the absolute temperature of melting, and $V = A/\rho$ ($\rho$ is the density). Therefore, by knowing molecular weight and density, the viscosity can be calculated. Figure 4.37 depicts the relationship between viscosity and composition of Sn/Pb alloys.[34]

## 4.11 MECHANICAL PROPERTIES OF SOLDER ALLOYS

For a given joint design, the mechanical properties vary with the alloy compositions, the soldering parameters, such as soldering tem-

**TABLE 4.8**

## Physical Properties of Common Solder Alloys

| Alloy Composition | Density (g/cm³) | Thermal Conduc- tivity (W/M·K)* | Electrical Conduc- tivity (IACS %)** | TEC*** X 10⁻⁶ (°C) | TEC*** X 10⁻⁶ (°F) | Hardness (HB)**** | Surface Tension (dyne/cm) |
|---|---|---|---|---|---|---|---|
| 100-Pb | 11.3 | 35.3 | 7.9 | 29.3 | 16.3 | 1.5 | 439 |
| 5-Sn/95-Pb | 10.8 | 35.2 | 8.1 | 28.4 | 15.8 | 8 | — |
| 10-Sn/90-Pb | 10.5 | 35.8 | 8.2 | 27.9 | 15.5 | 10 | 437 (500°C) |
| 20-Sn/80-Pb | 10.0 | 37.4 | 8.7 | 26.6 | 14.7 | 11 | 467 |
| 30-Sn/70-Pb | 9.7 | 40.5 | 9.3 | 25.6 | 14.7 | 12 | 420 |
| 40-Sn/60-Pb | 9.3 | 43.6 | 10.1 | 24.7 | 13.7 | 12 | 369 |
| 50-Sn/50-Pb | 8.9 | 47.8 | 10.9 | 23.6 | 13.1 | 14 | 476 |
| 60-Sn/40-Pb | 8.5 | 49.8 | 11.5 | 21.6 | 12.0 | 16 | 481 |
| 63-Sn/37-Pb | 8.3 | 50.9 | 11.8 | 21.4 | 11.8 | 17 | 490 |
| 100-Sn | 7.3 | 73.2 | 15.6 | 23.5 | 13.0 | 3.9 | 548 |
| 62-Sn/36-Pb/2-Ag | 8.4 | — | 11.5 | 23.6 | 13.1 | — | 376 |
| 1-Sn/97.5-Pb/1.5-Ag | 11.3 | 23.0 | 6.0 | 28.9 | 16.0 | 13 | — |
| 10-Sn/88-Pb/2-Ag | 10.4 | — | — | — | — | — | 272 |
| 42-Sn/58-Bi | 8.6 | — | 4.5 | — | — | 22 | 391 (800°C) |
| 96.5-Sn/3.5-Ag | 7.4 | 33 | 16.0 | 28.7 | 15.9 | 14.8 | 480 |

(continued)

115

Table 4.8 (continued)

| | * | **** | *** | | ** | | |
|---|---|---|---|---|---|---|---|
| 95-Sn/5-Sb | 7.3 | 28 | 11.9 | 29.6 | 16.4 | 15.0 | — |
| 48-Sn/52-In | 7.3 | 28 | 11.7 | 20.0 | 11.1 | — | — |
| 70-Sn/18-Pb/12-In | 7.9 | 45 | 12.2 | 24 | 13.3 | — | — |
| 30-Pb/70-In | 8.2 | 38 | 8.8 | 28 | 15.6 | — | — |
| 40-Pb/60-In | 8.5 | 29 | 7.0 | 27 | 15.0 | — | — |
| 60-Pb/40-In | 9.3 | — | — | — | — | — | — |
| 92.5-Pb/2.5-Ag/5-In | 11.0 | 25 | 5.5 | 25 | 13.9 | — | — |
| 90-Pb/5-Ag/5-In | 10.9 | 25 | 5.6 | 27 | 15.0 | — | — |

*     watt per meter °Kelvin
**     International Annealed Copper Standard %
***     thermal expansion coefficient
****     hardness of Brinell

116

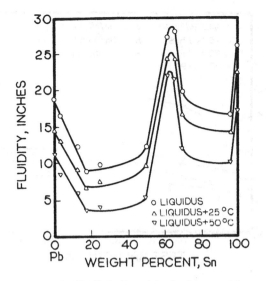

**Figure 4.37** Relationship between viscosity and composition of Sn/Pb alloys.[34]

perature, soldering time, the temperature of testing, and strain rate of testing.

Figure 4.38 is a plot of data of tensile strength measured at two temperatures, 20°C and 100°C, respectively for several bulk solder

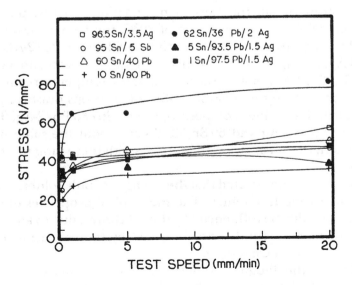

**Figure 4.38(a)** Tensile strengths of bulk solders tested at (a) 20°C.

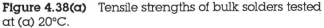

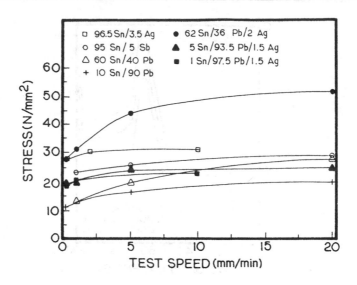

**Figure 4.38(b)**  Tensile strengths of bulk solders tested at (b) 100°C.

alloys.[35] As can be seen, the tensile strength generally increases with increasing test speed at both temperatures.  Among the alloys tested, the 96.5-Sn/3.5-Ag and 63-Sn/36-Pb/2-Ag solders exhibit the highest tensile strength, and the high lead-containing solder 10-Sn/90-Pb has the lowest, over the range of test rate (1-20 mm/min).

Shear strength data for copper ring and plug joints are plotted in Figure 4.39.[35] At the temperature of 20°C, the 58-Bi/42-Sn alloy is ranked the highest in shear strength with 62-Sn/36-Pb/2-AG, 96.5-Sn/3.5-Ag, 95-Sn/ 5-Sb, and 60-Sn/40-Pb ranked as second, and the 1-Sn/97.5-Pb/1.5-Ag solder the lowest. As test temperature increases to 100°C, the strength of 58-Bi/42-Sn solder drops drastically. The shear strength of solder compositions of 62-Sn/36-Pb/2-Ag, 95-Sn/5-Sb, 96.5-Sn/3.5-Ag and 60-Sn/40-Pb are about the same magnitude. Again, the shear strength increases with increasing test rate in general.

It has been demonstrated that the strength of bulk solders may not reflect the strength of joints.  The mechanical properties of solder joints may readily be influenced by the substrates, due to alloying or intermetallic compound formation, and by the soldering process that is discussed in Chapter 9.

Reviewing the fatigue strength data for various solder alloys indicates that the increasing magnitude of cyclic stress generally

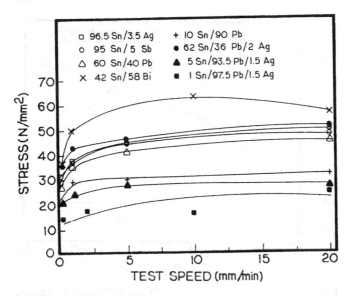

**Figure 4.39(a)** Shear strengths of copper ring and plug joint (a) at 20°C.

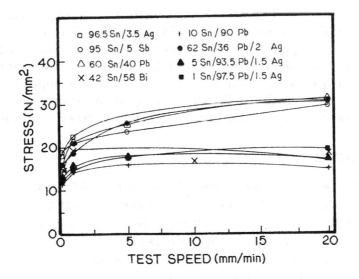

**Figure 4.39(b)** Shear strengths of copper ring and plug joint (b) at 100°C.

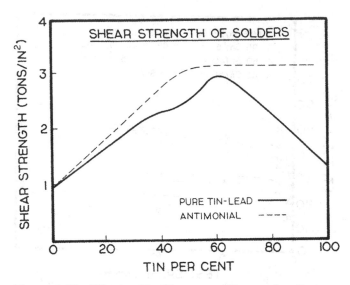

**Figure 4.40** Effects of Sn/Pb composition and antimony on shear strength.[35]

reduces the fatigue life, and 62-Sn/36-Pb/2-Ag and 96.5-Sn/3.5-Ag have somewhat better fatigue strength among the common solder alloys tested.[29]

It is reported that the creep strength may be improved by the addition of antimony up to 6% of tin content for tin-lead alloys, or by copper alloying at 0.1% concentration.[36] The antimonial hardening effect as shown in Figure 4.40 also indicates that the maximum shear strength occurs at the eutectic or near-eutectic composition. In general, 95-Sn/5-Sb and 96.5-Sn/3.5-Ag are more suitable for high temperature applications, due to their higher creep resistance.

## REFERENCES

1. Max Hansen, *Constitution of Binary Alloys* (New York: McGraw-Hill, 1958).

2. H. Udin, "In Metal Interfaces," *American Society for Metals* (1952): 114.

3. H. Udin, A. J. Shaler, and John Wulff, *J. Metals* 1-2 (1949): Trans. 186.

4. H. Udin et al., *J. Metals* 3 (1951): 1206, 1209.

5. D. V. Atterton and T. P. Hoar, *Nature* (1951): 167, 602.

6. F. Sauerwald, *2–Metallkunde* 35 (1943): 105.

7. G. L. J. Bailey and H. C. Watkins, "The Flow of Liquid Metals on Solid Metal Surfaces and Its Relation to Soldering, Brazing and Hot-Dip Coating," *Journal of the Institute of Metals* 80 (1951–52): 57.

8. C. S. Smith, Trans. *Am. Inst. Mech. Engrs.*, 175(1984): 15.

9. G. J. Davies, *Solidification and Casting* (London: Applied Science Publ., 1973).

10. Bruce Chalmers, Trans. AIME 200, 519 (1954).

11. F. Weinberg and B. Chalmers, *Canadian Journal of Physics* 29: 382.

12. F. Weinberg and B. Chalmers, *Canadian Journal of Physics* 30 (1952): 488.

13. J. W. Rutter and B. Chalmers, *Canadian Journal of Physics* 31 (1953): 15.

14. D. Walton, W. A. Tiller, J. W. Rutter, and W. C. Winegard, "Instability of a Smooth Solid-Liquid Interface During Solidification," *Journal of Metals*, Transactions AIME, 1023, (1955).

15. Merton C. Flemings, *Solidification Processing* (New York: McGraw-Hill, 1964).

16. Bruce Chalmer, *Principles of Solidification* (Krieger Publ., 1964).

17. Michael Romberg. Leybold Technologies, Inc., Enfield, Connecticut.

18. E. Peissker, "Production and Properties of Electrolytic Copper Powder," *International Journal of Powder Metallurgy and Powder Technology* 20-2 (1984): 87.

19. D. J. Hodkin, P. W. Sutcliffe, P. G. Hardon, and L. E. Russel, "Centrifugal Shot Casting: A New Atomization Process for the Preparation of High-Density Alloy Powders," *Powder Met.* (1973): 278–313.

20. S. Abbowitz, "A New Way to Make Titanium Alloys and Composites," *Metal Progress* 89 (1966): 62–61.

21. B. Champagne and R. Angers, "Size Distribution of Powders Atomized by the Rotating Electrode Process," *Modern Development in Powder Metallurgy* 12 (1980): 83–104.

22. B. Champagne and R. Angers, "Fabrication of Powder by the Rotating Electrode Process," *The International Journal of Powder Metallurgy and Powder Technology* 16-4 (1980): 319–367.

23. B. Champagne and R. Angers, "Rotating Electrode Process Atomization Mechanisms," *Powder Metallurgy International* 16-3 (1984): 125.

24. T. Takeda. Japan National Research Institute for Metals, Tokyo, Japan.

25. Ulf Backmark, Nils Backstrom, and Lars Arnberg, "Production of Metal Powder by Ultrasonic Gas Atomization," *Powder Metallurgy International* 18-5 (1986).

26. D. L. Erick. Battelle, Columbus Division, Columbus, Ohio.

27. Instrument Corporation, Norcross, Georgia.

28. "Solder Manual" (American Welding Society).

29. Publication 155 (International Tin Research Institute, 1947).

30. Publication 175 (International Tin Research Institute, 1967).

31. D. V. Atterton and T. P. Hoar, "Surface Tension of Liquid Metals," *Nature* 167 (1951): 602.

32. E. N. C. Andrade, "The Viscosity of Liquids," *Proc. Roy. Soc.* A-215 (1952): 356.

33. E. N. C. Andrade, "The Viscosity of Liquids," *Proc. Roy. Soc.* A-215 (1952): 36.

34. D. V. Ragone, C. M. Adams, and H. F. Taylor, Transactions of the American Foundrymen's Society, 64, 650 (1956).

35. Publication 656 (International Tin Research Institute, 1986).

36. Publication 93 (International Tin Research Institute, 1959).

# Rheology of Solder Pastes

Rheology is the science of deformation and flow of materials. The ultimate goal of rheology is to understand and anticipate the force needed to cause a given deformation or flow in a material under a set of specified conditions. It is broad in scope in its own field. This chapter is only intended to cover selected areas which are directly related to solder paste materials in terms of phenomena, characterization, practical measurements, and applications. Some fundamentals underlying the applications are briefly introduced. In addition, the single point viscosity measurement and its potential variations are discussed. Since this chapter discusses viscosity frequently, the units of viscosity used are either CGS-system (centimeter-gram-second) centipoises and poises, or SI (International System) Pascal-seconds (1 pa.s = 10 poises).

## 5.1  REQUIREMENTS AND DRIVING FORCES

Solder paste, as already described is composed of multiple ingredients of basically organic as well as metallic nature. The organic portion constitutes the matrix for metallic particles. The particles are homogeneously distributed in the matrix with or without the aid of dispersion agents. In conventional coating systems containing an organic carrier phase and dispersed oxide filler particles, the organic carrier phase is usually designed to have a yield value high enough to counteract gravitational force exerted on particles, in order to keep the solid particles suspended. There are similarities between coating products and solder paste in many aspects. Yet, in solder paste, different inherent factors are involved, making the solder

123

paste rheology more of a challenge. Factors such as the following are significant:

- drastic difference in density between organic matrix and metallic particles
- suspended particle of metallic nature
- relatively large particle size
- spherical particles
- high load of particles

The drastic difference in density between the organic matrix (0.7-1.2 g/cm³) and the metallic particles (7-12 g/cm³) facilitates the settling of particles, and the large size (up to 75 microns) of particles with small surface area reduces the interactions between the particle and matrix which further facilitates settling. The metallic nature of particles renders the mixing or dispersion technique more delicate. The high load of particles, commonly more than 40 volume percentage, contributes to the complexity of the rheological behavior.

The ideal rheology of solder paste, from an application point of view, is expected to deliver the following properties:

- specified shelf stability
- no separation or settling
- no dripping
- no stringiness, but with enough tackiness to hold components
- no slumping/sagging, but with capability of leveling
- perfect transfer through printing pattern
- clean release from the screen
- clean breakoff through fine needles
- nearly instant recovery after transfer
- temperature insensitivity

In practice, instant recovery after paste transfer and temperature insensitivity are difficult to come by. It is thus obvious that trade-off and balance are essential to the design of pastes having a rheology able to accommodate the application processing parameters. (The common paste application methods are covered in Chapter 6.)

The primary driving forces underlying the rheology of solder paste include both kinetic and thermodynamic contribution. In principle these cover thermal motion, with magnitude determined by the mean free path of molecules/particles, and by temperature, in conjunction with intermolecular/interparticular forces. The forces can be viewed as electrostatic repulsion and attraction between particles, and as the affinity between the chemical matrix and the particle surface. The particles will not aggregate when the affinity of particle surface for the chemical matrix exceeds the attractive forces between particles, or when steric effect through adsorption hinders the aggregation. Therefore, the rheology of solder paste may be affected by the following factors:

* composition, shape, and size of suspended particles
* chemical composition of suspending matrix
* relative concentration of effective ingredients in matrix
* structure of ingredients in matrix
* interactions between matrix and suspended particles either in physical or chemical nature, including wetting and solvation
* volume fraction occupied by the suspended particles—usually, the higher amount of particles, the more deviation from viscous flow
* internal structure and its response to external forces
* interactions among particles and resulting aggregates and flocculants
* temperature

The difficulty of predicting the rheology of such a system is apparent, due to the lack of knowledge of the structure and the nature of forces exerted by molecules/particles. However, behavior can be characterized. Section 5.4 devotes itself to the characterization of solder paste. It is also apparent that solder paste is not an elastic material, nor a pure viscous material. Viscoelasticity normally describes the behavior of solder paste. Sections 5.2 and 5.3 cover some basic theoretical background and introduce the relevant parameters for describing viscoelasticity.

## 5.2  FLOW BEHAVIOR: THEORETICAL BACKGROUND

When applying a force to a viscous material contained between two parallel plates as shown in the simple model in Figure 5.1, one plate

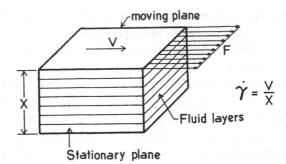

**Figure 5.1** Laminar flow of a viscous material.

is made to move at a constant velocity while the other plate is stationary. A laminar flow in the material results. The relationship between the force ($F$) per unit area (shear stress), and the change in velocity ($v$) along the distance ($x$) between two plates (shear rate) was first defined by Newton:

$$\tau = \eta \, \dot{\gamma},$$

where $\tau$ is the shear stress, $\dot{\gamma}$ is the shear rate, and $\eta$ is the apparent viscosity. The viscous flow described by this relationship is called Newtonian flow. Figure 5.2 indicates the flow curves for shear stress versus shear rate and viscosity versus shear rate, respectively.

For most materials, however, Newtonian behavior is not observed. Instead the behavior represented by a more general relation applies, where material flow behavior depends on both shear rate and time ($t$), as follows:

$$\eta \, (\dot{\gamma}, t) = \frac{\tau \, (t)}{\dot{\gamma} \, (t)}$$

Rheologically, materials which do not flow until the applied force exceeds a minimum value are called plastic. Plastic flow is well represented by the Bingham model as

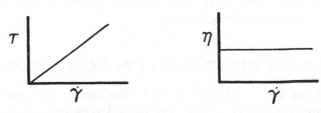

**Figure 5.2** Newtonian flow.

$$\tau = \tau_y + \eta_p \, \dot\gamma,$$

where $\tau_y$ is the yield value, and $\eta_p$ is the plastic viscosity.[1] The yield value is normally a measure of cohesive strength related to flocculation or internal structure of suspended particles. Plastic flow is illustrated in Figure 5.3. With change in shear rate, materials which behave as in Figure 5.4 are called pseudoplastic. In contrast, materials shown in Figure 5.5 are dilatant.

Many materials decrease in viscosity when subjected to increasing rate of shear, and as the shear is removed the viscosity builds up. The time required to build up the viscosity depends on the material, particularly on its internal structure. This time-dependent flow behavior, defined as thixotropy, is generally illustrated by Figure 5.6. The area within the ramping-up and ramping-down curves repre-

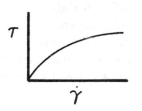 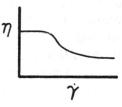

**Figure 5.3**  Plastic flow.

**Figure 5.4**  Pseudolastic flow.

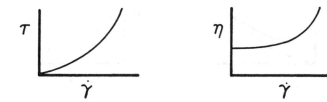

**Figure 5.5**  Dilatant flow.

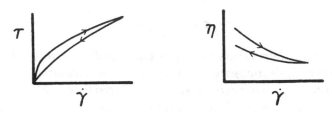

**Figure 5.6** Thixotropic flow.

senting the increasing shear rate and decreasing shear rate, provides
the qualitative information about the extent of thixotropy.

Another occurrence of flow phenomenon as material is subjected
to shearing is called rheopectic flow, as represented in Figure 5.7.

Although rheopectic flow is not well understood, it is postulated
that a perfect structural orientation is associated with this phenomenon, assuming *in situ* chemical reaction and evaporation are absent.

In addition to idealized Newtonian flow, non-Newtonian types are
exemplified by plastic, pseudoplastic, dilatant, thixotropic and rheopectic flows. As illustrated in Figures 5.3-5.7, the viscosity of pseudoplastic material decreases with increasing shear rate due primarily to
the alignment of molecules/particles along with direction of shearing. The dilatant material displays increased viscosity as shear rate
increases. It is believed that the flow resistance is attributed to the
formation of close-packed structure. It often occurs in the systems
containing very small solid particles being suspended in a highly
defluocculated manner.[2] The plasticity of materials is characterized
by the presence of a yield value which is the threshold shear stress in
order to overcome the forces of internal structure to start flow. In
addition to shear thinning effect, the thixotropic material exhibiting
a time-dependent reestablishment in viscosity results in a hysteresis
loop in the ramping up and down cycle.

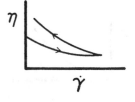

**Figure 5.7** Rheopectic flow.

## 5.3 ELASTIC BEHAVIOR: THEORETICAL BACKGROUND

Parallel to Newton's Law for viscous fluids is Hooke's Law for solids. Hooke's Law relates shear stress ($\tau$) and shear strain ($\gamma$) in the following equation:

$$\tau = G\gamma,$$

where $G$ is the shear modulus.

Under stress, the deformation of most materials differs substantially from the idealized case described by Hooke's Law. The theoretical and mathematical treatment of viscoelastic properties are thoroughly covered by J. D. Ferry.[3] Because of the time factor involved in the viscoelastic phenomenon, dynamic mechanical measurements are needed for characterizing the viscoelasticity. The relevant parameters which are commonly used for characterization are defined as follows. Under dynamic measurements, the stress is applied sinusoidally at a given frequency ($\omega$), as represented by Figure 5.8.

The strain also alternates sinusoidally but is out of phase with the stress at phase angle ($\delta$). Therefore, the mathematical equations for stress and strain, respectively, are

$$\tau = \tau_0 \sin \omega t$$
$$\gamma = \gamma_0 \sin (\omega t + \delta).$$

The dynamic shear modulus may be written as a complex modulus, $G^*$, with real and imaginary parts, as shown in the vectorial components

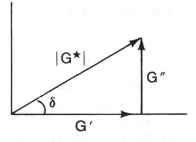

$$G^* = G' + iG''$$

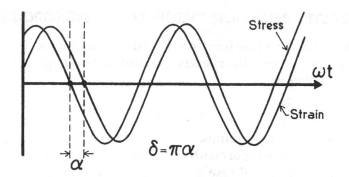

**Figure 5.8** Sinusoidal stress and strain.

The imaginary part of the complex modulus is the damping term which determines the dissipation of energy when material is deformed, and is called loss modulus $G''$. The real part of the complex modulus represents the amount of recoverable energy stored as elastic energy, and is called storage modulus ($G$). The ratio of loss modulus to storage modulus is called loss tangent, (tan δ).

$$\tan \delta = \frac{G'}{G}$$

The preceding equation is a convenient parameter in practical measurement, which indicates the relative portion of viscous and elastic properties of a material.

The complex moduli are related to complex viscosity by the following equations:

$$G'' = \omega \eta'$$
$$G' = \omega \eta''$$

From these equations, the loss tangent can be expressed by

$$\tan \delta = \frac{\eta'}{\eta''}.$$

The preceding mathematical equations signify that under dynamic cyclic deformation, the two components $G'$ and $G''$ represent the energy portion stored or recovered elastically and the energy lost or dissipated within the material, which in turn represent elastic and viscous components of a material.

Stress relaxation and creep are another two phenomena observed in viscoelastic materials.[2,3] When measuring stress relaxation, a constant strain is applied to the material and the stress is measured as a function of time. The decay of stress follows the equation

$$\tau = G\gamma_0 \, e^{-t/\lambda},$$

where $\lambda$ is the stress relaxation time and $\gamma_0$ is the constant strain. In many actual situations $G$ and $\lambda$ are not constants; the simple exponential relation as just expressed may not be sufficient to describe the phenomenon. In measuring creep, a constant stress is applied and the time dependent deformation is followed.

## 5.4  MECHANICAL MODELS

The viscoelastic behavior of materials has been described by using a combination of the linear viscous element, dashpot, and the linear elastic element, spring. J. C. Maxwell proposes to have a simple series combination of an elastic spring with shear modulus G and a dashpot containing liquid with viscosity $\eta$, as shown in Figure 5.9.

In the Maxwell model, both spring and dashpot are subjected to the same stress, and the overall strain is the sum of the strain in the spring and the dashpot:

$$\gamma = \gamma_{\text{spring}} + \gamma_{\text{dashpot}}$$

Upon sudden application of a stress, the spring experiences an instantaneous strain, followed by a linear viscous creep. As time proceeds, the spring retracts and the dashpot extends to maintain a

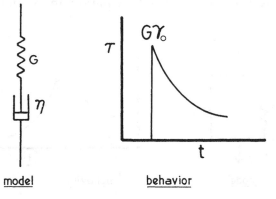

Figure 5.9  Maxwell mechanical model and behavior.

constant strain of the whole system. As more of the spring retracts, the rate of retraction declines. This model is able to describe stress relaxation phenomenon, as shown in the stress versus time curve of Figure 5.9.

The Voigt model consists of an elastic spring in parallel with a dashpot as shown in Figure 5.10. The strain in the spring and dashpot is the same and the total stress supported is the sum of the stresses in each element,

$$\tau = \tau_{spring} + \tau_{dashpot}$$

Upon application of stress, the dashpot does not yield instantaneously, and it supports the stress entirely. As the system is extended, the dashpot flows and creep takes place. The spring then sustains the stress, and the dashpot reduces flow as illustrated in Figure 5.10. The strain follows:[4]

$$\gamma = \frac{\tau_0}{G}(1 - e^{-t/\lambda})$$

After the stress is removed, strain decays exponentially as follows:

$$\gamma = \frac{\tau_0}{G} e^{-t/\lambda}$$

Based on this, the Voigt model describes creep response but lacks the instantaneous elastic response, while the Maxwell model accounts for stress relaxation but does not display the recoverable creep characteristics. Three- and four-parameter models are derived

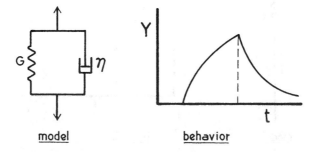

model                          behavior

**Figure 5.10**  Voigt mechanical model and behavior.

in order to provide a representation of all phenomena observed with viscoelastic materials. The interested reader is referred to the Reference 5 for more in-depth treatment.

## 5.5  FLOW CHARACTERISTICS

With the basic viscoelastic parameters in mind, the characterization of a solder paste can be based on its flow behavior and elastic properties. This section discusses the flow behavior; Section 5.6 covers the elastic behavior.

Responses of paste to shear stress, shear rate, and time are the areas to be studied to understand flow. The basic relationships to be measured include shear stress versus shear rate; yield point; viscosity versus shear rate; viscosity versus time; and shear stress versus time. There are different techniques available to measure the viscosity or flow behavior—namely, capillary, cup, falling ball, and rotational viscometry. Rotational viscometry is the most prevalent technique for measurement of solder paste rheological behavior. The sensors adopted are cone-and-plate or plate-and-plate design as shown in Figure 5.11. The measurements of the torque required to drive the cone or plate, together with the corresponding rotational speed, generate the shear stress (viscosity)–shear rate relationship of the paste. The shear stress as transformed from measured torque, is related to the cone-and-plate operation by

$$\text{shear stress} \propto \frac{\text{torque}}{R^3} \, ,$$

where $R$ is the radius of cone.

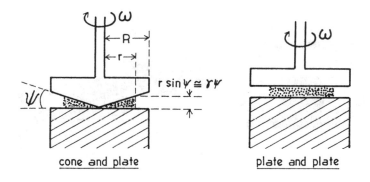

cone and plate                          plate and plate

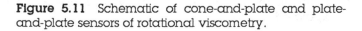

**Figure 5.11** Schematic of cone-and-plate and plate-and-plate sensors of rotational viscometry.

The viscosity is related to the radius of the cone, the cone-plate angle ($\Psi$), and angular velocity ($\omega$), as follows:

$$\text{viscosity} \propto \frac{\text{torque} \cdot \Psi}{R^3 \omega}$$

Viscometry equipment is shown in Figure 5.12.

### Shear Stress Versus Shear Rate

As examples, Figure 5.13 shows the flow profiles of Paste A and Paste B. Paste A is slightly shear-rate dependent. The shear rate ramping up and down curves indicate a small amount of thixotropy as represented by the area between the two curves. Paste B displays a relatively large yield point and moderate thixotropy.

### Yield Point

There is more than one way to determine the yield point. For convenience, the yield point may be obtained by the extrapolation of

**Figure 5.12** Rotational viscometer. (*Courtesy of Haake Buehler Instruments, Incorporated.*)

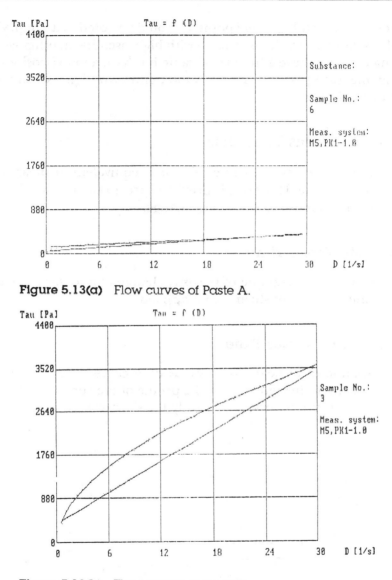

**Figure 5.13(a)**   Flow curves of Paste A.

**Figure 5.13(b)**   Flow curves of Paste B.

the flow curve to intercept the shear stress axis, or is taken as the
shoulder point (inflection) of the curve, if present. These techniques
are appropriate only for comparison purposes. When using these
techniques to derive yield point, it is more reliable to measure flow
at low shear rate conditions (<50 sec $^{-1}$).

Paste A as shown in Figure 5.13 has a low yield point, and Paste B shows a relatively high yield point. It should be noted that a high yield value is not necessarily associated with high viscosity. In other words, a material may have a high yield value but have a low viscosity—for example, catsup—and have low yield point with high viscosity—for example, honey.

## Viscosity Versus Shear Rate

The viscosity versus shear rate relation is equivalent to shear stress versus shear rate. However, it provides direct readings on viscosity change in response to shear rate change.

## Viscosity Versus Time

Figure 5.14 is an example of viscosity change of a solder paste with time after a constant shear rate is applied.

## Shear Stress Versus Time

An example of a shear stress versus time curve (stress relaxation curve) is shown in Figure 5.15. The profile of the curve and the area under the curve reflect qualitatively the characteristics of a solder

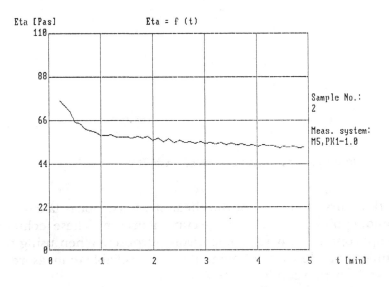

**Figure 5.14**   Viscosity versus time of a solder paste.

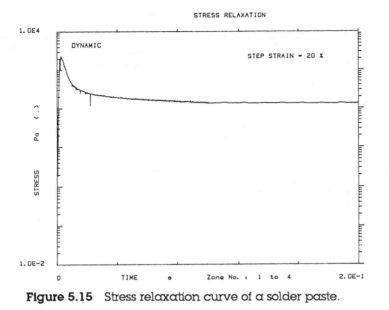

**Figure 5.15** Stress relaxation curve of a solder paste.

paste such as yield value, internal structure changes, and leveling viscosity.

## 5.6  ELASTIC CHARACTERISTICS

At a constant temperature the dynamic mechanical measurements that characterize the elastic properties include three parameters: storage modulus, loss modulus, and complex viscosity in response to operating parameters of frequency, strain amplitude, and time. (The definitions of the three properties are given in Section 5.3.)

### Storage Modulus, Loss Modulus, Complex Viscosity Versus Frequency

Figure 5.16 exemplifies the three subject properties of a solder paste under the influence of frequency sweep. The characteristics may be defined by analyzing the following features:

- magnitude of the three properties
- relative magnitude of loss and storage moduli

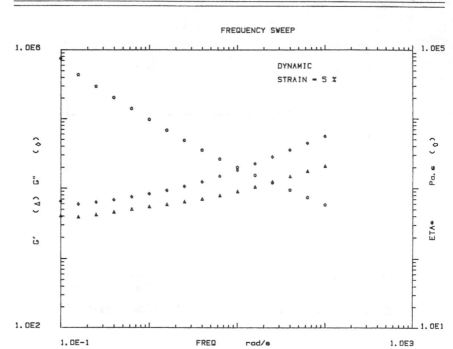

**Figure 5.16** Measurements of storage modulus, loss modulus, and complex viscosity of a solder paste in relation to frequency.

- relative changes of loss and storage moduli in relation to frequency change or strain change, or time
- relative rate of change in complex viscosity versus frequency or strain or time

## Storage Modulus, Loss Modulus, Complex Viscosity Versus Strain

This material may be characterized by the same examination as in the frequency sweep test.

## Storage Modulus, Loss Modulus, Complex Viscosity Versus Time

The time sweep illustrates the change of the three properties with time under constant frequency and strain. This test is particularly useful to examine the potential or expected reactions in the material.

## Loss Tangent Value

Equipment is available to directly measure loss tangent over a range of frequency. This is a direct way to provide relative weight of viscous and elastic portions in the material. Figure 5.17 is a plot of loss tangent versus frequency.

## 5.7  CHARACTERIZATION VERSUS PERFORMANCE

Combining flow measurement and elastic measurement, a solder paste can be adequately characterized. The characterization of the paste has to be correlated with the processing parameters as well as with the paste performance under a specific set of conditions. In view of the complexity of solder paste rheology and numerous processing variables, the following provides some general remarks. (Detailed information for a specific system has to be derived under a specified set of conditions.)

Under two main paste application methods—fine dot dispensing and pattern printing, which are discussed in Chapter 6—a paste possessing low yield point and slight plastic behavior is found most

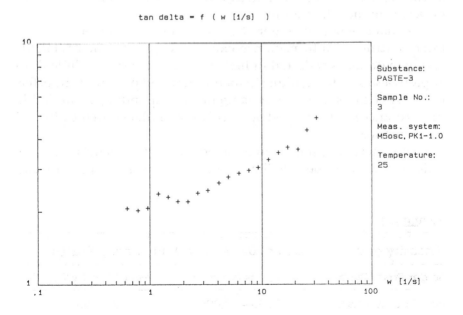

**Figure 5.17**  Loss tangent of a solder paste in relation to frequency.

suitable for dispensing applications. As shown in Figure 5.13, Paste
A, demonstrating good dispensing properties is relatively less sensitive to shear rate, and the recovery as indicated in the ramping down
curve is nearly instantaneous. This is consistent with the formulation
requirement that a dispensing paste normally has a lower upper limit
in volume fraction of metal load as well as a lower viscosity range, in
comparison with pastes designed for stencil/screen printing applications. Table 5.1 lists typical viscosity and metal load percentages for
the dispensing and printing application techniques, although some
exceptions may exist.

Low yield value physically signifies that paste will start to flow
under low shear stress. Too high a yield point may result in poor
leveling which leaves unsmooth surface, and too low a yield point
may cause running and slumping.

Printing paste normally demands a higher viscosity range. A
moderate yield and thixotropy are found to be suitable for printing
mechanics as demonstrated by Paste B of Figure 5.13. Too high a
degree of thixotropy may cause slumping and sagging. However,
shear thinning is desirable, especially at high shear rate, as encountered while a paste is being transferred through the opening during
a printing process. Dilatant behavior observed at high shear rate will,
however, hinder the transfer.

The data obtained under dynamic mode testing provide the
internal structure information as expressed by $G'$, $G''$ and $\eta^*$. Highly
elastic material as reflected in high $G'$ value is normally difficult to
apply to a substrate, resulting in incomplete and poor paste transfer-
ral. A specific range of loss tangent corresponding to desirable
performance can be derived by correlating the data with the physical
performance.

In the literature, some empirical equations for viscoelastic materials are available, but are limited to specific conditions. Unfortu-

**TABLE 5.1**

### Viscosity and Metal Load of Dispensing and Printing Pastes

| Solder Paste Type | Viscosity cps* | Metal Load Wt % |
| --- | --- | --- |
| Fine Dot Dispensing | 200,000–450,000 | up to 88 |
| Screen/Stencil | 450,000–1,200,000 | up to 92 |

*Centipoises

nately, there is no simple equation that can be used to predict the viscoelastic behavior of solder paste. The requirements, therefore, are to select appropriate testing conditions—particularly shear rate, strain amplitude, and frequency—that are representative of the actual applications, and to correlate the data obtained under such conditions with the actual performance. As is well established in paint and coating applications, the curves of viscosity recovery at a constant shear rate, as shown in Figure 5.18, correlate well with sag and leveling, which are two of critical performance parameters of solder pastes.[6] In this test, the viscosity recovery rate is plotted immediately after the application of high shear rate. With the establishment of data and performance correlation, the product demonstrating good performance can be readily identified by the curve obtained.

The relationship between the application parameters as expressed in shear rate and the viscosity requirement in coating application is illustrated in Figure 5.19.

Some studies have been carried out on rheology of thick film paste in relation to screen printing.[7, 8, 9, 10, 11, 12] A hypothetical relationship of viscosity as a function of time during screen printing and corresponding shear rates of major printing steps (stirring, squeegeeing, screening, and leveling) is illustrated in Figure 5.20.

These characterization measurements will serve as not only a research and development means, but also as a quality control tool from both users' and suppliers' viewpoints. Data in this area are sparse, so efforts in obtaining the data and information are warranted.

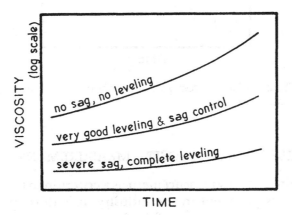

**Figure 5.18**  Viscosity recovery rate of three coating products.[6]

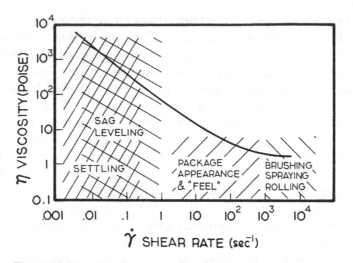

**Figure 5.19** Application performance of coating products in relation to viscosity and shear rate.[6]

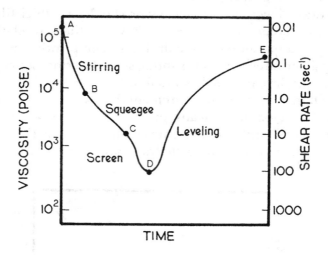

**Figure 5.20** Viscosity as a function of time during screen printing.[11]

## 5.8 SINGLE POINT VISCOSITY MEASUREMENT

A single point viscosity measurement as a criterion to the rheology of solder paste is prevalent in the industry. It is thus appropriate to discuss some practical aspects of single point viscosity measurement. One of the common techniques adopted is to use the Brookfield viscometer with a helipath spindle TF at 5 rpm in accordance with QQ-S-571E (Appendix I).

Measurement obtained by this technique is simple and convenient. However, single point viscosity is valid only when the following conditions are met:

1.  The rheology of the paste is known.
2.  The paste handling technique immediately prior to measurement is standardized.
3.  The measurement technique of the paste is standardized.

To illustrate the point, Table 5.2 is a summary of viscosity data measured with Brookfield Model RTV viscometer with helipath spindle TF at 5 rpm for Paste A and Paste B. The variables in this table are paste handling prior to descending spindle as expressed by the mixing time, and the measuring time—that is, the time interval from the start of the spindle descending to obtaining dial reading. This time interval also indicates the spindle position in the paste container.

**TABLE 5.2**

**Single Point Viscosity of Pastes A and B in Relation to Pre-mixing Time and Time-to-Read**

| | Viscosity (cps x $10^3$) | | | | | | |
|---|---|---|---|---|---|---|---|
| **Mixing time** | | | **Time-to-Read** | | | | |
| **(min)** | **30 sec** | **45 sec** | **60 sec** | **90 sec** | **120 sec** | **100 sec** | **180 sec** |
| **Paste B** | | | | | | | |
| 0 | — | — | 1,440 | 1,550 | 1,700 | — | — |
| 1 | — | — | 840 | 950 | 1,100 | — | — |
| 2 | — | 600 | 660 | 720 | 800 | 900 | — |
| 3 | — | 580 | 640 | 700 | 800 | 880 | 900 |
| **Paste A** | | | | | | | |
| 0 | 200 | 220 | 220 | 280 | 400 | 430 | — |
| 1 | 190 | 210 | 220 | 240 | 270 | 290 | — |
| 2 | 190 | 200 | 210 | 220 | 230 | 250 | — |
| 3 | 170 | 180 | 200 | 200 | 220 | 230 | — |

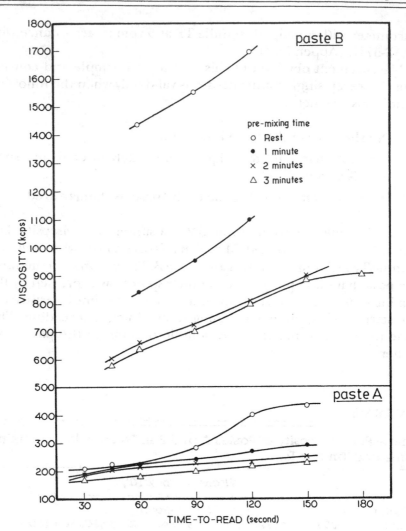

**Figure 5.21** Viscosity versus time-to-read at four different pre-mixing times for Paste A and Paste B.

Figure 5.21 plots the viscosity versus time-to-read dial for four different pre-mixing times of Paste B: no mixing, one-minute mixing, two-minute mixing, and three-minute mixing. The mixing technique for all measurements is kept the same with 1¹/4" diameter, three-blade propeller at a speed of 240 rpm. All measurements are in the same size container—4 fl oz jar with 2" diameter and 4" height. The data clearly indicate the effects of pre-mixing. Longer mixing generates lower viscosity numbers across the full range of reading time from half a minute to three minutes. Under the test conditions, after the paste is mixed thoroughly for two minutes, further mixing does not significantly affect viscosity. The data also indicate that the viscosity reading increases with increasing time-to-read, due to increased friction as the spindle is descending. An equivalent test is carried out for Paste A. The comparison between Paste A and Paste B indicates that Paste A has much less variation in response to mixing time and the time-to-read. Again, as paste is mixed for two minutes, further mixing has no significant effect and the viscosity variation with time-to-read becomes minimal. The comparative results demonstrate the relative sensitivity of pastes to handling (mixing) and measuring techniques. With these variations in mind, it is not surprising to see the inconsistency and lack of agreement from laboratory to laboratory and from operator to operator. Nonetheless, with the proper technique and control, viscosity data can be consistently obtained. Figure 5.21 consists of the good reproducibility of measurements conducted under the specified technique.

In practice, realizing such variations as well as being able to control them is important to the establishment of specification and communication between users and vendors, which in turn is important to overall quality and production yield. Single point viscosity can also be utilized to develop an indicator for shear thinning factor by measuring viscosities at two significantly different shear rates such as viscosities at 1 rpm and 20 rpm. The ratio of these two viscosities is the shear thinning factor. Again, the paste handling has to be kept consistent in order to obtain meaningful results.

## 5.9  VISCOSITY VERSUS TEMPERATURE AND OTHER FACTORS

Temperature is a well recognized factor which affects viscosity. Figure 5.22 exhibits the viscosity change with temperature for Paste A and Paste B over a temperature range of 15°C (59°F) to 45°C (113°F). It is apparent that the viscosity of paste can change significantly within this tested temperature range. The ambient temperature during paste application therefore affects paste performance. As temperature increases, pastes not possessing the designed flow control as discussed in Section 3.13, tend to slump. In order to evaluate this characteristic, the cold slump and hot slump tests are

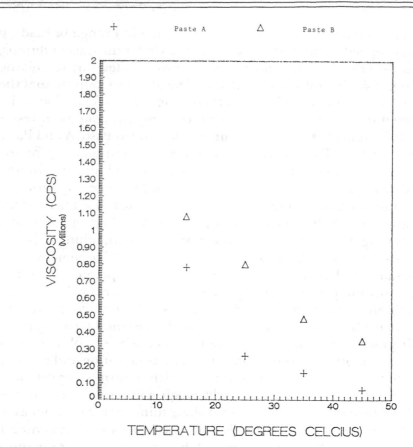

**Figure 5.22**  Viscosity of a solder paste in relation to temperature.

outlined in Sections 11.7 and 11.8, respectively. The cold slump is tested at ambient temperature against gravitational force, and the hot slump is tested against imposed heat.

Metal load and physical characteristics of metal powder affect the resulting viscosity and rheology of the paste. Figure 5.23 illustrates that paste viscosity varies with powder particle size, size distribution, and shape for flux/vehicle system A; Figure 5.24 illustrates for vehicle/flux system B over a range of metal load.

Viscosity may change with aging or storage if a chemical reaction between metal particles and organic matrix, and/or between the chemical ingredients is present, or if the state of flocculation changes as a result of physical adsorption or other interaction. Other contributors to viscosity change include the settling of particles and

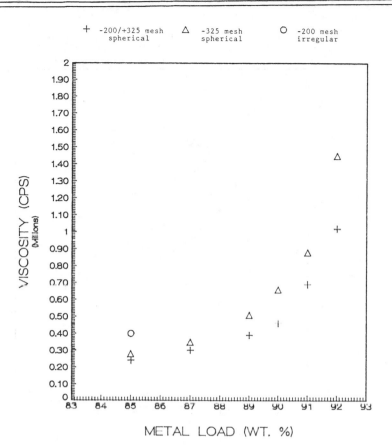

**Figure 5.23** Dependency of viscosity on physical characteristics of metal powder for flux-vehicle system A.

evaporation of volatiles in the paste. The paste containing high
volatiles is obviously prone to evaporation.

The entrapment of air or micro-air bubbles in the paste is quite
common during processing or handling. Viscosity is directly related
to the amount and the distribution of air bubbles. Normally, viscosity
increases with increasing air entrapment.

Question has risen about whether the viscosity of solder paste is
additive or subtractive for on-site adjustments when needed. A
majority of flux/vehicle systems in solder paste consists of mul-
tiphases with metal particles suspended in them, and the solder paste
is made kinetically stable rather than thermodynamically stable, in
contrast to a true "solution" or other multicomponent systems such
as microemulsion. It should not be assumed that viscosity is always

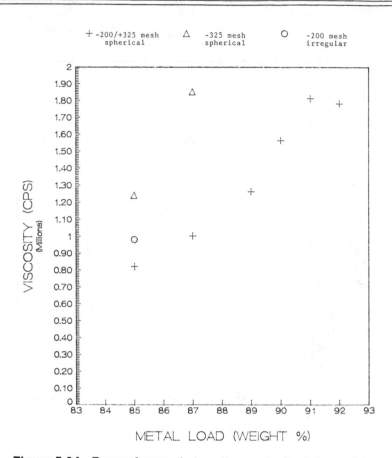

**Figure 5.24** Dependency of viscosity on physical character-istics of metal powder for flux/vehicle system B.

additive or subtractive among different pastes. Generally, a mixing technique of two pastes with low and high viscosities in order to achieve a medium viscosity is not recommended. In case of a practical need, the combined viscosity needs to be verified by actual tests to assure the compatibility and stability in practical conditions so that other performance characteristics are not jeopardized.

With respect to paste manufacturing, adjustments on a finished batch of paste in an attempt to reach a given viscosity should also be avoided to eliminate adverse potential. *Make it right the first time* is what to strive for. In addition, the processing parameters are vital to the consistency of a paste. Considering the viscoelastic nature result-ing from complex interactions present in the paste as discussed in the previous sections, it is not surprising to see the same paste recipe could produce different paste products under different processing parameters. Due to its proprietary nature, this is not further elaborated.

## REFERENCES:

1.  E. C. Bingham, *Fluidity and Plasticity* ( New York: McGraw-Hill, 1922).

2.  M. Reiner, *Deformation and Flow* (London: Lewis, 1949).

3.  John D. Ferry, *Viscoelastic Properties of Polymers* (New York: Wiley, 1970).

4.  J. C. Maxwell, Philosophical Transactions of the Royal Society, *157*, 49 (1867).

5.  Stephen L. Rosen, *Fundamental Principles of Polymer Materials* (New York: Wiley, 1982): 243.

6.  Robert E. Van Doren and Alfred J. Whitlon, "Rheological Control of Solvent-Based Coatings," *Modern Paint and Coatings* (Jun. 1981): 54.

7.  Dr. William A. Vitriol and Philip M. Hodge, "Sophisticated Techniques Solve Ink Manufacturing Problems," *Solid State Technology* (Mar. 1976).

8.  King F. Hsu, "The Measurement of High Speed Printing Thick Film Pastes (Inks) With a Mechanical Spectrometer," proceedings, 1981 International Hybrid Microelectronics Symposium: 8.

9.  D. A. Ross, T. T. Hitch, and D. C. McCarthy, "The Fundamentals of Rheology Measurement of Fluids Using the Ferranti-Shirley Viscometer," proceedings, 1981 International Hybrid Microelectronics Symposium: 454.

10.  Van S. Kandashian and S. J. R. Vellanki, "A Method for Characterizations of the Rheology of Thick Film Pastes," proceedings, 1978 International Hybrid Microelectronics Symposium: 44.

11.  R. E. Trease and R. L. Dietz, "Rheology of Pastes in Thick Film Printing," *Solid State Technology* (Jan. 1972).

12.  King F. Hsu, "The Viscoelastic Properties of Thick Film Pastes (Inks)," proceedings, 1985 International Hybrid Microelectronics Symposium: 67.

PART III

# Methodologies and Applications

# CHAPTER 6

# Application Techniques

## 6.1 INTRODUCTION

This chapter deals with the methodologies for consistently and accurately transferring solder paste onto the intended solder pads. Solder paste is endowed with characteristics that can be readily adapted to automation. The types of application methods include mesh screen printing, metal mask stencil printing, pneumatic dot and line dispensing, positive displacement dispensing, and pin transfer.

For printing methods, whether through mesh screen or etched metal mask, there are common traits between printing solder paste and printing other thick film pastes as used in making hybrid circuits. A lot of work has been performed on screen printing for creating thick film circuitry. The mesh screen printing of solder paste utilizes the same principle. However, major differences between solder paste and thick film pastes with respect to printing are present. They include the relatively coarser size of particles used in solder paste, the larger size of substrate and therefore large screen, and the geometry of printing patterns. Thick film pastes contain particle sizes from several microns to submicrons, and solder pastes broadly contain particles in the range of 75 microns to a few microns. The size of the substrate differs significantly, comparing hybrid board normally up to 6" x 6" (15.3 cm x 15.3 cm) with surface mount printed wiring board with size up to 24" x 24" (61.2 cm x 61.2 cm). The patterns of conductor traces, resistors, and also dielectrics are different in geometry and resolution from solder pads for chip carriers, transistors, diodes, passive chips, lead frame attachments, and other connections. With the advent of fine-pitch packages, the dimensions of solder pads continue to shrink, and the requirements of solder paste and its placement on solder pads change accordingly.

In addition to conventional mesh screen printing, metal mask stencil printing has been developed. Its prevalence increases due to its compatibility with solder paste as well as some performance and maintenance advantages, which are discussed in Section 6.3. For convenience, the metal mask stencil is also grouped as one of the screen types in the subsequent sections.

Although printing technique has been proven to be a viable method to economically produce accurate and reproducible transfer of paste onto the designated patterns, the technique essentially adapts to more or less flat surface. For the assemblies requiring solder interconnections on the irregular surface levels or in hard-to-reach areas, paste dispensing through needles becomes a necessary method.

## 6.2 PRINTER

The principal functional structure of a printer consists of a screen frame assembly, squeegee assembly, squeegee pressure control, screen registration system, substrate alignment system, and a substrate-holding system. The screen frame assembly of most commercial models has certain level of flexibility to accommodate different dimensions of screen. The screen registration system, normally utilizing three-point positioning, allows convenient setup. The squeegee assembly includes squeegee drive and squeegee holder for squeegee blade and flood bar. The squeegee can be driven by pneumatic, electromechanic, or hydraulic power. The squeegee drive capable of providing uniform and consistent speed is important to the printing process.

The squeegee pressure control is constructed with squeegee height control, which is called downstop and pressure regulation. Adjustments for downstop and pressure are separately monitored. The squeegee pressure control mechanically consists of coil spring, squeegee rod, and piston. The coil spring is designed to accommodate the cambers and variation encountered on the substrate surface. The squeegee pressure is generated by applying air pressure on the lower part to hold piston up, compressing the coil spring and preventing the movement of squeegee rod. In printing mode, air pressure enters the upper port and moves the piston downward and frees the squeegee rod, which allows the spring to force the squeegee rod down.

The substrate holding is usually accomplished with pressure differential above and beneath the substrate by means of vacuuming. The alignment of substrate to screen is carried out by adjusting micrometers in $x$, $y$ and $\theta$ (rotary adjustment) coordinates. Some new models of screen printer can be easily adapted to vision system for automatic screen-substrate alignment. The vision system provides accuracy repeatability and fast changeover for printer setup. In addition to setup, the vision system can also function as a post-inspection tool.

The microprocessor controlled system runs a calibration program to position $x$, $y$, and $\theta$ coordinates. The system then compares the difference between the known position and reference marks, and repositions the screen to align to the known position. Among the vision systems supplied by different manufacturers, the alignment accuracy falls in the range of $\pm 0.2$ mil ($\pm 5$ $\mu$m) to $\pm 1$ mil ($\pm 25$ $\mu$m). Varying with the model and manufacturer, the vision system is capable of delivering repeatability within $\pm 1$ mil (25 $\mu$m). The capability of a printer, whether it is manual, semi-automatic or automatic, to deliver accuracy and repeatability is vital to the yield and quality of the process.

For high volume operation, automatic load system in stack-type and partitioned-type magazines, and off-load conveyor are options for full automation. Figure 6.1 is an automatic printer system comprised of board loader, machine vision guided screen alignment, and board conveyor to a pick and place unit for component population.[1] The large board stack loader provides more accurate and consistent placement of the boards on the print carriage. It can be easily adapted to automatic vision screen alignment system through a standard interface.

For printer setup, although varying with specific models, some basic considerations should always be assured, including

the flatness of squeegee blade and its leveling to the substrate surface,

screen frame alignment and parallelism,

squeegee stroke length and speed,

screen snap-off distance,

**Figure 6.1** Overview of a screen printer. *(Courtesy of Affiliated Manufacturers, Incorporated–Presco.)*

squeegee downstop/squeegee pressure, and
substrate alignment.

In some models, the squeegee assembly is program-controlled to monitor and maintain the selected parameters from batch to batch and within a batch. The programmable parameters include snap-off, squeegee pressure, downstop, flood height and hop-over height.[1]
Most printer manufacturers provide the detailed procedures and demonstration service for printer setup and adjustment. It is highly recommended to follow the manufacturer's procedure to obtain best productivity and utilization of the specific equipment.

## 6.3  PRINTING MODE

Most printers provide options in printing mode such as print/flood, flood/print, print/hop-over/print, and double print. The print/flood mode contains both squeegee and flood blades. The paste is deposited between squeegee and flood blades, and the squeegee makes a print and the flood blade drags the paste back to the start position. In the flood/print mode, the flood blade of the squeegee carriage moves to flood the screen and the squeegee makes a print in the opposite direction. This flood/print mode allows the print cycle to end with a dry and clean screen. The flood blade is typically positioned about 8-15 mils above the screen. The print/hop-over/ print mode uses only a squeegee blade that makes a print and then jumps over the paste to make another print in the opposite direction. The single squeegee facilitates cleanup and provides a longer squeegee stroke. For metal mask screen, rigid or flexible, the flooding action is normally not needed due to the complete opening of print patterns.

## 6.4  SCREEN HARDWARE

There are basically three types of screen: mesh and emulsion screen, rigid metal mask stencil, and flexible metal mesh stencil. The mesh and emulsion screen is made of wire mesh patterned by polymer emulsion. Stainless steel, polyester, and nylon are common materials for wire. The inherent material characteristic differences between polymer and stainless steel wire cause different properties and performance in the screen. Table 6.1 is the summary comparison between stainless steel wire and polyester thread in resolution, registration, adaptability to substrate surface, deposit thickness, deposit uniformity, abrasion and chemical resistance, and production considerations.[2] Stainless steel in general can be drawn to a finer diameter than polyester. Table 6.2 compares screen made of stainless steel and polyester in design mesh counts and their corresponding wire diameter, mesh opening, and open area.[2] The angle of mesh wire to frame can be oriented between 22.5° and 90°.

The screen mesh is attached to an aluminum screen frame under tension (10–25 N/cm) with a bonding adhesive. The emulsion, commonly made of sensitized polyvinyl alcohol, is used to generate the pattern. The thickness of a screen is approximately equivalent to the sum of two wire diameters and the thickness of emulsion buildup, as shown in Figure 6.2.

The thickness of print deposit can be estimated as follows:

thickness of print deposit = (2 X wire diameter X % open area)
                                            + emulsion buildup

For example, for solder paste printing using 80-mesh stainless steel screen applied with 5 mils of emulsion buildup,

**TABLE 6.1**

## Comparison of Stainless Steel Wire and Polyester Thread[2]

| Characteristics | Stainless Steel Wire | Polyester Thread |
|---|---|---|
| Resolution | Depending on mesh count and wire diameter available in diameter, to 20 microns | Depending on mesh count and thread available in diameter, to 30 microns |
| Registration | Excellent for new and poor for used or aged | Poor for new and excellent for used or aged |
| Adaptability | Poor | Excellent |
| Abrasion and Chemical Resistance | Excellent | Good to fair |
| Production Consideration | Accuracy diminishes with use, easily damaged, not affected by temperature and humidity changes | Accuracy improves with use, highly resilient to damage, but affected by temperature and humidity changes |

**TABLE 6.2**

## Mesh Screen Parameters[2]

| Material Type | Mesh Count per in. | per cm | Thread Diameter mils | microns | Mesh Opening mils | microns | Open Area % |
|---|---|---|---|---|---|---|---|
| Stainless Steel | 80 | 31.5 | 3.70 | 94 | 8.80 | 224 | 49.6 |
| Stainless Steel | 80 | 31.5 | 1.97 | 50 | 10.53 | 268 | 71.0 |
| Polyester | 81.3 | 32.0 | 4.7 | 120 | 7.6 | 193 | 37.9 |
| Polyester | 81.3 | 32.0 | 3.9 | 100 | 8.4 | 213 | 46.2 |
| Stainless Steel | 105.0 | 41.3 | 3.15 | 80 | 6.37 | 162 | 44.8 |
| Polyester | 101.6 | 40.0 | 3.5 | 90 | 6.3 | 160 | 41.0 |
| Stainless Steel | 150.0 | 59.1 | 2.48 | 63 | 4.19 | 106 | 30.4 |
| Polyester | 149.9 | 59.0 | 2.5 | 63 | 4.2 | 106 | 39.5 |

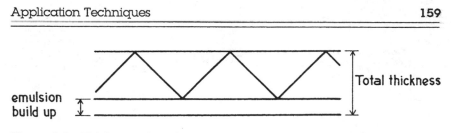

**Figure 6.2**   Thickness of mesh screen showing wires and emulsion buildup.

thickness of screen          = (2 x 3.7 mils) + 5 mils
                            = 12.4 mils, and
thickness of print deposit = ( 2 x 3.7 mils  x 49.6%) + 5 mils
                            = 8.7 mils

The metal mask stencil is made of solid metal foil which is chemically etched to the desired pattern, and then bonded to the aluminum frame. The metal foils are usually brass, stainless steel, copper, beryllium copper, and nickel. For flexible metal mask stencil, the foil is not directly bonded to the frame. A flexible mesh structure bridges the foil and the frame by bonding adhesives. The thickness of the screen as well as the thickness of printing deposit for both rigid and flexible stencil is essentially equal to the thickness of the foil used. The screens are shown in Figure 6.3 and a comparison of the three types is summarized in Table 6.3.

## 6.5   SCREEN ARTWORK AND PATTERNS

Another important aspect of screen is artwork. The quality of printed pattern starts from the precision and quality of the artwork. The artwork involves engineering design and drawing, transferring magnified drawing to a Mylar™ material (ruby), reducing the scale of the ruby, and generating a photopositive. During this process, practical aspects must be considered. The following are some of the particulars for generating artwork and patterns in real world practice, as provided by G. G. Petersen of Weltek International, Inc.

**Make the photocopy and notation of artwork**. Since most phototools contain more details than in that for solder pads, unless otherwise specified, all patterns on the phototool will appear as openings in the screen or stencil.

**Figure 6.3** Patterned screens. *(Courtesy of Utz Equipment, Incorporated.)*

**Specify how the solder pad openings will appear on the printed circuit board** to avoid "reverse image" on the final screen or stencil.

**Specify the requirement of off-center frame location;** otherwise the screen/stencil makers will center the image on the frame.

**Provide the exact $X$ and $Y$ dimensions between images** to prevent photographic exaggeration.

**Consider the "etch factor" at the design stage or at the photographic stage.** In the chemical etching process, the exposed metal areas which are not protected by photoresist material are etched away by acid spray. Because of the etching process and its undercutting at the edges of the resist pattern on the surface, all etched

## TABLE 6.3

### Summary of Characteristics of Three Types of Screens

|  | Screen Type | | |
|---|---|---|---|
|  | *Mesh Screen* | *Rigid Stencil* | *Flexible Stencil* |
| Advantages | low cost<br>fast turn-around<br>good gasketing<br>better surface con-<br>formance | complete open<br>area<br>low squeegee<br>wear<br>better print<br>geometry<br>easy cleaning<br>and mainte-<br>nance<br>easy setup | complete open<br>area<br>low squeegee<br>wear<br>better print<br>geometry<br>easy cleaning<br>and mainte-<br>nance<br>easy setup |
| Disadvantages | partial open area<br>of the image<br>easy to wear out<br>higher squeegee<br>wear<br>potential clogging<br>longer setup<br>more cleaning<br>and maintenace<br>susceptible to tem-<br>perature and<br>humidity<br>need a control<br>room | high manufac-<br>turing cost<br>less gasketing<br>etch factor<br>does not pro-<br>vide a flat<br>surface to the<br>substrate<br>on-contact print<br>only | high manufac-<br>turing cost<br>less gasketing<br>etch factor<br>need good bond<br>at metal mask<br>and flexible<br>mesh border |

dimensions, tolerances, and configurations are a function of material type and thickness. The openings are often extended beyond the required dimensions during etching, which is known as the etch factor. To compensate the etch factor, the guidelines for metal thickness greater than 0.005" are to reduce each dimension by 25% of metal thickness for square and rectangular patterns, and to reduce the radius by 25% of metal thickness for circles and similar patterns.

**Use line-width and hole opening size guideline;** the minimum line width should be equal to metal thickness for thinner than 0.005" metal thickness, and 1.25 times metal thickness for 0.005" or thicker metal mask. The diameter of an opening cannot be greater than the metal thickness for 0.005" and thinner metal mask, and 110% of metal thickness for greater than 0.005" metal mask.

**Make sure the screen frame fits the equipment.** For surface mount technology applications, the important frame specifications are size, configuration, and dimension.

## 6.6  SCREEN CLEANING AND MAINTENANCE

After print cycles, paste should be cleaned as soon as practical. Prolonged delay will jeopardize the cleanability. Screen should be cleaned by gently scraping and wiping the gross amount of paste off the screen, followed by rinsing the screen with proper solvents. Solvents such as trichloroethane/alcohol blend, Freon™–type, work, well with most pastes. Some screen manufacturers supply cleaning solvent specially formulated for the screen. Water and low-carbon chain alcohol may impair the emulsion in mesh screen and in the border of flexible metal mask, and should be avoided.

In order to avoid damaging the adhesives and emulsion, soaking in solvent should not be practiced during cleaning. Hot solvent should also be avoided. In cleaning screen, good industrial hygiene should be observed: Gloves, safety glasses, and adequate ventilation are mandatory. For screen storage, environment with moderate to low temperature (<100°F) and low humidity are conditions to prolong screen life.

Another important aspect of screen is its tension retention. All mesh screen and mesh border of metal stencil are stretched to a precise tension during manufacturing—for example, 10-20 Newton/cm or 20-40 mil deflection. The screen tension declines with usage. A regular check on screen tension is needed to assure its adequacy. A tensionmeter and tension gauge are commercially available. As one technique, the tension gauge measuring the screen deflection by placing the gauge on the front face of screen with added one-pound weight is commonly used.[3]

## 6.7  SQUEEGEE SYSTEM

The function of squeegee during printing is to render the screen in a uniform line contact with the substrate to be printed, and in conjunction with the squeegee system, to provide proper downstop and pressure for transferring the paste through the openings of the screen. The squeegee blade is usually made of polyurethane with different levels of hardness as expressed in a durometer, typically being 60-80. Commonly, the squeegee is formed in diamond shape or rectangular block. The edge of the squeegee usually makes a 45- or 60-degree angle with the surface of the screen.

The squeegee speed control is driven by a pneumatic or hydraulic system, providing a range of traveling speed from 1" to 20" per second. But practically, one uses 2"-8" per second to obtain print performance. The squeegee pressure control is commonly a spring system capable of pressure adjustment in response to any variations on the surface of substrate. The downstop determines the stopping position of the squeegee in its downward travel and is directly related to the true pressure on the screen. The squeegee downstop, squeegee pressure, and squeegee speed all affect the quality and thickness of the print in addition to the durometer value of the squeegee material and the angle between the edge of squeegee and substrate. These parameters function interdependently, and a simple equation to illustrate the relationship is not in existence. (Section 6.9 discusses some of the interrelationships.)

## 6.8  SNAP-OFF

The snap-off is the distance between the bottom of the screen and the surface of the substrate being set for off-contact print. For mesh screen printing, snap-off distance is often used, usually 20-50 mils. By having the snap-off distance, the screen and substrate are kept separate at all points except the transistory line contact between squeegee and the screen during printing. The setting of a snap-off distance brings in another parameter for printing process, which influences the desirable settings of squeegee parameters. (Its relationship to other parameters is discussed in Section 6.12.)

## 6.9  PRINTING PRINCIPLES

Several forces contribute to the paste transfer process: (1) the relative adhesion between the paste/screen interface and the paste/substrate interface, (2) the pushing force by the squeegee along $x$-direction, (3) the pressure (force per unit area) exerted by the squeegee along y-direction, (4) the hydraulic pressure created in the paste when the paste is compressed between the squeegee and the screen, and (5) the gravity force. Figure 6.4 is the schematic of paste transfer during printing. The squeegee makes a line contact with the screen and a close contact with a given amount of paste. The paste thus moves in front of the squeegee while the squeegee is in motion, and is released behind the squeegee throughout the openings to the substrate. During squeegee motion, the hydraulic pressure developed in the paste is believed to press against the screen and the squeegee, resulting in the lifting of the squeegee edge and the extrusion of paste through the screen openings.

D. E. Riemer has investigated the hydrodynamics of screen printing and derived the following relation between hydraulic pressure and printing parameters:[4]

$$P_{hydraulic} \propto V\eta \, \frac{\sqrt{v}}{(\sin \alpha)^2},$$

where $P$ is the hydraulic pressure in the paste, $\eta$ is the paste viscosity, $v$ is the squeegee speed, $V$ is the paste volume in front of squeegee, and $\alpha$ is the squeegee attack angle.

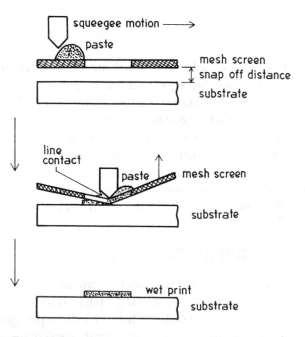

**Figure 6.4**  Schematic of paste transfer during printing.

As expressed in the equation, the hydraulic pressure in the paste increases with increasing paste viscosity, paste volume, and squeegee speed; however, it decreases with increasing squeegee angle. The hydraulic pressure is found to be directly related to the thickness of print, with higher hydraulic pressure resulting in a thicker print. In order to have sharp print resolution, the hydraulic pressure must be focused on the area near the squeegee edge. (The hydraulic pressure is not to be confused with squeegee pressure and downstop.) The squeegee angle is considered to affect the extent of area exerted by pressure. The smaller squeegee angle is associated with a pressure area farther out from the squeegee edge, causing poor print resolution.

The existence and extent of hydraulic pressure vary with different printing/screen systems. As external squeegee pressure increases, the effect of hydraulic pressure on the squeegee is reduced. The increased hardness of squeegee also reduces the effect of hydraulic pressure. As hydraulic pressure diminishes, other forces become prevalent. A theoretical treatment correlating the printing variables and the paste transfer provides further understanding of the interplay of some variables associated with the printing process.[4]

## 6.10  PRINTING VARIABLES

A large number of printing variables have been identified in the literature for mesh screen printing of thick film pastes.[6] Some variables can be directly translated to solder paste. For solder paste, variables may be grouped into squeegee system, substrate, screen systems, and paste-related areas, as shown by the following:

### Squeegee System

blade material type          squeegee size
blade hardness               squeegee speed
blade edge angle             downstop pressure
blade edge flatness          external pressure
blade configuration

### Substrate

material type                surface flatness
surface condition            substrate size

## Screen Systems

### Mesh Screen

wire material type
wire diameter
mesh tension
mesh count
mesh orientation
emulsion type
emulsion thickness
print pattern
screen size

### Metal Mask Stencil

material type
flatness
thickness
print pattern
stencil size

### Paste

paste consistency     chemical compatibility with the screen and
                          squeegee
paste rheology
paste stability       compatible thickness with printing pattern
paste tackiness

These variables, which may not be exhaustive, must all be considered when starting to set up a printing process.

## 6.11  PRINTING OPERATION

Among the large number of parameters affecting printing results, once the basic equipment parameters are properly selected, four operational parameters are to be monitored. They are the squeegee speed, squeegee downstop, squeegee pressure, and snap-off distance. These operational parameters can be readily adjusted in most printing equipment. The impact of each of these four parameters on the thickness and quality of the print pattern is relatively easy to comprehend. With other parameters being fixed, a one-to-one correlation can be found. For example, the print thickness normally increases with increasing squeegee speed. The print thickness in general increases with increasing snap-off distance. And the print thickness varies with squeegee pressure or downstop in such a manner that a minimum point is displayed with increasing either squeegee pressure or downstop. In addition, a larger squeegee attack angle is found to associate with an increase in print thickness.

However, in a real printing process, these parameters do not work in such a simple manner; they are interdependent.

Frequently, the one-to-one correlation is found to be violated. Therefore, empirical information needs to be obtained for a given paste and printer by generating the interrelation curves through systematic tests—for example, evaluating the print quality and thickness in relation to squeegee pressure at a given snap-off for different squeegee speeds, in conjunction with evaluating the print quality and thickness in relation to squeegee pressure at a given squeegee speed for different snap-offs. The information generated serves as a good baseline for real-world printing of the specific system.

The interpretation of the effect of paste viscosity for printing is facilitated with two equations. The shear rate imposed on the paste during printing is considered to be directly related to the speed of squeegee, and the thickness of the screen that paste has to pass through, and inversely proportional to the square of the opening dimension, as expressed by

$$\dot{\gamma} \propto \frac{vh}{D^2},$$

where $\dot{\gamma}$ is the shear rate, $v$ is the squeegee speed, $h$ is the thickness of the screen, and $D$ is the dimension of opening. For stencil printing, the shear rate that a paste is experiencing while passing through the opening is estimated to be in the range of $100 \ sec^{-1} \sim 500 \ sec^{-1}$.

The viscosity of paste ($\eta$) is inversely proportional to the shear rate which is imposed on the paste:

$$\eta \propto \frac{1}{\dot{\gamma}}$$

Thus, combining the two relations,

$$\eta \propto \frac{1}{\dot{\gamma}} \propto \frac{D^2}{vh}.$$

When other conditions are equal, the relation indicates that the thicker screen or higher squeegee speed demands lower viscosity of paste for good paste transfer, and a large opening requires a more viscous paste. This may also indicate the effect of screen size on printing. On a large screen, the paste with a given viscosity and with shear-thinning rheology tends to make heavier deposits through a large screen than through a small screen. Needless to say, the phenomenon naturally depends on the intrinsic rheology of the paste to be used.

## 6.12  CONSIDERATIONS FOR PRINTING PARAMETER SELECTION

For printing operation, two primary mechanisms dominate the quality of print: One is paste transfer through openings, and the other is screen release. The efficiency of paste transfer through pattern opening and the efficiency of screen release depend on various parameters as listed in Section 6.10. However, the maximization of screen release efficiency normally demands a different set of parameters as of paste transfer efficiency. It is therefore obvious that the selection of a set of parameters to balance these two mechanisms is the key to the quality of print.

No fixed number can be readily assigned to each parameter, since all parameters are interrelated. Changing one may change the effects of others in a nonproportional manner. Nonetheless, understanding the general relationship among the parameters and their effects on print in conjunction with systematic testing on the specific system including printer, board, and solder paste, parameters to achieve ultimate print quality can be established. The following illustrates some general relationships among parameters and their effects on print.

**For off-contact print, a proper snap-off distance is important.** If the snap-off distance is too high, squeegee pressure may focus on the screen but not be sufficient on the substrate, resulting in incomplete transfer. On the other hand, too small snap-off may cause paste smear and inefficient screen release. The properness of snap-off varies with screen size and squeegee length. Normally, the increased screen size requires increased snap-off, and increased squeegee length requires decreased snap-off. Since the increase in snap-off is normally applied to facilitate screen release, its adverse effect on paste transfer can be compensated by using a reduced squeegee speed and/or increased squeegee pressure.

**The fast squeegee speed favors screen release but makes the paste transfer unfavorable.** The squeegee speed also changes with print area: Larger area needs slower speed for good print, when other conditions are equal. In addition, faster speed can accommodate lower viscosity paste.

**Proper squeegee pressure is important.** While the squeegee speed and snap-off are maintained, high squeegee pressure can cause paste smear and bleeding, and low squeegee pressure can cause incomplete transfer.

**The hardness of squeegee affects the print quality.** When the squeegee is too soft, it tends to bend under pressure, resulting in bulldozing the paste. Thus minimal pressure should always be associated with soft squeegee. However, very hard squeegee is not always favorable due to its lack of the ability to conform to the surface variation.

The operating parameters have to be set up based on the requirements of the substrate to be printed in its size and pattern, the desired thickness of print, and the paste characteristics. Overall, the following lists some tips for printing solder paste.

**Select a proper solder paste** in terms of its viscosity, its response to mechanical shear, and its compatibility with the operation, such as paste open time.

**Make sure the paste is composed of compatible solder particle size** with the mesh opening of the mesh screen, or with the finest opening of the stencil. As a safe margin, the smallest dimension and the opening should be at least three times the diameter of the largest particles.

**Mesh screen requires a less viscous paste than stencil.** When measuring by a Brookfield RVT Model viscometer, TF/5 rpm, 3 min mixing/2 min reading, a viscosity of $700,000 \pm 150,000$ cps is a typical range for 80-mesh screen, and $900,000 \pm 150,000$ cps for stencil.

**Test chemical compatibility** among solder paste, squeegee material, emulsion (mesh screen), and bonding adhesives.

**Design for over-printing may lead to a soldering problem,** and is not recommended.

**Make sure the stencil opening matches the artwork.** A larger actual opening in stencil than in the original artwork is often encountered, as discussed in Section 6.5.

A squeegee with a diamond blade and high durometer reading ($\geq$70 ) are recommended for stencil print.

Excess squeegee pressure should be avoided. To start with low pressure and gradually raise the pressure until a clear sweep is achieved is a good technique by which to set the pressure level.

Make sure the squeegee drive provides a constant squeegee speed across the screen.

For flexible stencil, foil size should be smaller than the frame size so that a length of 1"–3" (2.5 cm-7.5 cm) is available for the flexible mesh.

Contact print has been successful with stencil, not being adopted for mesh screen. However, it may require less tacky paste.

Optimum snap-off distance varies with screen size. When board/screen changes, the snap-off may need to be adjusted accordingly.

Avoid the paste touching the substrate before the screen does by setting a proper set of parameters and selecting a right paste.

Optimum squeegee pressure often varies with squeegee speed. They often inversely relate to each other in that higher speed imposes lower pressure.

Pattern size should be kept small in relation to screen size. Meeting this condition provides more flexibility to optimize the printing parameters.

To maximize print resolution through facilitating the screen release, ultilize the relay function (if available) to delay substrate carrier return and/or squeegee lifting.

To balance the paste transfer and screen release in stencil printing, a good starting point is to set a high squeegee speed and bring squeegee pressure from low to adequate at a proper snap-off (5–20 mils).

To assure print thickness and uniformity, the wet print thickness can be measured through various techniques. The utilization of a laser sensor is one of the techniques to measure the height and height variations of a solder pad print.[1]

## 6.13 DIFFERENTIAL THICKNESS PRINTING

For surface mount printed circuit board containing a variety of solder pads requiring different thickness of print—for example, solder pads for 50 mil pitch pattern and fine pitch pattern (25 mil pitch and 20 mil pitch)—differential printing on multilevel stencil is developed. This technique can achieve one-pass printing to different thicknesses of paste deposit without having excess or starved print thickness for each group of solder pads. This can be done by using either single foil or two (multiple) layers of foil.

In single foil process, the metal mask stencil with a given foil thickness is recessed in areas for fine pitch pattern. The recessed areas with reduced foil thickness provide thinner print as required for fine pitch patterns. When the design is such that it requires thicker and thinner deposit patterns close together (typically, <200 mil), a two- or multiple-foil process is needed. The difference in thickness of paste print is obtained by adhering foils together through a lamination technique. The multiple foil portion provides the thicker print and single foil the thinner print.

Generally, a thickness of 10 mils of solder paste print is deposited for solder pads mounting a standard 50 mil pitch component, and 3–8 mils for fine pitch pads. To produce quality reproducible prints in multilevel stencil printing, the control of processing parameters and solder paste characteristics becomes more demanding.

## 6.14 PRINTING ENVIRONMENT AND PASTE HANDLING

Most solder pastes are quite responsive to temperature and mechanical shear. As temperature increases or shear stress is applied, paste tends to become more fluid (decrease in viscosity). The printing quality can thus be affected by ambient temperature fluctuation. Different pastes vary to different extent. Therefore a controlled environment is definitely a plus to assure consistent print. In addition to temperature, humidity is another variable which could affect paste performance.

Pastes also tend to "set" on shelf storage. For manually applying paste on screen, a gentle mixing prior to depositing on the screen is desirable to assure proper fluidity and homogeneity. To maintain consistent amount of paste on screen, paste dispensed from a cartridge through an automated dispensing unit is recommended.

The used paste (after being printed and exposed to atmosphere) is not recommended for reuse. The residual paste on the screen should be collected and disposed of through authorized metal scrap dealer with appropriate identification. (Confirm disposal of solder paste and associated wipe-off rags or cloths through local environmental requirements.)

## 6.15  PNEUMATIC DISPENSING

Basically, a pneumatic dispenser delivers paste by means of air pressure to push paste through a specified channel. The size of the paste deposit for attaching components or leads of electronic assemblies is usually very small. Table 6.4 provides the common needle gauge and corresponding diameters. A typical dispensing setup is illustrated in Figure 6.5. The paste reservoir is a syringe or equivalent container in different sizes, from 10cc to 600cc (20 fl. oz). The amount of paste packed in different sizes of syringes depends on the type of alloys and metal content. The common ranges are listed in Table 6.5. Air pressure applied on the top surface of the paste presses the paste out of the syringe reservoir, through the fine needle, and deposits it on the desired locations of the substrate. For discrete dot dispensing, pulsed air pressure regulated by a controller is used. For continuous line or bead of paste, continuous pressure is used. Operation can be manual or automated. Both types of equipment are available. The microprocessor, with numerically controlled *XYZ* table, directs the precise location of paste deposit on the designated locations.

For a reliable process, the key requirements are to maintain constant pressure and flow rate in the syringe and to have a pressure-

## TABLE 6.4

### Dispensing Needle Gauges and Inner Diameters

| Needle Gauge | Inner Diameter | |
|---|---|---|
| | in. | mm |
| 15 | 0.054 | 1.37 |
| 18 | 0.033 | 0.84 |
| 20 | 0.023 | 0.58 |
| 21 | 0.020 | 0.51 |
| 22 | 0.016 | 0.41 |
| 23 | 0.013 | 0.33 |
| 25 | 0.010 | 0.25 |
| 30 | 0.006 | 0.15 |

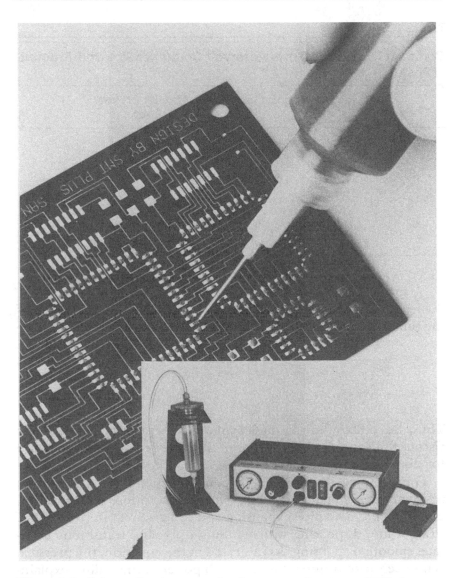

**Figure 6.5** Schematic of pneumatic dispenser. *(Courtesy of EFD, Incorporated.)*

stable paste. The pressure required to produce a constant flow rate depends on several parameters. Considering that the system is equivalent to pumping viscous fluid through pipe(s), the flow rate increases approximately with the square root of the applied pressure, and the pressure $(P)$ is related to the viscosity of the fluid $(\eta)$,

## TABLE 6.5

Common Dispensing Paste Reservoir Container Size and Amount of Paste

| Paste Reservoir Container (cc) | | Amount of Paste (g) |
| --- | --- | --- |
| 10 | | 35–45 |
| 30 | | 95–105 |
| 35 | | 100–110 |
| 60 | | 190–210 |
| 180 | (6 fl oz) | 450–650 |
| 360 | (12 fl oz) | 1000–1200 |
| 600 | (20 fl oz) | 1900–2100 |

the length of the tube ($L$), the volume of the fluid ($V$), the radius of the tube ($R$), and the time ($t$) by the following equation:[7]

$$P \cong \frac{\eta LV}{R^4 t}$$

Thus, the combination of smaller volume, short passing distance, less viscous fluid and the large radius of the tube demands less pressure for dispensing.

The pressure decay in the pipelines depends on the viscous resistance of the material and the pipeline transitions such as valves, elbows, bends, and pipe expansion and contraction.[8] In view of this, straight line dispensing with minimum pipeline transitions facilitates smooth dispensing. As indicated in the equation, the pressure is inversely proportional to the fourth power of the radius, explaining that one increment of the needle gauge can affect the paste dispensability drastically.

In addition to dispensing entirely driven by pressure, other modified systems have been developed. The dispenser with a syringe connected to the tubing where it is moved by a roller wheel as shown in Figure 6.6 is designed for versatile dispensing.[9] This equipment is basically a peristaltic pump without the pulsation. The amount of deposit is controlled by the motion of the roller wheel.

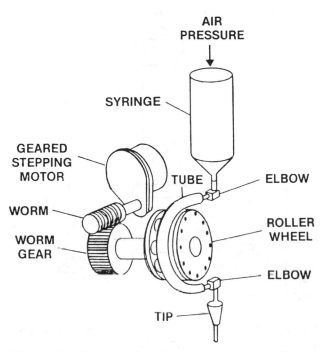

**Figure 6.6**   Pneumatic roller wheel dispenser.[9]

## 6.16   POSITIVE DISPLACEMENT DISPENSING

This type of dispensing is based on volume displacement through mechanical means, in contrast to that by air pressure. The equipment as shown in Figure 6.7 is designed with a lift station, a paste reservoir, pressure controllers, and dispensing head, which is the heart of the equipment.[10] The dispensing head is composed of a manifold block, pinch off block, tube block, and needle block. The manifold block directs paste into multiple channels to provide simultaneous multiple dot dispensing by passing through the pinch off block, tube block and needle block. The pinch off block consists of polymeric tubings, two pinch plates, and a displacement bar, which are driven by a piston pump. While the equipment is in operation, the top and bottom plates and the displacement bar are synchronized so that the paste first fills the tube between the top blade and bottom plate and then displaces the paste at the displacement bar area by closing the top plate and opening the bottom plate, as shown in Figure 6.8. The volume of paste being dispensed depends

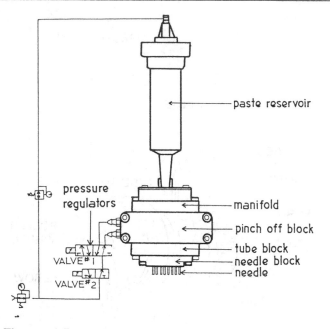

**Figure 6.7** Schematic of positive displacement dispenser. *(Courtesy of SCM–Dispensit.)*

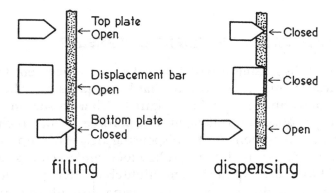

**Figure 6.8** Volumetric paste displacement.

on the position of the displacement bar, which is adjustable, and on the cross section of the tubing, as well as on the diameter of needle. As shown in Figure 6.7, the capability of dispensing multiple dots simultaneously is a unique and desirable feature.

Another positive displacement dispensing system, as shown in Figure 6.9, is designed with a rotary positive displacement pump which is able to deliver dot sizes as small as 0.010 in. diameter at a

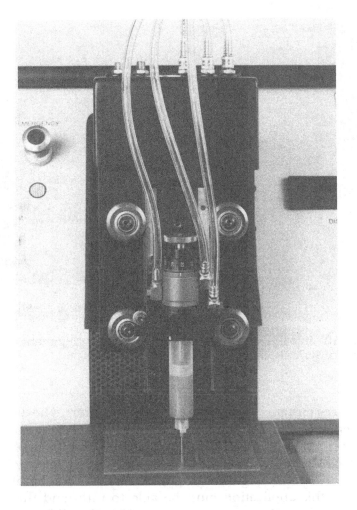

**Figure 6.9** Rotary positive displacement dispenser.
*(Courtesy of Camelot Systems, Incorporated.)*

speed of 3 dots per second. The equipment can be added to an intelligent vision system for boards or parts alignment as shown in Figure 6.10.

## 6.17  DISPENSING OPERATION

In deciding whether to use a pneumatic or a positive displacement system, an important factor is to select a proper solder paste which

**Figure 6.10** Intelligent vision system for dispensing alignment. *(Courtesy of Camelot Systems, Incorporated.)*

is pressure-resistant over a period of dispensing time. The dispensing time depends on the size of the syringe and the size of the dot placed. For an application using a 30 cc syringe containing 100 g of paste to dispense a single dot through a 20 gauge needle by 1-sec-on/1-sec-off pulsing, the dispensing time is approximately 15 h. Therefore, the paste for this application must be able to withstand the applied pressure at the ambient temperature for 15 h. Equally important are the quality of equipment and the setup of the equipment. Many practical points are to be considered for obtaining consistent and smooth dispensing, as the following indicates.

**Carefully select a proper dispensing system** whether for manual or robotic application.

**Thoroughly check the whole system** including the needle and connection tubes (if any) to assure the system is clean and clear.

**Always flush the system with a compatible "lubricant"** to wet the wall of the channel so that the paste will experience a minimum of friction to start.

**The paste syringes should be kept under cold storage** (or under conditions paste manufacturers recommend) to assure stability and freedom from physical separation.

**The paste syringe should be brought to a temperature in equilibrium with the ambient before starting dispensing.** As discussed in Section 6.10, the flow rate depends on viscosity, which can be altered by temperature change. Even a small temperature difference may cause wrong start, resulting in erratic results, particularly for very fine dots (e.g., 20 gauge or finer needle).

**Make sure that no foreign particles are introduced** and that no solder particles are flaked, since solder alloys are relatively "soft" and easily deformed.

**The pressure applied in the syringe should be kept at a minimum,** and the proper head pressure kept in the range of 15-25 psi. In case the paste requires much higher pressure (more than 40 psi) to dispense, the probability of having a consistent and continuous dispensing is very low.

**The external air pressure supply should be maintained constant.** An air regulator to assure constant pressure input is needed.

**The clearance (gap) between the needle and the substrate affects the shape and quality of the dot dispensed.** If the clearance is too little, the dot tends to be flattened out, and if too large, the dot tends to have long tailing. The "Hershey Kiss" shape is considered desirable.

**Location of the paste syringe should be considered.** For in-line production involving a reflow furnace, the dispensing system should be set up in a way that the paste syringe is not located near the reflow furnace.

**Select a "right" paste for the dispensing system.** The required characteristics of a dispensing paste include: a viscosity of 250,000 $\pm$50,000 cps (Brookfield RVT, Helipath TF/5 rpm, 3 min mixing/ 2 min reading); rheology displaying low thixotropy and low yield point, yet no drooping during dispensing and idle time; clean breakup without trailing tail (or stringiness); and physical stability in ambient temperature under applied pressure during the entire dispensing time. A viable dispensing paste should provide consistent and reproducible dot placement without any disruption until the whole syringe is completely depleted.

**Chemical compatibility is essential.** The materials used in a dispensing system, metal or polymer in nature, with which the paste is in direct contact must be checked for chemical compatibility with the paste to be used.

**For extensive downtime, the whole dispensing system should be flushed** without leaving any paste in any part of the system. This step may appear to be redundant, but in the long run it is time-saving and a quality assurance.

## 6.18  PIN TRANSFER

The pin transfer process is another available method to deposit paste. The potential advantage of this process is being able to deposit paste in small dot form to already populated boards where a printing process is not suitable, and where a large number of dots can be placed simultaneously. It is not particularly popular in the solder paste area, yet pin transfer is a viable process in principle. The process requires the design of a pin pattern which exactly corresponds to the board design as shown in Figure 6.11 and to the uniform wetting of the tip of the pin.[11] The basic parameters to be determined are the shape, the diameter and the length of the pin, the technique of making a reservoir layer and its control, and the characteristics of the paste.

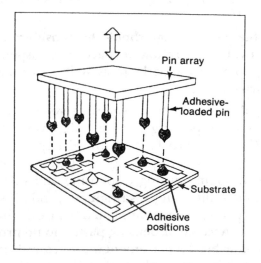

**Figure 6.11**  Schematic of pin transfer.
(*Courtesy of Hellers Industries.*)

A suitable paste for this process should possess the proper fluidity and wetting ability for the pins at the moment of dipping, but not be too fluid to stay on the pin for complete transfer to the designated location. The viscosity of a well constituted paste for pin transfer is in the range of 350,000 ± 100,000 cps (Brookfield RVT, Helipath TF/ 5 rpm, 3 min mixing/2 min reading). Since the paste reservoir is an open system, the paste also requires environmental stability.

## REFERENCES

1.  AMI–Presco, North Branch, New Jersey.
2.  Tamas Freeska, "Stainless Steel vs. Polyester: Fabric Comparison for Electronics Printing," *SITE* (Nov. 1986).
3.  Utz Engineering, Inc., Clifton, New Jersey.
4.  Dietrich E. Riemer, "Ink Hydrodynamics of Sheer Printing," proceedings, 1985 International Hybrid Microelectronics Symposium, ISHM: 52.
5.  Dietrich E. Riemer, "The Function and Performance of the Stainless Steel Screen During the Screen-Print Ink Transfer Process," proceedings, 1986 International Hybrid Microelectronics Symposium, ISHM: 826.
6.  Charles A. Harper, *Handbook of Hybrid Microelectronics* (New York: McGraw-Hill, 1974).
7.  W. H. Herschel, "Experimental Investigations Upon the    Flow of Liquid in Tubes of Very Small Diameter by J. L. M. Poiseuille," *Rheol. Met. Soc. of Rheology* (1940).
8.  W. H. McAdams, *Heat Transmission* (New York: McGraw-Hill, 1954).
9.  Noel Peterson, "A Solder Paste Dispenser for SMD Assembly," Surface Mount Technology, International Electronics Packaging Society 3-2: 989.
10. G. Blumb and K. Miller. SCM-Dispensit, Indianapolis, Indiana.
11. Robert M. Barto and Marc Peo, "Pin Transfer - Motions," *Circuits Manufacturing* (Jun. 1987).

# CHAPTER 7

# Soldering Methodologies

## 7.1 INTRODUCTION

This chapter covers the reflow methods currently available for reflowing solder paste. Literally, reflow is a misnomer; in the context of solder paste technology it means to flow or melt the paste, rather than to re-flow. Since it has been an accepted term in the solder paste field, "reflow" will be used in this text. The reflow methods discussed here do not deal with wave soldering or other soldering techniques unrelated to the paste form.

To reflow solder paste, three fundamental questions to ask are: What is the proper heating source for the specific assembly? How is the heat transferred to the intended soldering area? What level of control is needed for the process? To answer these questions can sometimes be difficult. But with good understanding of materials and engineering principles in conjunction with a thorough plan, the questions can be handled successfully. There are three basic types of heat transfer: conduction, radiation, and convection. The following provides a brief illustration of the basic principles of heat transfer, the equipment and technique, and the temperature monitoring.

## 7.2 THERMAL CONDUCTION

The heat of thermal conduction is transferred by molecular motion which occurs within the same body or between two bodies which are in physical contact with each other. Mechanistically, heat in solids is mainly conducted by elastic vibration of lattice movement in the crystal transmitted in the form of waves (phonons) and by the movement of conduction electrons. The heat flow per unit area ($Q$)

is related to thermal conductivity ($k$) and the temperature ($T$) gradient in a given distance ($x$), as expressed by Fourier's Law:

$$Q = -k\frac{\partial T}{\partial x}$$

Note that this is equivalent to Newton's Law of viscosity, relating shear stress between any two thin layers of fluid and a constant velocity gradient. The thermal conductivity of solids which conduct heat only by phonons is inversely proportional to the temperature above the Debye temperature. As the temperature is raised to a sufficiently high level, the mean free path decreases to a value near the lattice spacing, and the thermal conductivity is expected to be independent of temperature. At very low temperature, the phonon mean free path becomes of the same magnitude as the sample size and the conductivity decreases to zero at $0°K$. However, if conduction is electronic in nature, the thermal conductivity is expected to increase with increasing temperature. In all crystalline materials, both phonons and electrons contribute to the conductivity. Figure 7.1 shows the temperature effect on the thermal conductivity of oxides—alumina ($Al_2O_3$), Silica ($SiO_2$), beryllia (BeO), magnesia (MgO), Titanium dioxide ($TiO_2$), fused quartz, $MgO\cdot Al_2O_3$, and zirconia ($ZrO_2$). The thermal conductivity of some metals—copper (Cu), nickel (Ni), tungsten (W), tin (Sn), lead (Pb), antimony (Sb), and iron (Fe)—is shown in Figure 7.2.

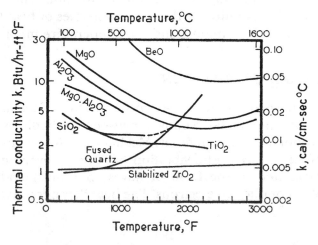

**Figure 7.1** Temperature effect on thermal conductivity of oxides.

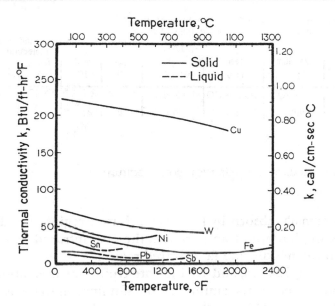

**Figure 7.2** Temperature effect on thermal conductivity of metals.

## 7.3  THERMAL RADIATION

The energy of thermal radiation is transferred by electromagnetic waves which originate from a substance at its elevated temperature. The thermal energy or emission power, $e$, is proportional to the fourth power of its absolute temperature, $T(°K)$.

$$e = bT^4,$$

where $b$ is the Stefan Boltzman constant.

The maximum wavelength of the radiation, $\lambda_{max}$, is inversely proportional to the absolute temperature (°K).

$$\lambda_{max} \propto \frac{1}{T}$$

These relations indicate that increasing temperature not only increases the emission power of an emitter but also shortens its maximum wavelength emitted. Figure 7.3 is the electromagnetic radiation spectrum and corresponding energy states. The radiation energy incident on the surface of a substance may be absorbed, reflected, or transmitted. In the range of wavelength of interest,

| X ray | Vacuum U.V. | near U.V. | Visible | near IR | mid IR | far IR | Microwave |
|---|---|---|---|---|---|---|---|
| less than $10^3$ Å | 100 - 200 mμ | 200 - 400 mμ | 400 - 800 mμ | 1 - 2.5 μ | 2.5 - 25 μ (4000 -400 $cm^1$) | 25 - 100 μ | > 100 μ |

displacement of nucleons and inner electrons of an atom
displacement of outer electrons of an atom
molecular vibrations and rotations
molecular rotations

**Figure 7.3** Electromagnetic radiation spectrum.

metals normally absorb in the near infrared region. Oxides and organics are, however, transparent in the near infrared region, and absorb in the middle and far infrared region.

Figures 7.4, 7.5, 7.6, and 7.7 indicate typical infrared absorption (absorbance versus wavenumber) of alumina, epoxy-glass, poly-imide-glass, and solder paste, respectively. The spectra were obtained by using Digilab FT-IR spectrometer. Absorbance instead of transmittance was directly measured in relation to wavenumber.

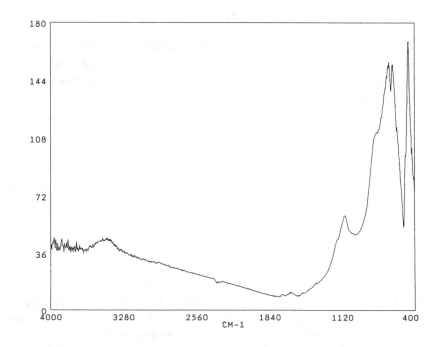

**Figure 7.4** Infrared spectrum of 96% alumina.

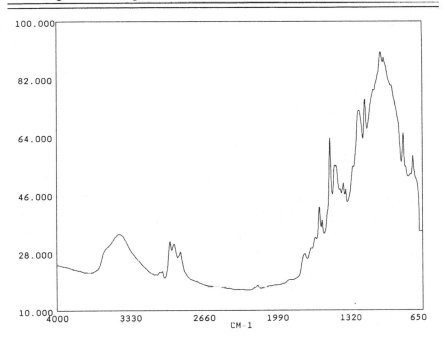

**Figure 7.5**  Infrared spectrum of typical epoxy-glass.

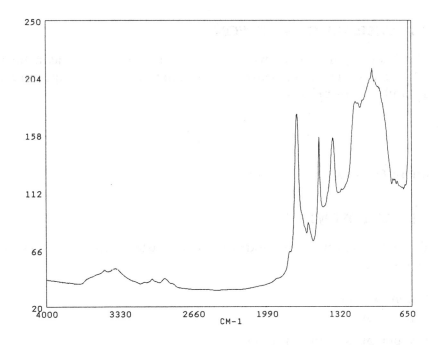

**Figure 7.6**  Infrared spectrum of typical polyimide-glass.

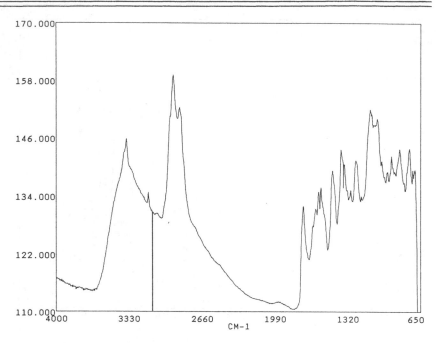

## 7.4   THERMAL CONVECTION

The heat transfer by convection is normally expressed as a heat transfer coefficient ($h$) between two temperatures (heating source, $T_2$, and workpiece, $T_1$).

$$h = \frac{Q}{T_2 - T_1} \; ,$$

where $Q$ is heat transfer rate.

## 7.5   REFLOW METHODS

The available reflow methods for solder paste utilize the following heat sources:

- conduction
- infrared
- vapor phase condensation

- hot gas
- convection
- induction
- resistance
- laser

Each of these methods possesses features which are particularly suitable for specific assemblies, although there may be a preference, depending on the volume of production, cost, type of components involved, type of materials involved, and other processing parameters.

## 7.6  CONDUCTION REFLOW

Due to the nature of heat transfer, this process is most suitable for the assemblies with flat surfaces, composed of thermal conductive material as the substrate, and with single-side component/device populations. Therefore, hybrid assemblies having small, flat surfaces and relatively high thermally conductive alumina substrate meet all of the criteria. Some limitations exist, yet this method provides a fast heating rate and operational simplicity. In addition, the visibility during the whole reflow process is a benefit and convenience in some cases. To solder less thermally conductive substrates such as FR-4 board or its equivalents, an infrared lamp on top of the conveyored heat platen aids the heat transfer efficiency. Assemblies have been successfully produced by using this conduction/IR lamp system.

## 7.7  INFRARED REFLOW: FURNACE TYPE

Performance of infrared furnaces has made a lot of progress in the past several years, in order to accommodate the demanding requirements of surface mount technology and component manufacturing.

The key design in furnaces is the heating source and heating control. Categorically, the availability of heating design can be grouped into three types: quartz-enveloped tungsten filament, nonfocused panel emitters, and honeycomb/downflow heating. The quartz/tungsten filament infrared source emits near to middle infrared radiation, with wavelengths ranging from 0.7–3.0 um, as shown in Figure 7.8.[1] The wavelength of emitted energy is controlled

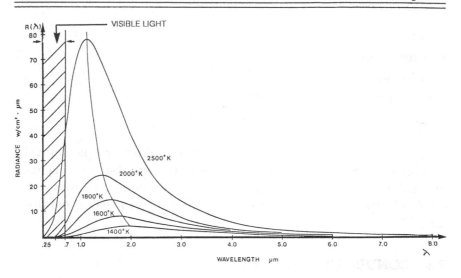

**Figure 7.8** Spectral response of tungsten filament infrared source. *(Courtesy of Radiant Technology Corporation.)*

by the voltage applied to the filament. In this range of wavelength, organic materials are essentially transparent, and metals are absorptive or reflective depending on the surface condition. However, longer wavelength in the range of 3–5 μm may also be present due to the re-emission from the backplane. The nonfocused panel emitter consists of an aluminized steel housing which encloses backup insulation, a resistance element and a fused quartz cloth emitter face.[2] This design operates on the "secondary emission principle"— that is, energy radiated from the element transfers to the ceramic face and emits evenly throughout its entire face. This allows the emission of a wide range of infrared wavelengths (2.5–7.0 μm) with predominant wavelengths between 1.5–5.5 μm. The honeycomb/downflow feature consisting of an aircraft-type honeycomb filter is reported to provide uniform temperature distribution by converting the turbulence at the top of a clamshell-type furnace into laminar flow as well as by stabilizing the atmosphere through the warm air layer of the honeycomb.[3]

Figure 7.9 shows a model of a surface mount infrared reflow system. The infrared furnaces are normally constructed with up to 12 individually controlled heating emitters. The emitters are commonly located in the top and bottom of the mesh belt.[1] Figure 7.10 is a schematic of a surface mount infrared furnace. The system is con-

**Figure 7.9** Infrared solder reflow furnace. *(Courtesy of Radiant Technology Corporation.)*

structed with 28 tungsten filament lamp, 24" belt-width, and 6 heat zones, wherein each zone has individually controlled top and bottom adjustment. The atmosphere, such as hydrogen, nitrogen, and synthetic gas, can be introduced into both heating and cooling sections of the furnace chamber. The furnace also is built with Venturi assisted exhaust. These individually controlled top and bottom zones add maximum flexibility to the temperature profile arrangement, therefore accommodating a variety of assembly designs and processes. In some furnaces, the distance between emitters and workpiece is also adjustable. In such cases the intensity of radiation is inversely proportional to the square of the distance between the emitter and workpiece. The potential heating rate of different designs and heating sources is rated as:

quartz/tungsten>nichrome>panel heater>convection furnace.

In infrared furnaces, regardless of the type of heating element and emitter design, the heat transfer involves more or less all three types—radiation, conduction, and convection—but different heating elements and designs impart different levels of radiation and

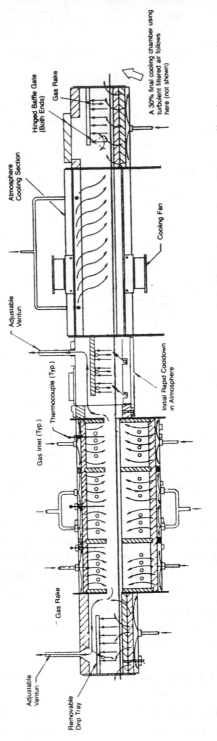

Adjustable Venturi

Removable Drip Tray

Gas Rake

Gas Inlet (Typ.)

Thermocouple (Typ.)

Adjustable Venturi

Atmosphere Cooling Section

Cooling Fan

Initial Rapid Cooldown in Atmosphere

Hinged Baffle Gate (Both Ends)

Gas Rake

A 30% final cooling chamber using turbulent filtered air follows here (not shown)

**Figure 7.10** Chamber schematic of infrared reflow furnace. (*Courtesy of Radiant Technology Corporation.*)

convection. The infrared character and penetrating ability in general follows in the order of:

quartz/tungsten>nichrome>panel heater>convection furnace,

which is in the same order as the heating rate.

## 7.8  INFRARED REFLOW DYNAMICS

It should be noted that the infrared reflow process is a dynamic process, and that conditions of the workpiece are in constant change as it travels through the furnace in a relatively short time. The momentary temperature that the workpiece experiences is the condition determining the reflow condition, therefore the reflow results.

In using infrared soldering, with respect to board and components, geometry, mass, thermal conductivity, and absorptivity are major factors to be considered. The color effect in relation to absorptivity and resulting temperature differentials, and mass effect on resulting temperature by using a quartz/tungsten furnace have been studied.[4] Figure 7.11 shows the results of temperature profiling

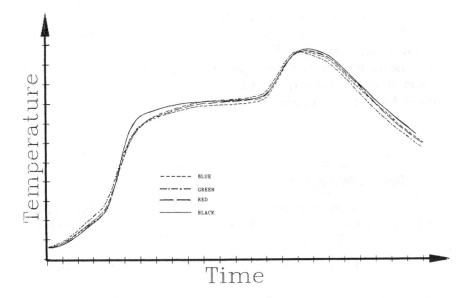

**Figure 7.11**  Color effect on temperature differential.[4]

for different colors, indicating that the colors blue, green, red, and
black make insignificant contribution to temperature differentials
under the specified testing conditions. In reality, the mass per unit
area affects the resulting temperature, and an increase in mass results in lower momentary temperature on the workpiece. Table 7.1
confirms that the mass per unit area affects the temperature. As thermal conductivity increases, the temperature differentials are expected to decline correspondingly. (Section 7.10 will illustrate further on the dynamic nature of infrared reflow.)

## 7.9  INFRARED REFLOW PROFILE

One controlling factor in operating an infrared furnace that needs
special attention is the "true" temperature measurement of the
solder area or other specified areas on the workpiece. The setting
temperature of emitters as displayed on the control panel does not
represent the true temperature of the workpiece. Actually, the true
temperature is significantly different from the setting temperature.
When measuring true temperature is impractical, a knowledge of
setting temperature in relation to reflow performance is the minimum requirement in order to have the reflow process under control.
(Sections 7.10 and 7.11 will discuss further the practical aspects of
temperature measurement and temperature profiling.)

The following two examples are to illustrate that temperature of
workpiece significantly differs from the setting temperatures. It
should be noted that the panel type furnace generally has slower heat
transfer than the lamp type, and there is always a convective heat
transfer between zones and to the workpiece.

**TABLE 7.1**

**Mass Effect on Workpiece Temperature[4]**

| Spacing | Mass Per Square Inch | | | Temperature (°C) | |
| | PLCC | Board | Total | Peak | Net Increase |
|---|---|---|---|---|---|
| 0.2" | 3.76g | 1.86g | 5.62g | 207 | 182 |
| 0.1" | 4.23g | 1.86g | 6.00g | 194 | 169 |
| | % Mass Increase = 8% | | | % Temp Decrease = 8% | |

For a nonfocus panel heating furnace with seven heating controls, the temperature settings are:

|  | Preheat | Zone 1 | Zone 2 | Zone 3 |
|---|---|---|---|---|
| *Top* |  | Control 1 | Control 3 | Control 5 |
|  | 500°C | 350°C | 250°C | 450°C |
| *Bottom* |  | Control 2 | Control 4 | Control 6 |
|  | — | 350°C | 250°C | 450°C |

The corresponding temperature profile on the workpiece is shown in Figure 7.12.

For a lamp-type furnace with six heating controls, the temperature settings (RT = room temperature) are:

|  | Zone 1 | Zone 2 | Zone 3 |
|---|---|---|---|
| *Top* | Control 1 | Control 3 | Control 5 |
|  | RT | RT | RT |
| *Bottom* | Control 2 | Control 4 | Control 6 |
|  | RT | RT | 350°C |

The resulting temperatures in three zones as monitored by thermocouple are:

| Zone 1 | Zone 2 | Zone 3 |
|---|---|---|
| 65°C | 157°C | 355°C |

The corresponding temperature profile on the workpiece is shown in Figure 7.13.

In general, three types of temperature profiles are considered, as shown in Figure 7.14: fast heating and moderate peak temperature (profile A), slow heating and moderate peak temperature (profile B), and fast heating and high peak temperature (profile C). There are unlimited intermediate profiles which can be realized. The precise profile depends on the solder alloy composition, the properties of the paste, and assembly involved. Profile A is suitable for surface mount eutectic paste which does not require predrying; profile B indicates the requirement of a slow heating stage which is most suitable for assemblies containing small ceramic chip capaci-

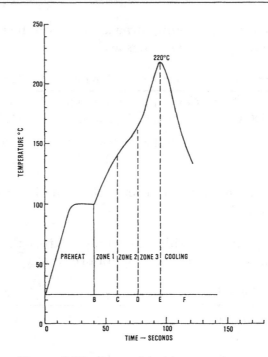

**Figure 7.12** Example of temperature profile obtained from middle-far infrared reflow system.[5]

tors or thermal shockprone components (the maximum heating rate for the components can be obtained from the component manufacturers) and/or for paste which needs predrying; profile C is a sharp profile with a high peak temperature (the peak temperature varies with the specific solder alloy used) that minimizes the preheating. The latter is found to be more compatible with the high lead-containing solders with liquidus temperature higher than 280°C, such as 10-Sn/88-Pb/2-Ag, 1-Sn/97.5-Pb/1.5-Ag, and 5-Sn/92.5-Pb/2.5-Ag, 10-Sn/90-Pb, and 5-Sn/95-Pb.

Many other factors are involved to obtain a precise temperature profile. The profile has to be optimized for a specific assembly and production environment. The repeatability of profile must be confirmed. Most equipment interfaced with a microprocessor is able to store the profile and other operating parameters which can be recalled as needed.

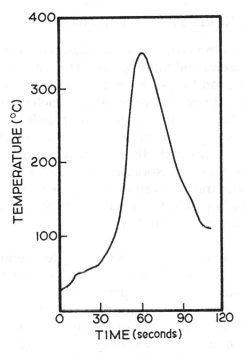

**Figure 7.13** Example of temperature profile obtained from near infrared reflow system.

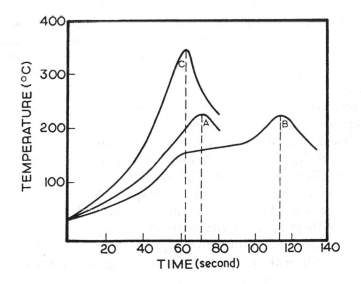

**Figure 7.14**  Three types of infrared reflow profile.

## 7.10  INFRARED REFLOW OPERATION

At this point, it is obvious that there are differences in characteristics among reflow systems and solder pastes. The user has to understand the characteristics and make a proper selection. With the proper furnace selection, the major operating parameters to be monitored include emitter height, temperature setting, belt speed, and atmosphere. The operating parameters depend on the specific assembly (workpiece). For a properly fixed distance between emitter and workpiece and for a given workpiece, the following outlines some information about infrared reflow systems with emphasis on steps and tips to "know" the selected infrared furnace and thus be able to set up an optimum operational process.

**Establish the correlation between setting temperature and measured temperature** of the workpiece at each of specified belt speeds and their relations with reflow performance. This is an important starting point to "know" the latitude of the furnace in relation to the workpiece of interest. As an example, the profiles and peak temperatures of a variety of settings in relation to solder reflow performance are monitored. The resulting correlation between solder performance and temperature provides a "working range" for the assembly as shown in Figure 7.15. The established graph serves as an important tool for the controlling process.

**Establish the correlation between measured temperature and belt speed.** That the peak temperature of the workpiece depends on belt speed, even when the setting temperatures are maintained the same, may seem obvious, but it is important to understand the reasons of the dependency. As the belt speed increases, the residence time of the workpiece in the heat zone decreases and the workpiece only has enough time to reach a lower temperature than it would at a slower speed. This is to be distinguished from the factors resulting from high load or larger total mass in heat chamber. For the purpose of illustrating the point, the response of the peak temperature on the workpiece to belt speed is conducted in a single zone Vitronic furnace at different settings. Figure 7.16 indicates quite consistent trends among different settings. The peak temperature depends on belt speed at all tested settings with peak temperature decreasing with increasing belt speed.

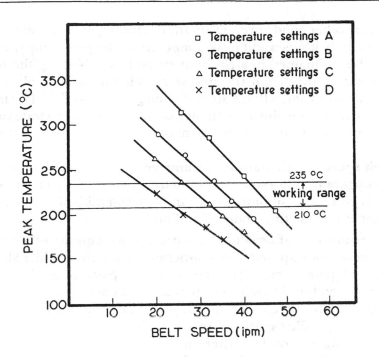

**Figure 7.15** Development of working range of temperature profiles.

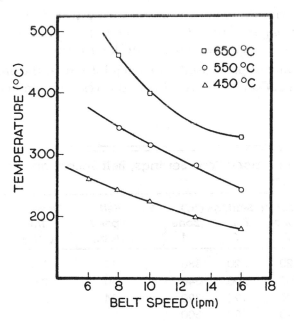

**Figure 7.16** Dependency of peak temperature on belt speed.

On the other hand, through an understanding of the relationship between belt speed and peak temperature, the peak temperature can be maintained in a constant range by establishing the temperature settings and the necessary belt speed. Data obtained from a four-zone reflow system are shown in Table 7.2.[1] The peak temperature is maintained in the constant range over a wide variation of belt speed and corresponding zone temperature settings.

**Belt speed is a dominating parameter in order to maximize the throughput,** from a volume production point of view. The desired high belt speed in most cases can be accomplished by the zone temperature settings, but within limits.

**Determination of peak temperature is a prerequisite of a reflow process.** The required peak temperature for a given solder alloy is a fixed parameter, regardless of which paste is used. It is a misconception that different pastes may demand different peak temperatures. The difference in paste can only change the temperature profile (temperature versus. time) requirement, not the peak temperature. The criterion to determine the peak temperature for reflowing a solder paste is to add 25°–50°C to the liquidus temperature of the solder alloy.

**Varying with the temperature profile, infrared reflow generally demands higher flux activity** than conduction, vapor phase condensation, hot gas, resistance, and laser reflow.

**The amount of paste used for each joint affects the time required to reach reflow** and also affects the residue characteristics. Its

**TABLE 7.2**

Relationship Among Zone Settings, Belt Speed, and Peak
Temperature[1]

| Temperature Settings (°C) | | | | Belt Speed (ipm) | Dewell Time (min) | Peak Temperature (°C) |
|---|---|---|---|---|---|---|
| Zone 1 | Zone 2 | Zone 3 | Zone 4 | | | |
| 360 | 220 | 220 | 380 | 40 | 1.5 | 211–221 |
| 320 | 200 | 200 | 340 | 30 | 2.0 | 210–220 |
| 300 | 200 | 190 | 320 | 24 | 2.5 | 216–224 |
| 290 | 190 | 180 | 300 | 20 | 3.0 | 214–222 |

sensitivity in this respect varies with the types of heating source and the workpiece.

**The material of the workpiece influences the heat transfer rate.** This is reflected in the soldering rate. For infrared furnaces generating wavelength in the middle-to-far infrared region, a printed circuit board generally absorbs heat faster than does hybrid board, with metallic substrate being the slowest. Heat conduction and convection in the infrared furnace may offset some differences, and the extent of conduction and convection varies with the assembly/furnace system.

**A sharp profile facilitates cleanability of solder residue and minimizes the metallurgical reactions,** if potentially present. For relatively high temperature soldering, furnace selection is particularly important in its temperature capability. Operating a furnace at its upper end of temperature capability does not provide a reliable temperature profile, and it also shortens furnace life.

**A near-middle infrared reflow system provides better temperature capability** for high temperature soldering (alloys with liquidus temperature > 280°C).

**For reflowing two-sided SMT board, a combination Edge and Belt Conveyor has been recently introduced.**[1] The edge conveyor allows SMT boards to be supported in minimum contact, with only the outside 0.5 cm (0.200 in.) of the board edge resting on support tabs spaced at 1.9 cm (0.75 in.) intervals, as shown in Figure 7.17(a). The front track remains stationary and the rear track is adjustable from side to side to accommodate any width of boards, as shown in Figure 7.17(b).

In the case of mixed technology that boards contain surface mount components on two sides and through-hole components on one side, equipment and process to solder both sides in one pass are available.[6] Figure 7.18 provides the schematic of the process. The process involves attaching bottom-side surface mount components with adhesive and after-adhesive curing, stenciling the top side with paste and inserting through-hole components on the top side. The reflow soldering utilizes both wave soldering and infrared quartz lamps.

**Atmosphere in the furnace affects the performance of solder paste to a certain degree.** Its impact is most outstanding when using high temperature alloy solder, particularly in residue char-

**Figure 7.17(a)** Combination of edge and belt conveyor. *(Courtesy of Radiant Technology Corporation.)*

**Figure 7.17(b)** Adjustable width of edge and belt conveyor.

acteristics and fluxing ability. Inert atmosphere (e.g., $N_2$) or reducing atmosphere (e.g., $H_2$ or $N_2/H_2$, synthetic gas) facilitates fluxing and residue cleaning. A proper gas flow, including air, in heating chambers, is important in order to minimize flux buildup in the interior of the chamber.

(More information regarding infrared reflow equipment and operation is available in the literature.[4, 7, 8, 9])

In summary, a logical starting point in conjunction with trial-and-confirming experimentation is an effective way to obtain an optimum infrared temperature profile, and thus a reliable and productive reflow process for a specific assembly.

**Figure 7.18(a)** Single pass soldering process for mixed technology boards (a) equipment. *(Courtesy of Hollis Automation Company.)*

## SINGLE PASS SOLDERING SYSTEM
### (SPS)

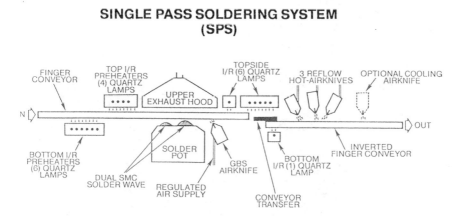

**Figure 7.18(b)** Single pass soldering process for mixed technology boards (b) schematic. *(Courtesy of Hollis Automation Company.)*

## 7.11   TEMPERATURE MONITORING

As mentioned in previous sections, capability of monitoring the "true" momentary temperatures on the workpiece is important to assure the reliability and quality of the process. Available temperature measurement techniques are reviewed.[10] Until now, most techniques involve the use of thermocouples. The recently developed temperature profiler is able to sense temperature at five locations simultaneously through thermocouples.[1] Figure 7.19 illustrates the temperature profiler and a printed circuit board (test board). During reflow process, the profiler moves along with the board and records data which are then processed by the designed software to generate a temperature profile.

Since most techniques of measuring furnace temperature use thermocouples, a few words about thermocouples are in order. At a steady state, a thermocouple always reads true temperatures. However, in a dynamic environment, such as infrared reflow, where the temperature on the workpiece is changing momentarily with the travelling distance, the response efficiency of a thermocouple needs to be considered. The response efficiency normally depends on the

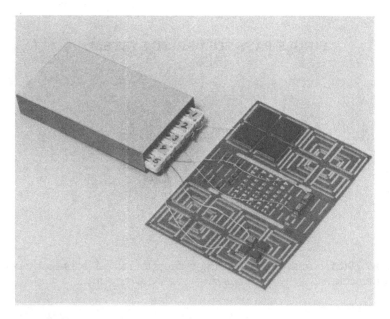

**Figure 7.19**   Temperature profiler. *(Courtesy of Radiant Technology Corporation.)*

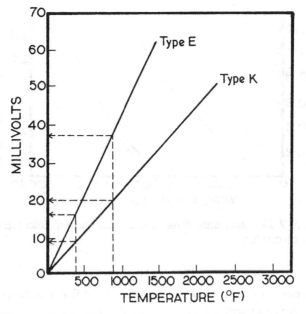

**Figure 7.20** Temperature response of thermo-couple type E and type K.

composition of thermocouple wire and the diameter of the wire. Figure 7.20 indicates the relative temperature response of two thermocouple types (Type K: nickel-chromium/nickel-aluminum, and Type E: nickel-chromium/copper-nickel), and Figure 7.21 shows the response time in relation to the diameter of thermocouple wire.

Another factor demanding attention is the attachment of thermo-couple to the workpiece. The way the thermocouple wire is attached to the workpiece should be kept constant. It is the intimate physical contact between thermocouple bead and the workpiece that deter-mines the workpiece temperature measured.

## 7.12  VAPOR PHASE REFLOW

Since vapor condensation soldering was invented by Western Elec-tric in 1975, fluid development and equipment development have made vapor phase soldering commercially feasible. Chemicals such as perfluorocarbon, fluoropolyethers, perfluorotriamylamine and perfluorophenanthrenes have been used as vapor phase fluids. The perfluorinated hydrocarbon has been the predominate fluid chemi-

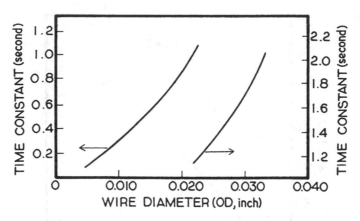

**Figure 7.21**  Response time of thermocouple in relation to wire diameter.

cal and is prepared by either electrochemical fluorination or fluorination of hydrocarbon expressed as follows:

$$H + \left( \begin{matrix} H & H \\ | & | \\ C - C \\ | & | \\ H & H \end{matrix} \right)_n H \; + \; 6n \; HF \; \xrightarrow{\epsilon} \; F + \left( \begin{matrix} F & F \\ | & | \\ C - C \\ | & | \\ F & F \end{matrix} \right)_n F \; + \; 6n \; H_2$$

$$H + \left( \begin{matrix} H & H \\ | & | \\ C - C \\ | & | \\ H & H \end{matrix} \right)_n H \; + \; 6n \; F_2 \; \longrightarrow \; F + \left( \begin{matrix} F & F \\ | & | \\ C - C \\ | & | \\ F & F \end{matrix} \right)_n F \; + \; 6n \; HF$$

Fluids with a wide range of boiling temperature (56°–253°C) are currently available. The fluid is generally characterized by thermal stability, chemical inertness, nonflammability, low vapor pressure, low surface tension, and high density. Table 7.3 summarizes typical properties of common reflow fluids.[11, 12] Under continuous, long-time operation, decomposed chemicals such as perfluoroisobutylene (PFIB) and hydrofluoric acid in the presence of moisture have been found. These chemicals are considered to be toxic or corrosive. Lately, the stability of fluid has been further improved to alleviate potential formation of decomposition products.

**TABLE 7.3**

**Typical Properties of Vapor Phase Reflow Fluids**

| Product | Boiling Point (°C) | Kinematic Viscosity (CS) | Vapor Pressure (torr, 25°C) | Surface Tension (dynes/cm; 25°C) | Molecular Weight |
|---|---|---|---|---|---|
| Fluorinert FC-43 | 174 | 2.8 | 1.3 | 16 | 670 |
| Fluorinert FC-5311 | 215 | 8 | <0.1 | 19 | 624 |
| Fluorinert FC-70 | 215 | 14 | <0.1 | 18 | 820 |
| Flourinert FC-5312 | 215 | 12.6 | <0.1 | 18 | 820 |
| Fluorinert FC-71 | 253 | 73 | <0.02 | 18 | 970 |
| Multifluor APF-175 | 180 | 3.5 | 0.67 | 25 | 570 |
| Multifluor APF-200 | 200 | 5.3 | 0.229 | 19.8 | 620 |
| Multifluor APF-215 | 215 | 8.0 | 0.12 | 21.6 | 630 |
| Multifluor APF-240 | 240 | 18.4 | 0.009 | 21.1 | 770 |
| Galden LS/215 | 215 | 3.8 | <0.001 | 20 | 950 |
| Galden LS/260 | 260 | 7.0 | <0.01 | 20 | 1210 |
| Galden LS/265 | 265 | 8.0 | <<0.01 | 20 | 1250 |

Note: Fluorinert is registered trademark of 3M Corporation. Multifluor is registered trademark of Air Products, Incorporated. Galden is the product of Ausimont.

The principle of the heat transfer in vapor phase method to achieve solder reflow is straightforward. The latent heat ($H$), as a result of phase transition from vapor to liquid, is the heat source. The latent heat stored in the vapor is transferred to the workpiece through the condensation process to reflow the solder.

$$F \left(\!\! \begin{array}{cc} F & F \\ | & | \\ C\!-\!C \\ | & | \\ F & F \end{array} \!\!\right)_{\!n}\!\! F \ \text{(gas)} \longrightarrow F \left(\!\! \begin{array}{cc} F & F \\ | & | \\ C\!-\!C \\ | & | \\ F & F \end{array} \!\!\right)_{\!n}\!\! F \ \text{(liquid)} \ + \Delta H$$

The whole workpiece is heated up to the temperature corresponding to the boiling point of the fluid used.

The selection of fluid is based on the solder alloy employed in the paste. A minimum of 25°C above the liquidus temperature of the alloy is required. For example, to reflow 63-Sn/37-Pb, the proper fluid has a boiling point of 215°C, and for 96.5-Sn/3.5-Ag, the fluid with boiling point 258° ± 7°C is needed, which is also the highest boiling point among the commercially available fluids.

All vapor phase reflow equipment is based on the same principle, yet is designed with different features by manufacturers, namely, immersion heater system[13] or thermal mass system,[14] batch unit or in-line unit, single fluid or dual fluid, and with or without the incorporation of preheating/predrying zone. Figure 7.22 is the schematic of an in-line system.

The rate of heat transfer in vapor phase reflow is relatively high in comparison with that of infrared radiation or convection. The rate of heat transfer decreases as the vapor and workpiece temperature

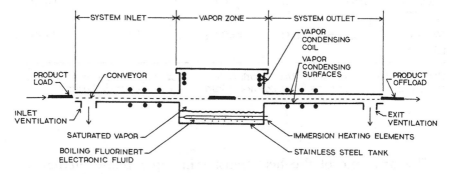

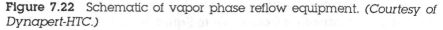

**Figure 7.22** Schematic of vapor phase reflow equipment. *(Courtesy of Dynapert-HTC.)*

approach equilibrium in an asymptotical manner, as shown in Figure 7.23.

The merits of vapor phase process include uniform temperature distribution, precise temperature on workpiece, nonoxidizing atmosphere, and fast heat transfer as shown in Figure 7.23 in comparison with hot air at equivalent temperature for heating steel bar. In addition, it is geometry-independent, providing uniform temperature to the workpiece. On the other hand, the fluid is costly, the equipment demands maintenance, and the operating temperature relies on the availability and quality of fluids.

Some practical points in using vapor phase reflow are as follows:

- The boiling point of fluid used should be at least 25°C higher than the liquidus temperature of solder alloy
- The materials of components and boards should be compatible with the fluid.
- A dual vapor is less prone to tombstoning problem encountered with small passive chips.

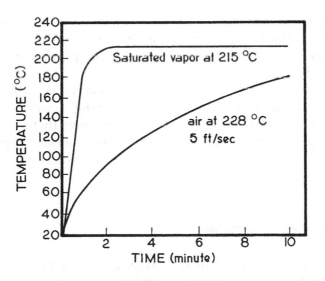

**Figure 7.23**  Heating rate of vapor phase condensation and hot air heat transfer to steel bar. *(Courtesy of 3M Corporation.)*

- The predrying of a paste eliminates skewing or tombstoning problem; common drying conditions are recommended to be 50°–100°C for 15–30 minutes depending on the type of paste. (More details are discussed in Section 10.4.)

- The preheating of the workpiece to offset the transient temperature gradient between J leads of PLCC and solder pads of printed wiring board reduces the potential starved or open joints due to molten solder wicking up to the leads. The problem is attributed to many sources and is discussed in Section 10.1.

- The purity and cleanness of the fluid in the sump should be maintained.

In this text, predrying and preheating are intended to have different meanings. Predrying denotes that heat is applied prior to reflow in order to expel volatiles in the paste, and preheating is to apply heat prior to reflow in order to "warm up" the workpiece.

## 7.13  CONVECTION REFLOW

Convection furnaces normally are capable of handling high volume production. The heating process is slow and needs more time to reach equilibrium conditions. Specific atmosphere can be easily introduced and controlled.

Focused convection reflow equipment is available, which is constructed with multiple preheating/predrying zones to provide forced hot air circulation preceding the air knives that direct necessary thermal energy to reflow solder.[15] The process is reported to complete the reflow process in much less time than could infrared and vapor phase technology.

## 7.14  HOT GAS REFLOW

With proper adjustment in flow rate, gas temperature, and the distance between the nozzle and workpiece, soldering by means of heat transfer from hot gas to the workpiece has been found to be convenient in many cases. The hot gas can solder within a relatively short time and is able to handle high temperature soldering with less vulnerability to residue charring.

## 7.15  RESISTANCE REFLOW

The heat in resistance reflow is transferred by the mechanical contact between the workpiece and the heater tip mounted in a resistance welding head which is heated by a power supply. It is applicable to both solder paste and prebumped solder. Fine pitch flatpacks (large scale integrated circuit devices) have been successfully soldered to the printed wiring board by using this technique. The process involves prebumping the solder pads with solder paste to produce 0.025–0.075 mm (0.001–0.003 in.) thickness of solder, and then re-flowing.[16]

## 7.16  LASER REFLOW

Two types of laser have been applied to solder reflow: carbon dioxide ($CO_2$) and neodymium-doped-yttrium-aluminum-garnet (Nd:YAG) lasers.[17, 18, 19, 20] Both generate radiation in the infrared region with wavelength around 10.6 µm from the $CO_2$ laser and 1.06 µm for the YAG laser. The wavelength of 1.06 µm is more effectively absorbed by metal than by ceramics and plastics. The wavelength of 10.6 µm is normally reflected by conductive surfaces (metals) and absorbed by organics. In addition to the wavelength difference, the Nd:YAG laser is capable of focusing down to 0.0500 mm (0.002 in.) diameter in comparison with $CO_2$ laser focusing to 0.100 mm (0.004 in.) diameter.

The main attributes of laser soldering are short duration heating and high intensity radiation which can be focused onto a spot as small as 0.050 mm (0.002 in.) diameter. With these inherent attributes, laser reflow is expected to

- provide highly localized heat to prevent damage to heat-sensitive components and to prevent crack of plastic IC packages,
- provide highly localized heat to serve as the second or third reflow tool for assemblies demanding multiple step reflow,
- require short reflow time,
- minimize intermetallic compound formation,
- minimize leaching problem,
- generate fine grain structure of solder,
- reduce stress buildup in the solder joint, and
- minimize undesirable voids in the solder joint.

With these attributes in mind, laser soldering is particularly beneficial to soldering densely packed regions where local solder joints can be made without affecting the adjacent parts, to soldering surface mount devices on printed circuit boards having heat sinks or heat pipes, and to soldering multilayer boards. In addition it also provides sequential flexibility of soldering different components and relieves the temperature requirement for adhesives used for mounting surface mount devices.

With respect to reflow time, laser soldering can be accomplished in less than a second, normally in the range of 100–800 msec. The laser can be applied to point-to-point connections through pulsation, as well as to line-to-line connections via continuous laser beam scan.

The fine pitch flatpack devices have been connected to printed wiring board by using YAG continuous laser beam scan on each side of the package. Prebumped solder pads and directly applied solder paste are feasible. In directly reflowing solder paste, although splattering and heat absorption problems have been observed, they are not incurable. To eliminate these problems, compatible properties of solder paste have to be designed to accommodate fast heating in relation to fluxing and paste consistency, coupled with the proper design in equipment and its settings.

In using the laser, another concern is energy absorption by the printed circuit boards which leads to board damage. This is considered to have been corrected by switching from $CO_2$ laser to YAG laser. Due to the wavelength difference, the energy absorption by polymers can be minimized. E. F. Lish has found that sometimes complications may occur.[21] In assembling multilayer polyimide boards by using laser as a second step reflow, burning was found in the board while it was moving under the laser. The burning is traced to the color pigments contained in the adhesive which is used for attaching heat sinks. The problem is eliminated by using noncolor pigment in this adhesive. This is another clear demonstration that consideration of all materials in reflow process is needed.

Regarding the resulting solder joint, fine grain microstructure and significantly reduced intermetallic compounds formation at copper and solder (63-Sn/37-Pb) interface have been observed when using laser soldering in comparison with other reflow methods.[19] Stress buildup in the solder joint due to the thermal expansion coefficient difference between the material on both sides of the solder joint is expected when the reflow method requires the whole assembly to be exposed to the soldering temperature. Localized heating and short duration of exposure by using laser is expected to generate less stress in the joint for the assemblies having thermal expansion coefficient difference among the materials. Further studies in this area are apparently warranted.

## 7.17  INTELLIGENT LASER SOLDERING

The intelligent laser soldering system manufactured by Vanzetti Systems synchronizes the laser reflow soldering and laser/infrared inspection, which is discussed in Section 9–13. The key functions of the equipment are fulfilled by a dual-shuttered Nd:YAG laser (typical C.W. 30 W at the source, which become approximately 20 W at the target, in a focal area of 0.019 in. diameter) to heat solder joint; a helium-neon laser for positioning; the optical fiber to homogenize and to precisely aim the laser beam; an infrared detector to control the solder joint formation; and a minicomputer to monitor and process data.

The infrared detector monitors and controls the reflow process by measuring the radiation emitted from the joint. As focused laser energy impinges on the solder by opening the control shutter, the infrared detector also views the solder. As soon as the detector senses the phase change from solid to liquid (melting), the emitted signal closes the shutter. By doing so, a correct amount of laser energy is supplied and therefore the overheating is avoided. The analog signal from infrared detector depicts the thermal history of the solder joint formation, which is called infrared signature. The computer then converts the analog signal to digital which is compared with standard expected signal, indicating the quality of the as-formed solder joint. Figure 7.24 depicts the radiation versus time profile during the formation of a solder joint between lead of a surface mount component and the solder pad. The solder joints produced through this system are inherited with the merits of laser soldering as listed in Section 7.15. Furthermore, the system provides real-time detection of any deteriorating phenomena to allow prompt correction at the source and eliminates the need for subsequent inspection.

By using this technique, components containing 280 leads with 0.010 in. (0.25 mm) pitch have been successfully mounted on a 40-layer printed circuit board within 34 sec.[22] The synchronized operation provided by the Intelligent Laser Soldering System is expected to yield savings both in capital investment and in labor manufacturing.

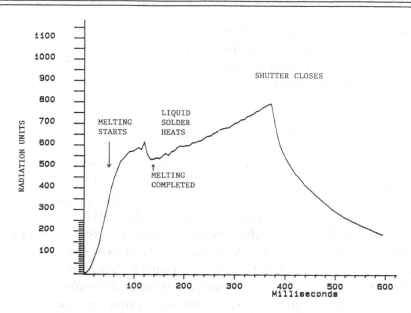

**Figure 7.24** Thermal signature during solder joint formation of an SMD lead using Intelligent Laser Soldering System. *(Courtesy of Vanzetti Systems, Incorporated.)*

## 7.18  INDUCTION REFLOW

The basic setup in induction heating is supplying an alternating current to the heating coil that is magnetically coupled with the metal to be heated. When the alternating current is passed through the heating coil, a magnetic field is produced. The direction change in the magnetic field as a result of the alternating current induces voltage in the metal placed in this field, which in turn generates the current in the closed electric circuit.

The heat produced is proportional to the resistance of the metal at the frequency of the current flow ($Rf$) and to the square of the current ($I$) passing through the metal, as shown by

$$Q \propto R_f I^2.$$

Induction heating is characterized by surface heating, extremely rapid heat transfer, and the capability of reaching high temperature quickly. For surface mount on printed circuit board, this method has its limits, but it is a viable technique for reflowing high temperature paste for lead attachment.

## 7.19  TEMPERATURE PROFILING VERSUS PASTE PERFORMANCE

When heat is applied to paste prior to solder melt, paste may respond in one or more ways such as volatile evaporation, chemical activation, chemical degradation, and paste spattering. The response depends upon the temperature and heating rate applied, and the specific type of paste—that is, its chemical constituents.

For tin-lead (tin-lead-silver) eutectic or near-eutectic alloys, the temperature in the range of room temperature to about 160°C is considered as preheating/predrying stage. Within this range, volatile evaporation and flux activation are expected to occur.

Volatile evaporation prior to solder melt can be an advantage to flux activity and to void reduction in solder joint if a dryable paste is used. The volatiles are also related to other soldering problems such as component skewing or tombstoning. Dryable paste is defined by its composition containing a certain amount of chemicals which are volatile in this temperature range in contrast to nondryable paste that does not contain volatiles, although most chemicals have vapor pressure to some extent.

For nondryable paste, preheating/predrying may not be a benefit. In some cases, excessive preheating/predrying may have adverse effect on paste, such as solder balling. On the other hand, to ramp up temperature too fast without preheating/predrying on dryable paste may cause paste spattering, resulting in solder balls and inadequate fluxing activity. Spattering from too fast heating is only one of several sources to solder balling. (Section 10.5 discusses the sources of solder balling.)

To establish a reflow temperature profile, the understanding of paste properties is obviously important. In addition, solder alloys play a role in paste performance in relation to profiling. For low temperature alloys such as bismuth- and indium-containing alloys, much lower temperature is required for reflow. The preheating/predrying temperature that can be used for flux activation and volatile evaporation is accordingly lower. Again, the knowledge of paste properties is crucial to proper reflow profile.

For high temperature alloys such as high-lead alloys requiring reflow temperature at 300°C or above, chemical degradation which may cause chemical charring and subsequent residue cleaning difficulty is the most concern. Avoiding the charring by minimizing soak time at moderate-high temperature during preheating/predrying stage is one of the major considerations to temperature profiling.

To complete profiling, cooling from melt is another part of the profile. Cooling rate has been recognized as a factor for microstructure formation of solder alloy, as shown in Figure 4.20, which is in turn a factor for solder joint integrity. In general, fine grain of Sn/Pb eutectic structure is developed when solder melt is cooled down at a rapid rate. For high-lead alloys, Figure 7.25 (a) and (b) indicate the microstructure development of 25-Sn/75-Pb solder with decreasing cooling rate from (a) → (b). The size of dendrites is increased significantly as cooling rate is decreased. The peak temperature setting is another factor which affects microstructure development when other conditions are equal. Figure 7.26 (a) shows the development of large size dendrites of 25-Sn/75-Pb solder due to slow cooling. Figure 7.26 (b) shows the existence of voids between dendrites and dendritic arms when the peak temperature is increased by 30°C. In this case, the shrinkage of eutectic phase is considered to be the source of voids.

## 7.20  TEMPERATURE PROFILING VERSUS COMPONENT

When using a reflow method of which the whole assembly needs to be heated, such as vapor phase system, convection furnace, and infrared furnace, several component-related concerns have emerged.

**Figure 7.25(a)**  SEM microstructure of 25-Sn/75-Pb at different cooling rates: (a) fast cooling.

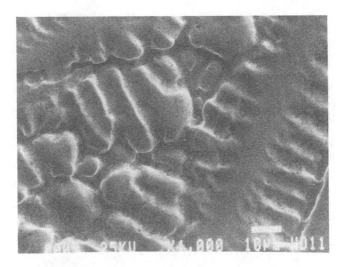

**Figure 7.25(b)**  SEM microstructure of 25-Sn/75-Pb at different cooling rates: (b) slow cooling.

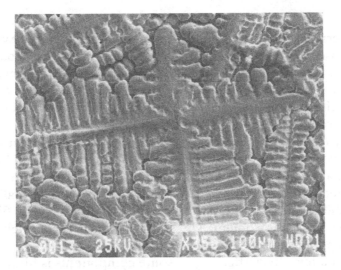

**Figure 7.26(a)**  SEM microstructure of 25-Sn/75-Pb melt at higher peak temperature.

These concerns have not been associated with wave soldering, or other reflow methods such as laser, hot air, and resistance, which do not require heating of the whole assembly.

Assembly is consisted of a variety of components and devices which are made by different materials and in different sizes and mass. Thus

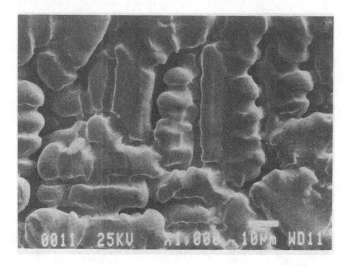

**Figure 7.26(b)**   SEM microsctucture of 25-Sn/75-Pb.

their thermal conductivity and thermal shock resistance are varying. The transient temperature at different components and board can therefore vary in the range of 5°–40°C. Infrared reflow systems which are built with independently controlled top and bottom emitters provide flexibility in temperature profiling, enabling to compensate the difference among board and components.

The capacitor and resistors have shown crack problem during reflow process when temperature profile is incompatible. The heating rate and cooling rate to the assembly containing these components need to be designed to accommodate the thermal shock vulnerability. It is found that the heating rate of slower than 6°C/sec can overcome the problem. The specific criteria recommended by the component manufacturers should be followed.

The shadow effect (heating to some components is hindered by adjacent components) has been a concern in near-infrared systems. This effect can be alleviated by a proper furnace setting, such as lowering power input and increasing convection heating. It is also suggested that the use of diffuse ceramic back plane is able to minimize the shadow effect.[23]

Crack problems have also been encountered in plastic IC packages, particularly in packages involving large IC chips. It is found that the crack developed during vapor phase reflow (or during any other heat excursion, not necessarily associated with vapor phase) is caused by the vapor pressure of moisture inside the package, which delaminates the resin and metal (chip substrate) and subsequently results in crack. To reduce the moisture absorption/desorption in the package, a protective packing system for plastic IC packages was developed and found to be effective.[24] In another study, the moisture content in the plastic flat pack packages was also identified to be the cause of the package crack during reflowing, and laser soldering was found to be able to prevent the crack of such flat pack IC packages by localizing the heat only at solder joint area.[25]

## 7.21   TEMPERATURE PROFILING VERSUS SOLDERING PROBLEMS

Temperature profiling in vapor phase (belt speed change) and in infrared reflow systems is directly related to some of the problems on production floor such as component skewing, tombstoning, and solder wicking on surface mount J leads. (These areas are discussed in Chapter 10.)

## 7.22   COMPARISON OF REFLOW METHODS

With the above discussion in mind, each reflow method has its uniqueness and beneficial features in cost, in performance, or in operational convenience. Table 7.4 summarizes the strengths and limitations of each method. For localized and fast heating, laser excels over other methods, with hot gas in second place. For uniform temperature of the whole assembly, vapor phase ranks first. For versatility, volume, and economy, infrared with proper wavelength is the choice. For low volume, hybrid assembly, conduction heating is a convenience. For conductive components requiring high temperature soldering, induction heating meets the requirements. The ideal equipment is the one which is most compatible with the specific assembly and production environment in yield, quality, and cost.

**TABLE 7.4**

**Reflow Methods—Feature Comparison**

| Reflow Method | Benefits | Limitations |
|---|---|---|
| Conduction | Low equipment capital<br>Rapid temperature changeover<br>Visibility during reflow | Planar surface and single-side attachment requirement<br>Limited surface area |
| Infrared | High throughput<br>Versatile temperature profiling and processing parameter | Mass, geometry dependence |
| Vapor Phase Condensation | Uniform temperature<br>Geometry independence<br>High throughput | Difficulty to change temperature<br>Temperature limitation<br>Relatively high operation cost |
| Hot Gas | Low cost<br>Fast heating rate<br>Localized heating | Temperature control<br>Low throughput |
| Convection | High throughput | Slow heating |
| Induction | Fast heating rate<br>High temperature capability | Applicability to non-magnetic metal parts only |
| Laser | Localized heating with high intensity<br>Short reflow time<br>Superior solder joint<br>Package crack prevention | High equipment capital<br>Specialized paste requirement<br>Limit in mass soldering |

# REFERENCES

1. Ahmet N. Arslancan. Radiant Technology Corporation, Cerritas, California.
2. Equipment Literature. Vitronics Corporation, Newburyport, Massachusetts.
3. Excel Company, Ltd., Index Corporation, San Jose, California.

4. Ahmet N. Arslancan and David K. Flattery, "Infrared Reflow for SMT: Thermal and Yield Considerations," proceedings, 1987 National Electronic Packaging and Production Convention–West: 357.

5. R. G. Spiecker and D. Haggis, "Soldering Techniques for Surface Mounted Leadless Chip Carriers," 1986 proceedings, International Symposium on Hybrid Microelectronics, 902.

6. A. V. Sedrick, "How to Solder Two Sides in One Pass," *Circuits Manufacturing* (Nov. 1986).

7. James E. Smorto, "Computer Controlled Infrared Solder Reflow: One Manufacturer's Approach," *SITE* (Apr. 1987).

8. David K. Flattery, "Infrared Reflow for Solder Attachment of Surface Mounted Devices," *Connection Technology* (Feb. 1987).

9. Norman R. Cox, "Infrared Solder Reflow of Surface Mounted Devices," *Hybrid Circuit Technology* (Mar. 1985).

10. J. Holloman and D. Rall, "Advances in Reflow Soldering Process Control," proceedings, 1987 National Electronic Packaging and Production Convention–West: 393.

11. "Fluorinert Liquids," product bulletin, 3M/Industrial Chemical Products Division.

12. "Multifluor Inert Fluids," product bulletin, Air Products and Chemicals, Incorporated.

13. Product bulletin, HTC, Concord, Massachusetts.

14. Michael J. Rickriegel, "The Benefits of Dual Vapor Inline Vapor Phase Soldering" (Corpane Industries, Incorporated).

15. Product data sheet, Hollis Automation, Nashua, New Hampshire.

16. "Catalog 501," Hughes Aircraft Company, Carlsbad, California.

17. E. Lish, "Applications of Laser Microsoldering to SMDs, proceedings, 1985 International Hybrid Microelectronics Symposium: 1.

18. F. Burns and C. Zyetz. *Electronic Packaging and Production* (May 1981).

19. Edgar A. Wright, "Laser Versus Vapor Phase Soldering," Society for the Advancement of Material and Processing Engineering (1985).

20. Carl B. Miller, "Lasers in Hybrid Circuit Manufacturing," *Hybrid Circuit Technology* (Jul. 1984).

21. Private communication: E. F. Lish. Martin Marietta Corporation.

22. Private communication: Riccardo Vanzetti. Vanzetti Systems, Incorporated, Stoughton, Massachusetts.

23. Norman R. Cox, "Lamp IR Soldering," *Circuits Manufacturing* (May 1988).

24. I. Anjoh et. al., "Analysis of the Package Cracking Problem With Vapor Phase Reflow Soldering and Corrective Actions," *Brazing and Soldering* 14 (Spring 1988).

25. Hiroshi Miura, "Overview of YAG Laser Soldering Systems," *Electronic Manufacturing* (Feb. 1988).

# CHAPTER 8

# Cleaning

## 8.1 SOLDER PASTE RESIDUE

The residue which is left on or around a solder joint after reflowing solder paste is expected to be the mixture of ingredients in the flux/vehicle portion of the paste, interreaction products among ingredients, and the reaction product between ingredients and substrate or solder alloy. Therefore, chemically, the residue is composed of polar organics, nonpolar organics, ionic salts, and metal salts of organics. For a typical RMA-type system, hypothetical constituents in the residue may contain the following:

- gum rosin or other rosins
- gum rosin derivatives or other rosin derivatives
- carboxylic acid
- amine and derivatives
- hydrocarbon
- alcohol
- ether
- ester
- fatty acid or fatty amine
- lead salt of rosin acid
- tin salt of rosin acid
- copper salt of rosin acid

## 8.2   WHAT IS CLEANING?

With all these chemicals in mind, how should one thoroughly remove them? One needs to answer four questions in dealing with a specific assembly:

1.   What is the most effective and practical solvent system?
2.   What is the most compatible cleaning technique?
3.   How is the cleanliness measured?
4.   What are the criteria of cleanliness?

As electronics packaging goes into more dense and intricate design, the questions not only involve solvent selection but also the cleaning technique which can provide adequate access of solvent to hard-to-reach areas, such as underneath the components and between densely packed components and devices. Due to the overall relative inaccessibility and versatility of the circuit board design, the establishment of a test to verify the cleanliness is another task one must accomplish. This may involve both direct measurement and functional tests.

This chapter is intended to provide such information that the above questions can be properly handled when selecting and operating the solvent/equipment system.

## 8.3   CLEANING PRINCIPLE

When an object having a completely exposed surface is to be cleaned, one should consider the intrinsic properties of a cleaning agent in its wetting ability, softening ability, and solubilizing ability. In addition, the environment of the object is also an important factor to consider in achieving a good cleaning. In other words, a cleaning agent must be able to be intimately in contact with the "debris" to be removed, to "blend" with the debris, and to carry away the debris from the object through flushing and adequate mobility. In dealing with the mix of exposed surface and hidden surface, additional factors such as accessibility and capillarity effect are involved.

Wetting ability of a liquid cleaning agent is known to be related to the viscosity, the density in relation to the debris, and the surface tension of the liquid. Generally, wetting is inversely related to viscosity and surface tension, and is directly related to density. (The solubilizing effect is covered in Section 8.5.) Softening ability is strongly dependent on temperature, although it is in most cases related to solubilization. It should be noted that the dependency of capillary effect on the physical properties of a liquid is a different relationship. For a two-plate system,

$$x \propto \frac{\gamma}{Wdg},$$

where $x$ is the distance which the liquid travels, $\gamma$ is the surface tension of the liquid, $W$ is the gap between plates, $d$ is the density of the liquid, and $g$ is the gravitational constant. Higher surface tension is therefore expected to promote the capillary effect. Cleaning efficiency thus lies in the net balance of these forces, which depends on the design of the assembly and the specific process.

## 8.4 FACTORS OF CLEANING EFFICIENCY

The factors affecting cleaning efficiency are outlined as follows:

**Physical and chemical properties of the flux/vehicle system of the solder paste.** The cleaning solvent selected and solder paste used have to be compatible with each other.

**Reflow time.** With respect to cleaning, the reflow time is considered as the time that heat is imposed on the residue. Thus the reflow time is expected to affect cleaning. Figure 8.1 exemplifies

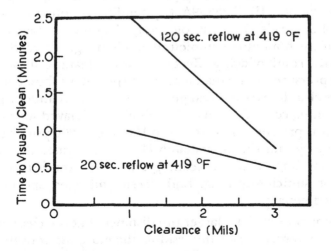

**Figure 8.1** Effect of reflow time on cleaning over a range of clearance.[1]

the effect of reflow time on cleaning at different levels of clearance.[1] Longer reflow time demands more cleaning time to achieve the same level of cleanliness, and excessive reflow may render the residue uncleanable.

**Reflow temperature.** In addition to time of heat exposure of the residue, temperature is another parameter affecting residue. Excessive temperature aggravates the chemical reactions producing undesirable reaction products that demand more vigorous cleaning. In an extreme case, the charred organics may never be able to be removed with the state-of-the-art cleaning techniques. For surface mounting components on board, the use of tin-lead eutectic alloy still keeps reflow temperature in the moderate temperature range with respect to thermal decomposition and chemical reaction. For most lead attaching applications using high temperature solder alloys, the required reflow temperature is aggressive to most organic chemicals. In such cases, control of reflow temperature becomes more critical to residue cleaning.

**Residue age.** After the solder paste is reflowed, the time elapsed between the completion of reflow and the commencement of cleaning contributes to the variation of cleaning results. The extent depends on the chemical, physical, and thermal properties of the specific paste used and on the reflow process. In some cases, the effect may not be significant enough to be detected. However, it is prudent to clean the residue immediately after the reflow. A complete in-line assembling process mandates this step. Military Specification MIL-P-28809A (Appendix III) on printed wiring assemblies calls for post-soldering cleaning to be conducted within one hour after completion of soldering. Figures 8.2 and 8.3 show the result of aging effect on residue cleaning at two levels of spray pressure.[2] Immediately cleaned specimens show consistently better cleanliness over a range of cleaning times under a given set of cleaning conditions than the 30-minute delayed specimens. As the spray pressure increases from 100 psig to 200 psig, cleaning is much easier for the immediately cleaned specimen than the one-hour specimen at short cleaning time. However, if this cleaning time is sufficiently long, both fresh and aged specimens are cleaned equally.

**Component size.** The larger the distance that the cleaning agent needs to travel to reach the residue, the more vigorous the level of cleaning process needed. Therefore, the vigor of cleaning condi-

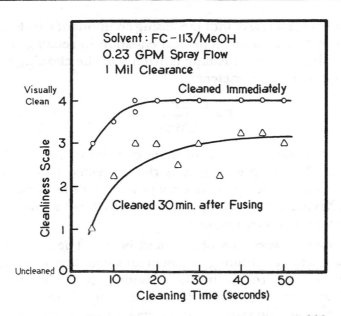

**Figure 8.2** Effect of residue age on cleaning at 100 psig spray pressure.[2]

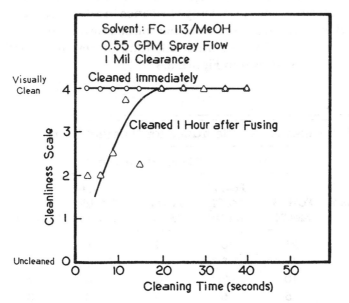

**Figure 8.3** Effect of residue age on cleaning at 200 psig spray pressure.[2]

tion needed increases with increasing component size. R. Musselman has derived a mathematical equation in relating the shear
force, which reflects cleaning efficiency to the cleaning parameters and design parameters:[3]

$$F = \frac{C\rho\Delta P^2 Y^4 A}{288\,\mu^2\,x^2},$$

where $C$ is the coefficient of drag, $\rho$ is the density of cleaning
solvent, $P$ is the pressure, $A$ is the projected frontal area of
contamination facing the velocity profile, $\mu$ is the viscosity of the
fluid, $X$ is the width of component, and $Y$ is the clearance between
the component and the board.

**Clearance between component and board.** The clearance is a
critical factor to cleaning. Musselman's equation indicates that
the shear force created during a cleaning operation depends on
the fourth power of the clearance distance. The demands on
cleaning conditions to achieve cleanliness in relation to the
clearance is shown in Figure 8.1. Diminishing clearance results in
prolonged cleaning time under the given testing condition to
achieve the same level of cleanliness. Table 8.1 further shows the
effect of different solvent systems. As can be seen, the cleaning
time increases with decreasing clearance in all cases for all solvents
tested. Another test also shows that the clearance effect diminishes
if the cleaning time is sufficiently long under the specified test
conditions, as shown in Figure 8.4.

**TABLE 8.1**

**Cleaning Times versus Clearance for Different Solvents[1]**

| | Cleaning Time (min) | | | | | |
|---|---|---|---|---|---|---|
| Clearance (mils) | FC-113 MeOH | FC-113 MeCl$_2$ MeOH | FC-113 MeCl$_2$ | 1,1,1- Trichlor- oethane | MeCl | IPA*/ H$_2$O |
| 1 | 21 | 16 | 14 | 18 | 17 | — |
| 3 | 14 | 8 | 9 | 11 | 9 | — |
| 6 | 10 | 7 | 8 | 8 | 7 | >30 |
| 15 | 5 | 5 | 6 | 3 | 3 | — |

*Isopropyl alcohol.

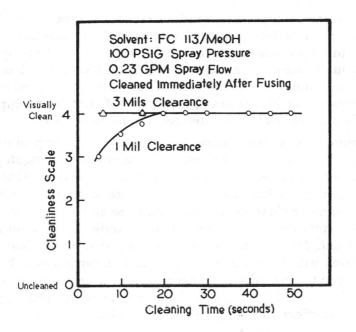

**Figure 8.4**  Effect of clearance on cleaning over a range of cleaning time.[2]

**Compatibility with substrate material and flux/vehicle system.** The chemicals in the paste in some cases may stain the alumina substrate when reflowing solder paste on hybrid board. Some chemicals may react with polymeric materials used on printed wiring board such as solder mask. When the problem is small, minimizing the reflow time and temperature may alleviate the problem. The incompatibility (reactivity) not only jeopardizes cleanability but also may lead to other soldering-related problems.

**Solvent selection.** The solvent should be effective for the specific solder paste. Generally, when used alone, neither fluorocarbon-based nor chlorocarbon-based is a safe approach. A polar solvent is needed to complement the nonpolar nature of fluorocarbons and chlorocarbons to achieve good cleanliness. (Section 8.5 covers the basic properties of different solvent system.) In addition to cleaning capability, the solvent has to be compatible with the materials involved in the packaging. Compatibility can also be obtained by compliance with the component and board manufacturing requirements. For example, Hewlett Packard has issued a

corporate component standardization document which states
that each surface mount device must have resistance to mild
cleaning processes, especially being capable of tolerating a minimum of four minutes exposure to Freon TMS at 40°C, including
a minimum of one minute exposure to ultrasonic immersion at a
frequency of 40 KHz and a power of 100 W/ft$^2$.[4] (The compatibility
of ultrasonic cleaning is discussed in Section 8.8.)

**Sequence of cleaning steps.** Depending on the composition of the
residue resulted from reflow, the sequence of cleaning step affects
the effectiveness of otherwise equivalent processes. One should
keep in mind that cleaning an aggregate of chemicals may not be
the same as cleaning the individual chemicals, and that cleaning
does not entirely rely on solubility of individual chemical in the
solvent. The "matrix" effect (meaning that a chemical or compound with insufficient solubility can be carried away by imbed-
ding in the grossly soluble matrix) contributes significantly to the
end results of cleaning. An improper sequence may negate the
matrix effect and thus jeopardize the cleanability.

**Spray parameters.** The interrelated spray parameters of pressure,
flow rate, and velocity affect cleaning efficiency. Table 8.2 lists the
different spray pressures and corresponding flow rates and veloci-
ties. Increasing pressure results in increasing efficiency. In the
meantime, increasing volume of spray increases cleaning efficiency. However, spray pressure and spray volume are operating
trade-offs.[6] It was found that a high-volume spray (1.1 GPM at 50
psi) provided equivalent cleaning efficiency to a high-pressure

**TABLE 8.2**

**Spray Pressure and Corresponding Flow Rate and Velocity[5]**

| Orifice diameter: 0.054" Dispersion: 15 Fan | | |
|---|---|---|
| Pressure (psig) | Flow (gpm) | Velocity (ft/sec) |
| 400 | 0.73 | 102.3 |
| 200 | 0.55 | 77.1 |
| 150 | 0.40 | 56.1 |
| 100 | 0.23 | 32.2 |
| 50 | 0.12 | 16.8 |

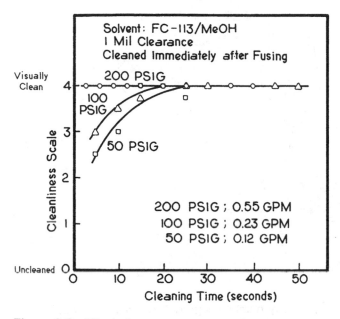

**Figure 8.5** Effect of spray pressure on cleaning over a range of cleaning time.[2]

spray (0.77 GPM at 150 psi). Figure 8.5 illustrates the dependence of cleanliness on the spray pressure applied. At a pressure of 200 psig, cleanliness becomes independent of the cleaning time under the testing conditions specified.[2]

**Spray nozzle design.** The design of the nozzle provides different patterns of spray and volume of spray, which results in different angles at which the spray stream impinges on the board.[6] Figure 8.6 illustrates the adjustable nozzles for delivering fan pattern and flat pattern spray.

**Spray nozzle location.** The distance between nozzle and board, and the position of the board in relation to the level of solvent in the sump also affect the cleaning efficiency, especially for the boards mounted with low clearance components.

**Solvent temperature.** The temperature of the solvent directly affects its solvency and the softening ability. With the exception of temperature effect on the cavitation intensity of ultrasonic sump, among the common cleaning solvents, higher temperature delivers better cleaning efficiency. The solvent temperature is limited by its boiling point.

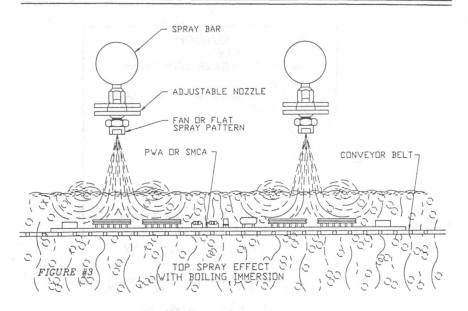

**Figure 8.6(a)** Spray nozzle design (A). *(Courtesy Detrex Corporation.)*

**Figure 8.6(b)** Spray nozzle design (B). *(Courtesy Detrex Corporation.)*

**Dwell time.** Cleanliness improves with increasing dwell time in each step to a certain extent and reaches a plateau, as well demonstrated in Figures 8.2–8.5.

**Ultrasonic aid.** Ultrasonic cleaning has been considered to be the most effective cleaning means. Lately the high pressure spray (50–200 psig) has been developed to enhance the cleanability of boards which contain low clearance components and cannot be cleaned with ultrasonics. Figure 8.7 compares the cleaning efficiency among different cleaning parameters. In this particular test, the ultrasonics is shown to be more effective than 50 psig and 100 psig spray, but not as effective as 200 psig spray.

**Component layout.** The layout of components on the board in relation to the direction of conveyor movement in the cleaner affects the flushability and accessibility. The direction of conveyor movement in line with the open end of components (SOIC, capacitors, resistors, transistors) and staggered components as shown in Figure 8.8 facilitate cleaning. However, for four-sided components (quad pack, PLCC), the effect may be minimal.

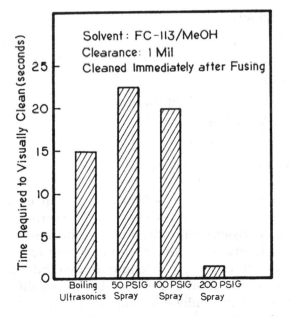

**Figure 8.7** Comparison of ultrasonic and high-pressure spray cleaning.[2]

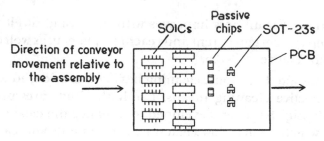

**Figure 8.8**  Component layout in relation to conveyor movement direction in a cleaner.[7]

## 8.5  SOLVENT

The major ingredients used in common cleaning solvents for electronic applications include the following:

- trichlorotrifluoroethane
- 1,1,1-trichloroethane
- trans -1,2 dichloroethylene
- acetone
- methylene chloride
- methanol
- ethanol
- isopropanol
- other aliphatic, low carbon-chain alcohols
- terpene hydrocarbon
- water
- water with surfactants/detergents

The commercially available solvents as listed in Table 8.3 are the azeotropes or blends of 1,1,2-trichlorotrifluoroethane with one of the alcohols, methylene chloride, and acetone, and the azeotropes or blends of 1,1,1-trichloroethane with one or several alcohols. The true azeotrope acts as a single substance in that the vapor produced has the same composition as the liquid, and has different boiling point from any of the individual substances. In addition, terpene hydrocarbon and aqueous media are other alternatives.

The solvents should be chemically inert toward the assembly.

## TABLE 8.3

### Composition of Common Commercial Solvents[9, 10, 11]

| Solvent | Nominal Composition |
|---|---|
| Freon TF | 100% trichlorotrifluoroethane |
| Freon TA | 11% acetone, balance Freon TF |
| Freon TES | 4% ethanol stabilized, balance Freon TF |
| Freon TMS | 6% methanol stabilized, balance Freon TF |
| Freon TMC | 50% methylene chloride, balance Freon TF |
| Freon TWD 602 | 6% water, 2% detergent, balance Freon TF |
| Freon SMT | 5.7% methanol, 24.9% trans 1,2–dichloroethylene, 0.3% stabilizer package, balance Freon TF |
| Prelete R | 94% inhibited 1,1,1 trichloroethane, 6% aliphatic alcohol |
| Genesolv D | 100% trichlorotrifluoroethane |
| Genesolv DE | 4.5% ethanol, balance Genesolv D |
| Genesolv DI | 3% isopropanol, balance Genesolv D |
| Genesolv DMS | 4.0% methanol, 2.0% ethanol, 1.0% isopropanol, 1.0% nitromethane, balance Genesolv D |
| Genesolv DA | 9.4% acetone, 0.3% nitromethane, balance Genesolv D |
| Genesolv DMC | 51.6% methylene chloride, 9.3% hydrocarbon, balance Genesolv D. |
| Genesolv DTA | 3.3% methanol, 41.7% methylene chloride, balance Genesolv D. |
| Genesolv DTA + | 4.3% methanol, 45% methylene chloride, 5.2% hydrocarbon, balance Genesolv D. |
| Bioact EC-7 | 65-95% terpene hydrocarbon |

Note: Freon is Du Pont's trademark for its fluorocarbon solvents; Prelete is Dow Chemical's trademark for its defluxer solvent; Genesolv is Allied Signal's trademark for its solvent systems; Bioact is Petroferm's trademark.

Several physical properties are important for the cleaning functions. They are:

- boiling point
- density
- viscosity
- surface tension

- solubility parameter
- solvency
- latent heat of vaporation

The boiling point controls the upper temperature of the cleaning process; viscosity and density are related to wetting and penetrating ability; surface tension balances the wetting and capillary phenomena; latent heat of evaporation is related to heat demand on equipment; solvency indicates the solvating power; and the solubility parameter provides compatibility nature for mutual dissolution, as discussed in Section 8.6. Table 8.4 lists these properties of a number of commercially available solvents. A new solvent system, Freon SMT has recently been introduced by the Du Pont Company. According to W. G. Kenyon, Freon SMT was especially designed to eliminate "white residues" as a result of overheated multilayer board and flux, to increase overall productivity by cleaning an assembly in one pass, to facilitate recycling by virtue of its azeotropic composition, and to comply with the relative ozone depletion potential restrictions per the Montreal Protocol.

A group of terpene hydrocarbon solvents has been introduced to the industry by Petroferm, Incorporated. The solvent can dissolve rosin appreciably as tested with abietic acid. It is readily emulsifiable. The solvent has a relatively low flash point and cannot be used for vapor degreasing. Recommended cleaning technique is room temperature spray/immersion with or without ultrasonic in conjunction with subsequent water cleaning.

## 8.6  SOLUBILITY PARAMETERS

Solubility parameter is a cohesive property of materials, also known as Hildebrand parameter, named after Joel H. Hildebrand for his pioneering fundamental concept on the solubility of nonelectrolytes.

The existence of the liquid and solid state over certain ranges of temperature and pressure indicates that there are interactions or forces among molecules to hold them together. This cohesive effect in the condensed phases is expressed in terms of cohesive pressure ($P$), which is equivalent to cohesive energy ($E$) per unit volume ($V$).

$$P = -\frac{E}{V}$$

TABLE 8.4

Physical Properties of Common Commercial Solvents[9, 10, 11]

| Solvent | Boiling Point (°C) | Boiling Point (°F) | Density (g/ml) | Viscosity (cps) | Surface Tension (dynes/cm) | Latent Heat of Vaporation (Btu/lb) | Latent Heat of Vaporation (cal/g) | Solvency (KB value) |
|---|---|---|---|---|---|---|---|---|
| Freon TF | 47.6 | 117.6 | 1.565 | 0.682 | 17.3 | 63.1 | 35.1 | 31 |
| Freon TA | 43.6 | 110.5 | 1.406 | 0.542 | 18.7 | 85.4 | 47.4 | 51 |
| Freon TES | 44.4 | 111.9 | 1.496 | — | 17.2 | 76.7 | 42.6 | 37 |
| Freon TMS | 39.7 | 103.5 | 1.477 | 0.70 | 17.4 | 90.7 | 50.4 | 45 |
| Freon TMC | 36.2 | 97.2 | 1.420 | 0.461 | 21.4 | 104.0 | 57.7 | 86 |
| Freon TWD 602 | 44.4 | 112.0 | 1.494 | 0.94 | 17.3 | — | — | 21 |
| Freon SMT | 38.4 | 101 | 1.376 | 0.52 | 22.9 | 100.6 | 55.9 | 96 |
| Prelete R | 73.3 | 164 | 1.28 | 0.618 | 25.2 | 103.7 | 57.6 | 124 |
| Genesolv DE | 44.6 | 112.3 | 1.49 | 0.70 | 19.0 | 77.3 | 42.9 | 38 |
| Genesolv DI | 46.7 | 116.1 | 1.53 | ~0.70 | 19.2 | 70 | 38.9 | 35 |
| Genesolv DMS | 40.3 | 104.5 | 1.46 | 0.77 | 18.4 | 89.5 | 49.7 | 49 |
| Genesolv DA | 44.5 | 112.1 | 1.42 | 0.55 | 18.8 | ~85 | 47.2 | 46 |
| Genesolv DTA | 34.2 | 93.6 | 1.31 | ~0.5 | 21.0 | 109.7 | 60.9 | 125 |

(continued)

*Table 8.4 (continued)*

| | | | | | | | |
|---|---|---|---|---|---|---|---|
| Genesolv DTA+ | 33.8 | 92.8 | 1.32 | 0.47 | 21.5 | 122.8 | 68.2 | 182 |
| Genesolv DMA | 36.2 | 97.2 | 1.29 | ~0.5 | 10.8 | 20.6 | 11.4 | 98 |
| Methylene Chloride | 39.8 | 104 | 1.32 | 0.43 | 28 | 142 | 78.8 | 115 |
| Bioact EC-7 | 171 | 340–372 | 0.84 | 1 | ~ | ~ | ~ | ~ |
| 1,1,1-Trichloroethane | 74.1 | 165 | 1.32 | 0.79 | 25.9 | 102 | 56.6 | 120 |
| Water | 100 | 212 | 1.0 | 1.0 | 73 | 972 | 540 | — |
| Water surfactant | ~100 | ~212 | ~1.0 | ~1.0 | ~30 | ~972 | ~540 | — |

Note: Freon is Du Pont's trademark for its fluorocarbon solvents; Prelete is Dow Chemical's trademark for its defluxer solvent; Genesolv is Allied Signal's trademark for its solvent systems; Bioact is Petroferm's trademark.

Hildebrand and Scott[8] defined that solubility parameter, related to the cohesive pressure by

$$\delta = P^{1/2} = -\left(\frac{E}{V}\right)^{1/2}.$$

The unit of solubility parameters is $cal^{1/2} \ cm^{-3/2}$. They later advanced the idea that a molecule will be attracted most effectively by other molecules when they have the same cohesive pressure. This concept suggests that when two molecules have the same internal pressure they will be completely miscible. On the other hand, if the cohesive pressures are significantly different among molecules, the molecules with the greater cohesive pressure will stick together and exclude the molecules with less cohesive pressure.

In organic materials, atoms of a molecule are held together by strong covalent bond whose energy are of the order of 50–100 kcal/mole. However, the intermolecular forces are a different kind of bonding. For nonpolar molecules, the forces arise from a momentary dipole that results from a positive nucleus and a negative electron cloud in each atom. The momentary dipole will induce an electron distribution in an oppositely oriented dipole in the neighboring molecule. These attractive forces between molecules are called van der Waals forces or dispersion forces. Even molecules that are nonpolar, like methane or benzene, exhibit momentary dipole moments, which holds molecules together as represented by:

The bond strength in this case is very weak, usually less than 1 kcal/mole.

For polar molecules, the intermolecular forces are the result of dipole-dipole interactions, dipole-induced dipole interactions, and hydrogen bonding. The dipole-dipole interactions occur between the permanent dipole moment between molecules, and the dipole-induced dipole force is the interaction between permanent dipole moment of one molecule and the polarizability of the other. Hydrogen bonding involves the interaction between a hydrogen atom covalently bonded to an electronegative atom of one molecule and the electronegative atom of another molecule. The hydrogen atom serves as a bridge between two electronegative atoms, as represented in the following by the dotted lines.

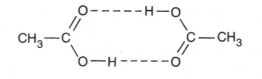

In contrast to covalent bonds in which each molecule supplies one electron to form the bond, hydrogen bonding has only one molecule (donor) supplying the pair of electrons. Its bond strength is much weaker than a covalent bond being about 2-10 kcal/mole, but it is stronger than that of van der Waals forces.

All the above-mentioned forces contribute to the cohesive energy, therefore to the solubility parameter. Assuming the total cohesive energy is made up of a linear combination of the contributors, dispersion interaction ($Ed$), polar interaction ($Ep$) and hydrogen bonding ($Eh$), then $-E = - (Ed + Ep + Eh)$, and

$$-\frac{E}{V} = -\frac{Ed}{V} - \frac{Ep}{V} - \frac{Eh}{V}.$$

Thus,

$$\delta^2 = \delta_d^2 + \delta_p^2 + \delta_h^2.$$

The solubility parameter and the contributing portions of some materials are listed in Table 8.5. The solubility parameter is a useful reference parameter to anticipate the miscibility between materials. In general, a similarity of solubility parameter is a necessary condi-

## TABLE 8.5

### Solubility Parameters of Selected Materials*

| Material | $\pm d$ | Solubility Parameter ($cal^{1/2} cm^{-3/2}$) $\delta_p$ | $\delta_h$ | $\delta$ |
|---|---|---|---|---|
| Water | 6.4 | 15.4 | 16.8 | 23.8 |
| Acetone | 7.6 | 5.0 | 3.4 | 9.8 |
| Methylene Chloride | 8.9 | 3.1 | 3.0 | 9.9 |
| Methanol | 7.4 | 6.0 | 10.9 | 14.5 |
| Ethanol | 7.7 | 4.3 | 9.5 | 12.9 |
| 2-Propanol | 7.8 | 3.0 | 8.0 | 11.5 |
| 1-Butanol | 7.8 | 2.8 | 7.7 | 11.3 |
| 2-Butanol | 7.7 | 2.8 | 7.1 | 10.8 |
| n-Hexane | 7.3 | 0 | 0 | 7.3 |
| n-Heptane | 7.5 | 0 | 0 | 7.5 |
| n-Octane | 7.6 | 0 | 0 | 7.6 |
| n-Nonane | 7.7 | 0 | 0 | 7.7 |
| n-Decane | 7.7 | 0 | 0 | 7.7 |
| n-Hexadecane | 8.3 | 0 | 0 | 8.0 |
| Cyclohexane | 8.2 | 0 | 0.1 | 8.2 |
| Benzene | 9.2 | 0 | 1.0 | 9.1 |
| Toluene | 8.8 | 0.7 | 1.0 | 8.9 |
| Biphenyl | 10.5 | 0.5 | 1.0 | 10.6 |
| 1,1,1-Trichloroethane | 8.3 | 2.1 | 1.0 | 8.7 |
| 1,1,2-Trichlorotrifluoroethane | 7.2 | 0.8 | 0 | 7.2 |
| Prelete | — | — | — | 8.6 |
| Freon TMS | — | — | — | 7.4 |
| Freon SMT | — | — | — | 7.9 |
| Genesolv DMS | — | — | — | 7.4 |
| Freon TA | — | — | — | 7.4 |
| Genesolv DA | — | — | — | 7.4 |
| Freon TE | — | — | — | 7.4 |
| Genesolve DE | — | — | — | 7.4 |
| Di-n-butyl phthalate | 8.7 | 4.2 | 2.0 | 9.9 |
| N, N-Dimethyl formamide | 8.5 | 6.7 | 5.5 | 12.1 |
| Ethylene glycol | 8.3 | 5.4 | 12.7 | 16.1 |
| Glycerol | 8.5 | 6.0 | 14.3 | 17.6 |
| Propylene glycol | 8.2 | 4.6 | 11.4 | 14.8 |

*Courtesy of Allan F. M. Barton, *Handbook of Solubility Parameters and Other Cohesion Parameters* (CRC Press, Inc., 1983).

tion to obtain good miscibility. However, it should be noted that a similarity may not always assure miscibility.

## 8.7 SOLVENT CLEANING TECHNIQUE AND EQUIPMENT

The key to designing the cleaning process is to effectively combine the available basic steps and to select parameters in each step so that the resulting process is a best fit to the specific assembling of interest, in delivering the cleanliness as well as meeting economical and practical issues. The basic techniques in cleaning include vapor degreasing, liquid spray, liquid immersion, high pressure spray, and liquid immersion with ultrasonic aid. It is obvious that the selection of combinations of steps and the sequence of steps are numerous. A few approaches in the order of increasing cleaning intensity are as follows:

vapor degreasing only
vapor → spray → vapor
vapor →immersion → vapor
vapor → immersion → spray → vapor
vapor → spray → immersion → spray → vapor
vapor → spray → immersion/spray → spray → vapor
vapor → spray → vapor/spray/high-pressure spray → vapor
vapor → spray → immersion/spray/ultrasonic → spray → vapor

The permutations can increase as one considers the choices of top and bottom spray, the alternative of spray impinging angle or inclining belt, the availability of different levels of spray pressure, the spray pattern, and the nozzle position.

The cleaning equipment manufacturers have made significant progress in the development of equipment capability and versatility over the last few years. Figures 8.9 and 8.10 are the section view and its corresponding operational schematic of a flux cleaner, respectively, manufactured by Electrovert, USA Corporation.[12] The equipment is composed of vapor prewash and immersion wash, then 35 psi spray wash with 18 nozzles (8 top and 10 bottom) and high-pressure spray with 18 fan jets at 250 psi, followed by spray rinse, immersion rinse, and vapor rinse, and dry.

In order to clean the less accessible areas, the design and location of the spray nozzle also influence the results. The cleaner can be designed to have high volume spray above the immersion zone with

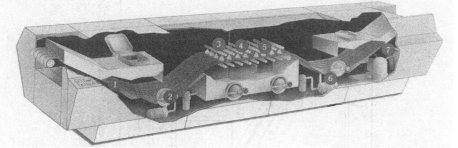

**Figure 8.9** Section view of a solvent cleaning system. *(Courtesy of Electrovert, USA Corporation.)*

the printed circuit board $1/2$ to 1 in. lowered into immersion zone. The cleaning for double-sided surface mount board can be accomplished by using spray from both above and below the board. A cleaner can also be designed to spray over the ultrasonic sump. It is reported that the agitation of the immersion zone can dampen the ultrasonic cavitation but still be able to ultilize ultrasonic efficiency.[6] Figure 8.11 illustrates the feature of designing the nozzle distance in relation to the level of hot sump by lowering the nozzle to within two inches to the board; this is expected to further enhance the "turbulence" in the immersion sump to clean less accessible areas or hard-to-clean residues.[6]

In designing and selecting cleaning processes for densely packed surface mount boards, the high-pressure spray or ultrasonic aid is necessary to clean the areas covered with components having less than six mils clearance between the board and the component. In addition, the solvent selection should be in congruence with the equipment construction—for example, the cleaner built for fluorocarbon-based solvent will not be suitable for use of chlorocarbon solvent or aqueous solution due to the differences in solvency or reactivity and in latent heat of vaporization, as indicated in Table 8.3. In comparison, the fluorocarbon-based solvents are characterized by high density, low surface tension, low temperature, low latent heat of vaporization and low solvency; and the chlorocarbon-based solvents, with high temperature, high latent heat of vaporation, and high solvency.

Based on a cleaning study program, Naval Weapons Center at China Lake in California endorses the use of fluorocarbon or chlorocarbon in combination with a high percentage of alcohol, and also endorses the addition of a rinse with deionized water after the solvent cleaning, as effective cleaning solvent systems.

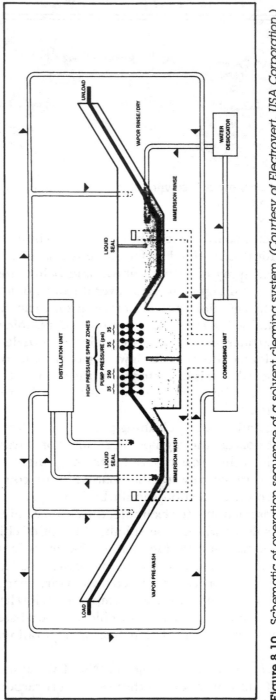

**Figure 8.10** Schematic of operation sequence of a solvent cleaning system. (*Courtesy of Electrovert, USA Corporation.*)

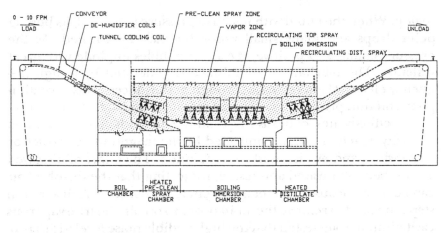

SPRAY – IMMERSION W/TOP SPRAY – SPRAY
CLEANING CYCLE

**Figure 8.11** Schematic of a cleaning system with combined high-pressure spray and immersion. *(Courtesy of Detrex Corporation.)*

## 8.8 ULTRASONIC CLEANING

Ultrasonic aid is a most effective step. However, the use of ultrasonic aid to clean electronics assemblies caused a controversy. The U.S. Military Specification MIL-P-28809A (Appendix III) explicitly prohibits the adoption of ultrasonic cleaning. The concern is not without basis. It has been found that the wire bonding inside the package has been weakened or damaged, and solder joints that connect small passive chips on the board exhibited cracks, and in some cases the small chips even fell off the board. These incidents, however, should not lead to the conclusion that ultrasonic use is an unsuitable means in all cases. Rather, it indicates the need of some more understanding and testing to identify workable parameters, specifically, frequency, power and the effect upon the components. An ultrasonic cleaning study is currently being conducted at the Electronic Manufacturing Productivity Facility at the Naval Weapons Center. Ultrasonic cleaning could be a successful process to clean solder paste residue under a qualified process condition, although this remains to be confirmed by forthcoming data.

The fundamental concept of ultrasonic cleaning is to utilize the cavitation defined as the implosion of microscopic vapor cavities within the solution, which is induced by the changing pressure differentials in the ultrasonic field. The pressure differentials occur by the interchange between negative and positive pressure in a liquid region. When the liquid with negative pressure is created, its boiling point drops and many small vapor bubbles are formed. As the pressure changes to positive, the small bubbles implode with great violence. The mechanical soundwave is generated by a high frequency electrical energy wave through a transducer. This cavitation phenomenon provides mechanical agitation and scrubbing effect.

The effectiveness of ultrasonic cleaning depends on the cavitation intensity which in turn is controlled by the magnitude of power or pulse width and by dissolved air in solution. The effect of dissolved air has been illustrated and tested, indicating that the dissolved air can act as an acoustical screen and energy absorber.[13] A deaeration step is needed to remove the air in order to obtain the true vaporous cavitation. It is suggested that the high audible noise level with a pronounced hissing sound and minimal visible bubbles in solution coupled with violent surface activity are the signs of ultrasonic efficiency. It is also suggested that the temperature of the solution is a factor in ultrasonic efficiency. The desirable temperature is approximately in the range of 80-98% of the boiling point of the solution.

With the superior cleaning efficiency that ultrasonics can provide, the accomplishment of identifying the compatible parameters and their correlation to a specific assembly will be a significant contribution to the electronics packaging area.

## 8.9 AQUEOUS CLEANING

With respect to cleaning medium, fluorocarbon-based and chlorocarbon-based mixtures have been adopted in the industry. Evaluating water as a cleaning medium has lately been resurged due mainly to the concerns and uncertainties of occupational health hazard, stratospherical variations, and environmental issues relating to other solvents.

Water is ionic in nature, therefore it is expected that aqueous cleaning is ideal to clean water-soluble flux systems as well as to deal with the ionic portion of an RMA-type residue. The study of aqueous cleaning of water-soluble residue (OA-grade flux) suggests that aqueous cleaning provides adequate cleanliness for boards containing SOIC-14, types 0805, 1206, and 1812 chip capacitors as reflected by the results of surface insulation resistance test and ionic contaminant test by determining the conductivity of test solution.[14] However, a typical RMA paste produces residue containing both polar and nonpolar compounds.

From a paste development point of view, in order to utilize aqueous cleaning, two types of solder paste are to be considered: water-soluble and water-insoluble. For a water-insoluble type, such as conventional RMA-grade, aqueous cleaning needs to incorporate additives which render the water-insoluble ingredients soluble or miscible. For example, a known additive is a saponsifier that changes the water-insoluble rosin into a soluble salt as represented by:

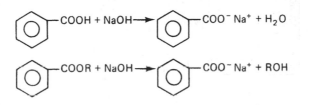

Based on the same principle, different additives and techniques are worthwhile studying. The objective is to convert the insoluble ingredients either as an individual or as an aggregate into water soluble or miscible forms through chemical reactions, with the aid of external temperature as supplied by water temperature, or pressure as provided by equipment design within a practical range of time and other parameters. In addition, the pH value of the medium should not be overlooked in the design to facilitate the residue removal.

Figure 8.12 is an example of available aqueous cleaning equipment constructed with pre-rinse, wash, and air-knife dry step followed by hot deionized water rinse and air knife/IR dry. Since water is ionic itself, the ability to thoroughly dry the assembly in the areas that are hard to reach, including crevices, is vital. A high volatile solvent vapor rinse may be an effective step to expel residual water by utilizing the water-miscible nature of the selected solvent such as low carbon alcohol.

The aqueous cleaning can also be used in conjunction with terpene hydrocarbon solvent cleaning for conventional RMA-type residue. In this design, the terpene hydrocarbon high impact energy spray followed with air-knife drying are preceding steps to the aqueous cleaning.

Although more availability of in-line aqueous cleaning equipment and water-compatible components in conjunction with safety and environmental issues related to solvents have increased the interest of using aqueous cleaning, aqueous cleaning today still lacks military approval.

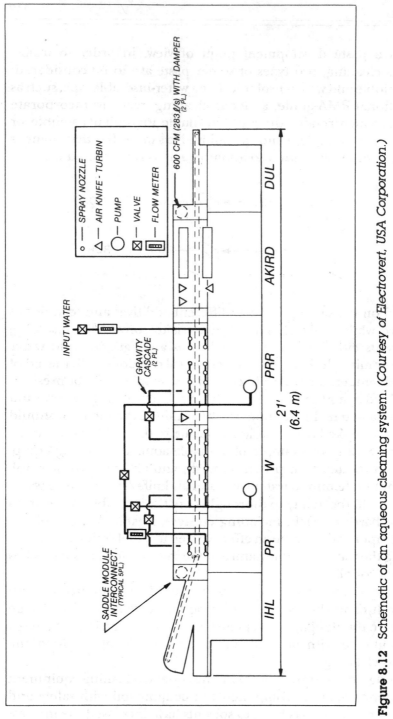

**Figure 8.12** Schematic of an aqueous cleaning system. (*Courtesy of Electrovert, USA Corporation.*)

With respect to waste water handling, waste water should not be discharged directly into public sewage systems without testing and/ or treatment. The sample of effluent should be tested to determine the treatment needed prior to discharging into sewage. Ion exchange and chemical treatment have been used to precipitate out heavy metals. The precipitates after filtration and drying are then packaged for separate disposal, and the remaining effluent can be discharged into sewage. The federal, state, and local environmental agencies should be consulted for implementing proper waste disposal.

## 8.10  WHITE RESIDUE

The occurrence of "white residue" after solvent cleaning has been recognized as a tenacious residue which is hard to remove under most established cleaning processes. The white residue has been often so described, yet it may arise from a variety of different origins and thus need different treatments. For a typical RMA-grade paste, the possible sources which constitute the white residue may be the formation of one or more of the following species:

- polymerized rosin/resin
- interaction products of flux/vehicle ingredient with strong chemical activity
- metal salt of rosin acid, mostly tin, lead salt
- other metal salts
- interaction product(s) between paste ingredients and board material

The interaction between paste ingredient and board material may be due to incompatibility, an improperly cured solder mask, or contaminated or uncleaned board. The apparent incompatibility between solder paste and board material or solder mask can be treated by using alternate material and a solder mask. The board condition problem can be corrected readily through better control and inspection. The polymerized rosin may demand a cleaning solvent with higher solvency. In some cases the polymerized resin can be softened or dissolved by the flux system from which it is formed. The tin and lead salt of rosin, and other metal salts may demand a stronger solvent and the "matrix effect" for removal.

W. L. Archer has studied the cleanability of tin and lead abietate (abietic acid is a primary constituent in most rosins) and found that a 1,1,1-trichloroethane/alcohol system is more effective than a fluorocarbon/alcohol system.[15] This finding is consistent with solubility results as shown in Table 8.6 and with the upper temperature capability provided by chlorocarbon.

In most cases, the tenacious residue is related to the reflow process and often formed from combined origins. The main reflow process parameters affecting the formation of a tenacious residue are heat exposure and atmosphere. As a rule of thumb, the formation of a tenacious residue can be decelerated by maintaining a minimum heat exposure (time and temperature) and by conducting reflow in an inert atmosphere (nitrogen or oxygen-free atmosphere). Reducing atmosphere may also be a help. Another effect frequently overlooked is the matrix effect as discussed previously—that is, insoluble ingredients in the residue can be removed by "riding" with a sufficient amount of the soluble portion of the residue, and being flushed away. Thus, it is obvious that the sequence of cleaning steps and the use of cleaning agent should be strategically selected.

## 8.11  SOLDER BALL REMOVAL

In addition to cleaning the organic residue, completely removing solder balls is a parallel objective. The smaller the solder balls are, the more difficult the cleaning is, due to the overwhelming force of binding of small particles to the surface of the board or substrate. Figure 8.13 illustrates that the binding force of particles to surfaces drastically increases with decreasing particle size.[17] The forces ap-

**TABLE 8.6**

### Solubility* of Metal Abietate[16]

| Solvent | Tin Abietate | Lead Abietate |
|---|---|---|
| Water | 3.2 | 0.7 |
| Dimethyl Formaldehyde | 207.9 | 99.0 |
| Chlorocarbon Blend | 275.5 | 17.1 |
| Chlorofluorocarbon Blend | 40.8 | 2.2 |

*Solubility in mg/g of saturated solution at room temperature.

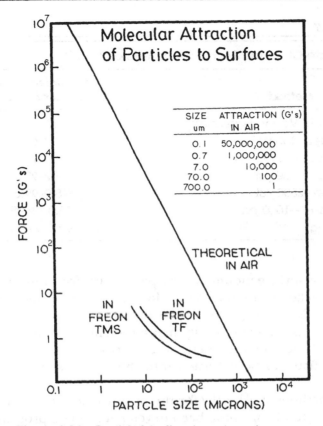

**Figure 8.13**  Particle binding force to surface versus particle size.[17]

plied to the solder balls from cleaning process need to exceed the attraction force of solder balls with surfaces in order to dislodge the solder balls. Table 8.7 shows the results of an investigation on the removal of 5-micron (0.2 mil) particles on glass substrate.

## 8.12  CLEANLINESS MEASUREMENT

Ionic contamination measurement, chemical and spectroscopic rosin analysis, functional tests, Vanzetti laser inspection, electron microscopic surface contaminant analysis (Auger, ESCA), as well as visual examination have been utilized to measure cleanliness.

The ionic contamination detection method as specified in U.S. Military Specification MIL-P-28809A (Appendix III) is designed for through-hole printed wiring board, and in some cases may not be

**TABLE 8.7**

**Particulate Cleaning Method Efficiencies[18]**

| Method | % Efficiency for Particles <5 microns |
|---|---|
| Freon TF Vapor degreasing | 11–20 |
| Gas Jet—100 psi | 52–61 |
| Freon TF Ultrasonics | 24–92 |
| Freon TF—50 psi | 92–97 |
| Water—2500 psi | 98.0–99.5 |
| Freon TF—1000 psi | 99.8–99.95 |
| Wiping | 99.6–99.98 |

suitable for surface mount boards, particularly for those containing densely packed components with low clearance. A mild extraction technique is not expected to be able to rinse out the residual contaminants for conductivity measurement. In other words, the residue left behind after a cleaning process may be equally, if not more, difficult to be extracted into the testing solution (Section 11.32).

Two methods specifically for the measurement of residual rosin on the printed board have been developed.[19] (The procedures are included in Appendix IV.) Basically, the first method utilizes the principle of light scattering by small particles of rosin precipitate produced by adding dilute aqueous acid to the isopropanol washing solution. The turbidity then is measured by a turbidimeter or nephelometric equipment. The second method is ultraviolet (UV) spectrometry by following the UV peak at 241 nm, which is the absorbancy maximum in the UV spectrum of the principal rosin ingredient, abietic acid, as shown in Figure 8.14.

The overall residue detection on the solder joint can be accomplished by Vanzetti laser/IR technique, which is discussed in Section 9.12.

Visual examination in conjunction with ionic contamination measurement has worked very well to judge cleanliness for the readily accessible assemblies. However, to visually examine underneath the component mandates a destructive test.

Functional tests such as accelerated surface insulation resistance test (Section 11.33) and other electromigration resistance tests under temperature, humidity, and applied bias reflect useful infor-

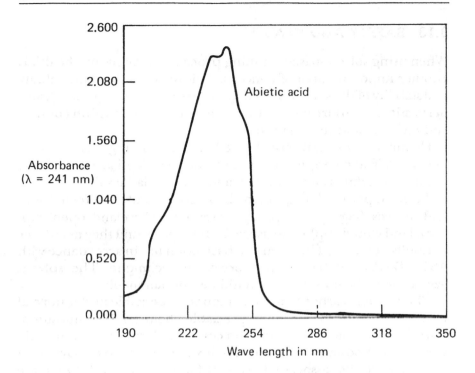

**Figure 8.14** UV spectrum of abietic acid.[19]

mation about cleanliness. Another part of accelerated functional tests is to correlate the test conditions to the service conditions that the assembly is to be exposed to and then to determine the acceptable level of cleanliness, such as the value of resistance and metal migration rate.

In studying the cleaning process for high voltage circuits with and without chip carrier, the cleaning measurements by using three methods—extraction/ionic contaminant technique, Auger surface contaminant analysis, and electric leakage test under accelerated aging and high voltage bias—indicate consistent results.[20]

A single absolute, direct, and nondestructive method to measure the cleanliness of an assembly is not in existence. However, for a known assembly, the proper combination of available techniques as just outlined can provide adequate information about the level of cleanliness. A functional test is always a viable choice for confirming the cleanliness. It cannot be overemphasized that the validity of the confirmation test relies on the test conducted on the specific assembly of interest, which is produced by the manufacturing process.

## 8.13  SAFETY AND HEALTH

When using solvent-based cleaning process, the safety and health is another important area of concern. This includes the flammability and stability of the solvent selected, the toxicity, carcinogenicity, and permissible exposure limit of the solvent, as well as emission control, and recovery and reclamation.

Flammable materials are defined as those having a flash point below 100°F and a vapor pressure of not over 40 psi at 100°F. Flash point is the temperature at which a liquid or volatile solid reaches a sufficient vapor level to ignite in air. There are different techniques and devices (tag open cup, tag closed cup, Cleveland open cup, Cleveland closed cup) to measure the flash point and they may differ in results obtained. The open cup flash point test in accordance with ASTM-D 1310-80 has been a prevalent technique. The solvent azeotrope or blend selected should be nonflammable.

Chlorofluorocarbons have been found to be stable under normal cleaning conditions without appreciable chemical decomposition. On the other hand, decomposition compounds such as dichloroethylene (low exposure limit) and 1,4-dioxane (flammable) have been identified to be associated with chlorocarbons. Stabilizers are needed in chlorocarbon in order to minimize the decomposition.

Each solvent or solvent system has its airborne permissible limit. The government and industrial agencies publish the threshold limits in TLV-TWA, TLV-STEL, and TLV-C. Threshold limit value–time-weighted average (TLV-TWA) is defined as the time-weighted average concentration for a normal 8-hour workday and a 40-hour work week, to which nearly all workers may be repeatedly exposed, day after day, without adverse effect. Threshold limit value–short-term exposure limit (TLV-STEL) is the maximum concentration to which workers can be exposed for a period up to 15 minutes continuously without suffering as specified. Threshold limit value–ceiling (TLV-C) is the concentration that should not be exceeded even instantaneously.

The threshold limit value for common cleaning solvents is in the range of 50-1000 ppm (parts per million in volume). It is used as a measure in the control of airborne concentration of the solvent and therefore the health hazard. Higher value is less critical with respect to emission control. The control guideline should comply with regulations issued by OSHA (Occupational Safety and Health Administration), EPA (Environmental Protection Agency), and ACGIH

(American Conference of Governmental and Industrial Hygienists), as well as by state and local government regulations.

The solvent emission (loss) is not only related to safety and health but also economics. The emission normally occurs through natural diffusion, dragout by parts, leakage, and improper handling. Most used solvents can be reclaimed by conventional distillation. Its efficiency, however, varies with different solvent systems.

For safe practice, the MSDS (Material Safety Data Sheet) supplied by solvent vendor must be consulted to understand the following aspects of the solvents:

- physical and chemical characteristics
- fire and explosion hazard
- reactivity
- health hazard
- precautions and protection for safe handling and use
- disposal information

In addition, regular air sampling should be conducted to confirm emission control. The information about the safety and health aspect of the solvent and the operation precautions of the equipment to minimize emission, as provided by the manufacturers, should always be observed. Avoiding inhalation, ingestion, and skin and eye contact is generally required for handling solvents.

## 8.14  OZONE DEPLETION

Scientific results have indicated that the chlorofluorocarbon (CFC) contributes to the depletion of ozone ($O_3$) in the stratosphere. The earlier findings have resulted in the 1970s ban of CFC use in aerosol. Recent data substantiate earlier findings and give rise to global concern about the use of CFC in any other applications.

It is theorized that the stratospheric chlorine (Cl·) is a major species depleting ozone through following catalytic chlorine oxide cycle:[21]

$$Cl\cdot + O_3 - ClO + O_2$$
$$ClO + O\cdot - Cl\cdot + O_2$$

The origin of chlorine is derived from the source such as CFC in addition to natural ozone removal mechanisms in the stratosphere. This is based upon the extraordinary stability of CFC compounds that are able to release to the troposphere and not be removed by tropospheric process. The accumulated CFC in the troposphere can then diffuse into the stratosphere where CFCs are photodissociated by ultraviolet radiation, producing chlorine.

The Montreal Protocol, pending final enactment, imposes the limitation of CFC production and the schedule of implementation. Continued revision of environmental regulations is expected as more scientific data are revealed. The fluorocarbon industry has responded by controlling production for the near term, and in the meantime developing alternative solvents. The cleaning equipment manufacturers are working on new technologies for aqueous cleaning of surface mount assemblies as well as on further improved tightness of solvent-cleaning systems. Needless to say, solder paste manufacturers have to respond to the changes and collaborate with other parts of the assembly line to achieve a working system.

## REFERENCES

1. J. E. Hale and W. R. Steinacker, "Complete Cleaning of Surface Mounted Assemblies," *Surface Mount Compendium* 3:2 (International Electronics Packaging Society): 1113.

2. David S. Lermond, "Model Studies in the Cleaning of Surface-Mounted Assemblies with High-Pressure Fluorosolvent Sprays," *Surface Mount Compendium* 3:2 (International Electronics Packaging Society): 920.

3. Robert P. Musselman. Quadrex HPS, Incorporated.

4. W. G. Kenyon and D. S. Lermond, "Post-Solder Cleaning by SMAs," *Printed Circuit Assembly* (Aug. 1987).

5. D. S. Lermond, "Key Process Design Factors for Efficient Fluorosolvent/Vent Spray Cleaning of SMAs," *Printed Circuit Assembly* (May 1987).

6. Detrex Chemical Industries, Bowling Green, Kentucky.

7. C. A. Capillo, "Surface Mounted Assemblies Create New Cleaning Challenges," *Electronic Packaging and Production* (Aug. 1984).

8. J. Hildebrand and R. Scott, *The Solubility of Nonelectrolytes*, 3 ed. (New York: Reinhold, 1949).

9. Du Pont Freon solvent product bulletin.

10. Dow Chemical Defluxer solvent bulletin.

11. Allied Signal Genesolv solvent system bulletin.

12. William Down. Electrovert, USA Corporation.

13. James B. Halbert, "Solvent Cleaning of SMDs with Boiling and Quiescent Ultrasonics," *Surface Mount Compendium* (International Electronics Packaging Society).

14. R. Aspandiar, A. Piyarali, and P. (Ray) Prasad, "Is OA OK?" *Circuits Manufacturing* 3:2 (Apr. 1986): 935.

15. Wesley L. Archer, "Analysis of Rosin Flux Residue After Soldering and Cleaning," *Electronic Packaging and Production* (Feb. 1988).

16. Private communication: Wesley L. Archer.

17. R. P. Musselman and T. W. Yarbrough, "The Fluid Dynamics of Cleaning under Surface Mounted PWAs and Hybrids," *Surface Mount Compendium* 4:2 (International Electronics Packaging Society): 878.

18. I. F. Stowers, "Advances in Cleaning Metal and Glass Surfaces to Micro-level Cleanliness," *Journal of Vacuum Sciences Technology* 15:2 (1978): 751.

19. Wesley L. Archer, Tim D. Cabelka, and Jeffrey J. Nalazek. Dow Chemical Company.

20. S. D. Schlough and E. F. O'Connell, "Cleaning Processes for High Voltage Circuits With Soldered Chip Carriers," proceedings, 1981 International Conference in Microelectronics:28.

21. R. S. Stolarski and R. J. Cicerone, *Can. J. Chem.* 52 (1974): 161.

# PART IV

# Reliability, Quality Control and Tests

# CHAPTER 9

# Solder Joint Reliability and Inspection

## 9.1 INTRODUCTION

As illustrated in Chapter 2 about solder paste technology, after reflowing the solder paste and cleaning the residue, another major performance area is solder joint integrity during service. With the functions as electrical, thermal, and mechanical linkage, solder joint integrity in a practical environment is apparently vital to the overall function of the assembly.

Studies on mechanical properties of bulk solders and conventional structural solder joints, including through-hole solder joints, have been carried out over several decades. Data and information obtained from such studies are useful for current solder joint construction, yet in most cases data cannot be directly transferred or extrapolated to the surface mount solder joint and/or "microjoint."

In the assessment of service performance or endurance of a solder joint structure, the stress distribution, the strain amplitude, the strain rate, the cyclic nature of stress (either mechanical or thermal), the temperature, and other environmental factors are all to be considered. It is worth noting that thermal environment can cause the modification of fatigue behavior, and that the service temperature, whether it is a result of environment or of assembly operation, is expected to have significant impact on the mechanical properties and failure mode. Since the solder joint is composed of relatively low temperature alloys, a temperature of 80°C in many cases is more than 40% of solidus temperature of some solder alloys. The temperature would be high enough to cause creep, and creep characteristics often interact with the basic fatigue process.

In view of the complexity in most failure modes, in-depth analysis and review will be beyond the scope of this book. Instead, this chapter discusses the factors which affect the solder joint integrity and phenomena which commonly are linked to the failure of solder or alloy in general. The basic processes or factors which are believed to be probable encounters in solder failure while in service are

- inferior or inadequate mechanical strength,
- creep,
- mechanical fatigue,
- thermal fatigue,
- intrinsic thermal expansion anisotropy,
- corrosion-enhanced fatigue,
- intermetallic compound formation,
- detrimental microstructure development,
- voids,
- electromigration, and
- leaching.

In reality, failure mode mostly involves more than one of these phenomena and operates in combination.

## 9.2  FACTORS OF SOLDER JOINT INTEGRITY

Solder joint integrity can be affected by the intrinsic nature of the solder alloy, the substrates in relation to the solder alloy, the joint design or structure, the joint-making process, as well as external environment to which the solder joint is exposed. Therefore, to assure the integrity of a solder joint, step-by-step evaluation on the following items is warranted.

- suitability of solder alloy for mechanical properties
- suitability of solder alloy for substrate compatibility
- adequacy of wetting of solder on substrates
- design of joint configuration in shape, thickness, and fillet area
- optimum reflow method and reflow process in temperature, heating time, and cooling rate

- conditions of storage in relation to aging effect on solder joint
- conditions of actual service in terms of upper temperature, lower temperature, temperature cycling frequency, vibration, or other mechanical disturbance
- performance requirements under conditions of actual service
- design of viable accelerated testing conditions which correlate to actual service conditions

Although each step takes diligent efforts, the last item—to design a viable accelerated test with good correlation to real-world application—is most challenging. By fulfilling this, an assembly would not be "overtested" or improperly tested leading to negative conclusion, or be "undertested" resulting in unexpected field failure.

## 9.3  CREEP

Creep is defined as a time-dependent deformation of solids under stress. For metals or alloys, when a stress is applied, the material responds with an instantaneous strain in the elastic and/or plastic region. As time proceeds, the material may continue to deform in a time-dependent manner and eventually fail. This time-dependent deformation can theoretically occur at any temperature above absolute zero. In the low temperature region, the time-dependent strain accumulates in a logarithmic manner:

$$\varepsilon_t = \varepsilon_o + A \log t,$$

where $\varepsilon_o$ is the initial strain, and $\varepsilon_t$ is the strain at time $t$, and $A$ is constant. Within this region the creep strain is very limited and deformation processes normally do not lead to eventual fracture.

In the high temperature region ($0.4-0.5\ T_m$, $T_m$ is the melting temperature of the material), the creep curve departs from the logarithmic relationship, as shown in Figure 9.1, which represents a typical creep phenomenon. The actual creep curve for a material varies with the characteristics of the material, temperature, and stress conditions.

Dorn's approach assumes that thermal activation is fundamentally involved in creep.[1] The rate of steady-state creep depends on temperature and applied stress and follows the relation as

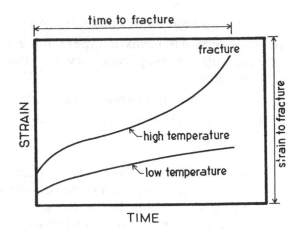

**Figure 9.1** Typical creep phenomenon at low and high temperature regions.

$$\varepsilon_t = f(\sigma, t) \exp(-Q/RT)$$

where $\sigma$ is the applied stress, $t$ is time, $R$ is the gas constant, and $Q$ is activation energy. For pure metals including Fe, Au, Cu, Al, Pb, Zn, In, and Cd, the value of the activation energy is found to be in good agreement with that for self-diffusion.[2, 3, 4] Subsequently a study of self-diffusion in polycrystalline tin also indicated in essential agreement with the activation energy for creep, although some discrepancy was found in earlier experiments.[2, 5, 6]

With respect to mechanisms, at temperatures and stresses where dislocation processes are insignificant, the time-dependent deformation is believed to occur by stress-directed vacancy flow.[7] Under such circumstances, the creep strain accumulation is attributed to diffusional transport of material and vacancies flow between sources and sinks.[8] The sources and sinks are normally external surfaces or grain boundaries, and the diffusion occurs through the crystal lattice or grain boundaries. At high stresses where dislocations become significant, the generation and movement of dislocations are considered to be predominant steps. It is well-established that under stress, dislocations will move, some passing out of crystals at the surface, others piling up at barriers, and some intersecting. Mott suggests that the important softening process at high temperature is dislocation climb, which is the rate-controlling process for high temperature creep.[9]

Therefore, the mechanism of creep depends not only on the material but also upon the range of stress and temperature encountered. For solid solution alloys, like most solders, their creep behavior may be different from that of pure metals.

Cannon and Sherby illustrated the division of alloys into two categories: Class I alloys whose behavior is different from pure metals and Class II alloys which behave similarly to pure metals.[10]

In Class I solid solution alloys, the steady state rate is given by:

$$\dot{\varepsilon} = K \frac{\sigma^3}{G^4} \qquad\qquad [9.1]$$

$$\dot{\varepsilon} = K \frac{kTD_s\sigma^3}{b^5 e^2 C G^4} ,$$

where $\sigma$ is the applied stress, $G$ is the shear modulus, $k$ is the Boltzmann constant, $T$ is the absolute temperature, $D_s$ is the diffusion coefficient of the solute, $b$ is the length of the Bergers vector, $C$ is the concentration of the solute atom, and $e$ is the fractional size difference between the solute and solvent atoms.[11] The creep rate of this category of alloys is proposed to be dislocation glide controlling. For Class II alloys where the solute atoms do not impede the dislocation movement, the creep is suggested to be dislocation climb controlling. Thus the same equation for pure metals is valid:

$$\dot{\varepsilon} = A' \left(\frac{\sigma}{E}\right)^5 \qquad\qquad [9.2]$$

$$A' = ADS^{3.5}$$

where $E$ is the elastic modulus, $D$ is the diffusion coefficient, and $S$ is the stacking fault energy.[12]

It is very informative that Cannon and Sherby illustrated a semi-quantitative analysis on when dislocation glide or dislocation climb is creep-rate controlling.[10] In their analysis, assuming all deformations take place during either glide process or climb process, it is expressed as

$$\dot{\varepsilon}_{\text{glide}} = \frac{\rho \bar{b} L}{t_{\text{glide}}}$$

$$\dot{\varepsilon}_{\text{climb}} = \frac{\rho \bar{b} L}{t_{\text{climb}}},$$

where $\rho$ is the mobile dislocation density, $b$ is the length of the Bergers vector, $t_{\text{glide}}$ is the average time a dislocation spends in gliding, $t_{\text{climb}}$ is the average time a dislocation spends in climbing, and $L$ is the average distance a dislocation moves parallel to the slip plane. Thus, the average total time from dislocation creation to annihilation is the sum of the average time a dislocation spends in gliding plus the average time it spends in climbing. That is,

$$t_{\text{total}} = t_{\text{glide}} + t_{\text{climb}}.$$

$$\frac{1}{\dot{\varepsilon}_{\text{total}}} = \frac{1}{\dot{\varepsilon}_{\text{glide}}} + \frac{1}{\dot{\varepsilon}_{\text{climb}}} \qquad [9.3]$$

Thus, by substituting Equations 9.1 and 9.2 into Equation 9.3,

$$\dot{\varepsilon}_{\text{total}} = \frac{A'K\sigma^5}{E^4 (EK + A'\sigma^2)}. \qquad [9.4]$$

When $EK \gg A'\sigma^2$, Equation 9.4 reduces to

$$\dot{\varepsilon}_{\text{total}} = \frac{A'\sigma^5}{E^5},$$

which is equivalent to Equation 9.2 describing Class II alloy creep behavior.

When $EK \ll A'\sigma^2$—that is an alloy with large atomic size difference between solvent and solute—then

$$\dot{\varepsilon}_{\text{total}} = \frac{K\sigma^3}{E^4},$$

which is the equation describing Class I alloy creep behavior. This analysis provides the significant fundamental knowledge in understanding the difference in creep behavior between pure metals and alloys. The two primary criteria to separate the two classes are the size difference between the solute and solvent atoms and the elastic modulus of the solvent. Alloys exhibiting high moduli and little size difference between solute and solvent follow the pure metal creep behavior, and alloys with low moduli and large solute-solvent size difference deviate from pure metal behavior. For common solder alloys, In/Pb, Pb/In, In/Sn, and Pb/Sn fall into Class I alloys, and Pb/Bi falls into Class II alloys.

To further understand the plastic deformation which occurs by dislocation movement or by the sliding of one plane of atoms over another (slip planes), the reader is referred to the dislocation theory and phenomena.[13]

## 9.4  FATIGUE

Fatigue, as one of the most common metal failure phenomena, occurs under alternating stresses. When a metal is subject to repeated applications of the same load, fracture occurs at much lower stresses than would be required for a single application of load. As a consequence, the metal can be damaged in many ways, ranging from minor misorientation of atoms to catastrophic cracking or failure. The basic material fatigue as illustrated by the Wohler curve, or S/N curve, indicates the dependence of the stress amplitude (S) on the number of loading cycles to complete fracture (Nf). Figure 9.2 shows a typical S/N curve. The stress limit at which the material can be cycled without fracture is termed fatigue limit, which, however, may not exist in some materials.

It is well known that the general responses of metals to cyclic loading involve hardening/softening and the development of plastic strains. On the atomic scale, when a metal is subjected to cycling loading, atomic motion and rearrangement by plastic flow will occur, resulting in hardening or softening, depending on the material. The atomic imperfection can form larger scale nuclei. In this phase, nuclei are formed in the slip planes and are highly concentrated. Nucleation sites where microcracks are usually originated include surface inclusions, fatigue slip bands, and grain boundaries. In this stage, microcrack nucleation takes place in a small part of the total volume and mostly at the surface layer, due to stress concentration

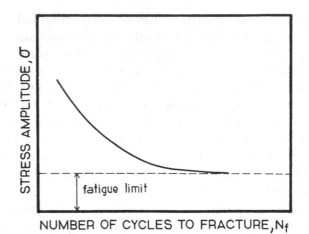

**Figure 9.2**  Typical fatigue curve (S/N curve).

in the surface layer. In the final stage, the crack propagates and ends in material fracture. Slip and twinning are believed to be prevalent in fatigue fracture mechanism. The stages of the fatigue process are generally proceeded in the sequence of hardening/softening, nucleation, propagation, and fracture, as the number of cyclic loading increases.

Macroscopically, the fracture surface as a result of fatigue shows combined regions of smooth appearance and rough appearance due to the initial slow crack spreading and the final rapid crack propagation.[14] The fatigue failures always start at the free surface and are very sensitive to stress raisers, such as surface defects and inclusions. In contrast to creep, fatigue failures do not involve macroscopic plastic flow, resembling brittle fractures in this regard. The movement of crack involves very little plastic deformation of the metal adjacent to the crack, and fatigue fractures can occur at very low temperature.

At higher temperature, more factors are involved. One of the most important considerations is creep effect. Other factors such as precipitation of new phases; grain growth; solution of precipitates; and surface compound formation due to corrosion, oxidation, and thermal stresses are all probable associated phenomena. It is important to recognize that small plastic strain, applied only once, does not cause any substantial changes in the substructure of materials, particularly of ductile materials. However, the repetition of this very small plastic deformation may lead to fracture failure.

## 9.5  THERMAL FATIGUE

The origin of stresses distinguishes thermal fatigue from mechanical fatigue. Thermal fatigue occurs under repeated applications of thermal stress, as a result of temperature changes. As temperature fluctuates, the creation of temperature gradients may cause thermal strains and associated thermal stresses across the section of material or assembly, resulting in the initiation of local plastic deformation. The thermal stresses may be introduced to the assembly when there is no temperature gradient if external constraints exist and the assembly consists of component materials with different expansion coefficients.

In the area of electronics packaging, different materials and/or material structures have been developed to alleviate the thermal stress caused by the differential thermal expansion coefficients among the component materials. Table 9.1 lists the thermal expansion coefficients for the materials which are commonly encountered in the electronics assemblies.

Effectiveness of solid metal core (Cu-Mo-Cu), heat pipe core, and copperclad-Invar incorporated in printed circuit boards to control the mismatch of thermal expansion coefficient between leadless chip carrier and epoxy-glass or polyimide-glass printed wiring boards has been well established. (Section 9.9 will discuss the effect of matched thermal expansion on the solder joint integrity in comparison with mismatched systems.) In addition, graphite material has been developed to control the thermal expansion coefficients of epoxy, polyimide, and Teflon board.[15]

Graphite has a negative thermal expansion coefficient, which is expected to counteract expansion of the materials within the package that have thermal expansion coeffficients greater than that of alumina, which is the housing material for leadless chip carriers. The location of the graphite layer in the printed wiring board is reported to be important. The structure with graphite layers in the outer layers of multilayer epoxy board enhances the rigidity of the structure, which is believed to improve the reliability of the package by minimizing the stress imposed on solder joint. The combined properties in negative thermal expansion coefficient, rigidity, and high thermal conductivity make the graphite enhanced printed wiring boards attractive, if cost is justifiable.

In addition to thermal expansion coefficient, the magnitude of thermal stress development depends on the amplitude of tempera-

## TABLE 9.1

### Thermal Properties of Common Electronic Packaging Materials

| Material | TEC* ($10^{-6}/°C$) | Thermal Conductivity (W/m °K) |
|---|---|---|
| Aluminum | 23.6 | 190 |
| Copper | 16–17 | 390 |
| Nickel | 13.3 | — |
| Silver | 19.7 | 422 |
| Silicon | 2.5 | — |
| Copper-Invar-copper** | 2.5–6.5 | 131 |
| Alloy 42 | 5.3 | 15.2 |
| Kevlar Fiber | –2 to –4 | — |
| Quartz Fiber | 0.5 | — |
| Glass Fiber | 4–5 | — |
| Graphite Fiber | –0.9 to 1.6 | 100–519 |
| Epoxy-Glass (x-y plane) | 12–18 | 0.3 |
| Epoxy-Glass (Z-axis) | 50–60 | — |
| Epoxy-Kevlar (x-y plane) | 5–8 | 0.2 |
| Epoxy-Quartz (x-y plane) | 4–6 | — |
| Epoxy-Quartz (z-axis) | 25–30 | — |
| Polyimide-Kevlar (x-y plane) | 3–7 | — |
| Polyimide-Glass (x-y plane) | 11–14 | — |
| Polyimide-Quartz ((x-y plane) | 6–9 | — |
| Teflon-Glass | 9.6 | — |
| Thermoplastics | 13–20 | — |
| Porcelainized Low-carbon Steel | 10 | 46.7 |
| 94% Alumina | 6.4 | 17.9 |
| 99.5% Beryllium Oxide | 6.4 | 207 |
| Alumina Chip Carrier | 5.5–6.5 | — |
| Sn/Pb Solder | 25–32 | 20–40 |
| Aluminum Nitride | 4.6 | 84–250 |
| Silicon Nitride | 3.2 | 33 |

*Thermal expansion coefficient
**Depending upon copper/Invar ratio

ture change, on heating rate, and on the material properties such as thermal conductivity, heat capacity, density, mechanical properties such as yield point and elastic modulus, and geometrical factors.

Therefore, it is apparent that a precise thermal analysis for an assembly composed of different materials is immensely difficult. In practice, the complexity is further enhanced by the temperature dependence of the fatigue process and creep process activated by the same temperature effect, in addition to thermal stresses induced by temperature cycles.

It has been misconceived that "ductile" materials provide better thermal fatigue life expectancy. In this connection one should note that material will behave differently under thermal cycling than under thermal shock, due to the difference in transient temperature distribution and the rate of stress application. Under a suddenly applied temperature change, ductile materials demonstrate the capability to withstand the conditions of thermal shock to the extent that stress relief can be brought about by plastic deformation. Brittle materials, with a limited plastic range, are more vulnerable to this form of stress, especially when their thermal conductivity is low as well. However, under repeated thermal cycling, the cycles of re-versed plastic strain impose high-temperature fatigue conditions on the material. Ductile materials may eventually fracture after being exposed to repeated cycles of stress.

Coffin and Manson have derived the relationship that the total strain range ($\varepsilon_t$) is the sum of plastic component ($\varepsilon_p$) and elastic component ($\varepsilon_e$) per cycle.[16, 17] By using Hook's Law—$\varepsilon = \sigma/E$, where $E$ is the elastic modulus—the elastic component can be obtained by measuring the strain amplitude of the specimen under loading. With

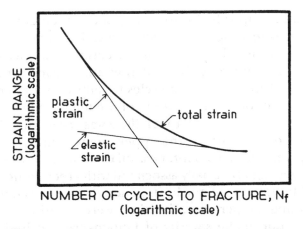

NUMBER OF CYCLES TO FRACTURE, $N_f$
(logarithmic scale)

**Figure 9.3** Strain range in relation to cyclic life.[17]

known elastic strain, plastic strain is calculated. An empirical relationship is established as

$$\varepsilon_t = \varepsilon_p + \varepsilon_e$$

$$= MN_f^Z + \frac{G}{E}N_f^\gamma ,$$

where $N_f$ is the number of cycles to failure (cyclic life), and $M$, $Z$, $G$, and $\gamma$ are material constants. Figure 9.3 indicates the relationship of the cyclic life and the total strain range as the sum of plastic and elastic components.

As shown in Figure 9.3, the elastic component is negligible in the low-cycle range and the total strain coincides with the plastic component. At the high-cycle range, the plastic component becomes negligible and the elastic strain accounts for the total strain.

At low-cycle range, the relationship is reduced to

$$\varepsilon_t = \varepsilon_p = MN_f^Z;$$

and with $Z$ equal to $-\frac{1}{2}$,

$$N_f = M\frac{1}{\varepsilon_p^2} .$$

This relationship is found to be in good agreement with fatigue results of a wide range of metals, indicating that fatigue life is inversely proportional to the square of plastic strain per cycle.

Under temperature cycling, Glenny and Taylor indicated the dependence of the number of cycles to failure on upper temperature, rather than on the magnitude of the temperature change.[18, 19] Experiments further showed that failure was accelerated by increasing the time of residence at the upper temperature, and the crack generally propagated in an intergranular manner. These characteristics as evidenced are usually associated with creep rather than with fatigue. Although the fatigue life is generally considered to decrease with increasing temperature, situations were found where localized creep has reduced the severity of fatigue stress at high temperatures.[20]

## 9.6  INTRINSIC THERMAL EXPANSION ANISTROPY

Thermal stresses may also be introduced even when there is no temperature gradient, when anisotropic thermal expansion is induced by anisotropic lattice structure.

Boas and Honeycombe studied the properties of tin-base and lead-base alloys under cyclic temperature treatment and found a marked difference between the two alloys.[21] The carefully polished specimens were alternately immersed in a hot oil bath at 150°C and a cold oil bath at 30°C, and held in each bath for different duration. Three types of phenomena that were observed in some specimens after such cyclic temperature treatment were deformation marks, rough surface, and migration of grain boundaries. The deformation was found to be affected by the duration of cycle, the number of cycles, the temperature, the metal employed, and the orientation of crystals. Tin specimens showed signs of plastic deformation after a small number of cycles, and these became more pronounced as the number of cycles increased. On the other hand, surface roughing and cracking were not observed in lead specimens. This difference is attributed to the anisotropy of thermal expansion in tin due to its noncubic crystal structure (tetragonal), in contrast to lead with cubic crystal structure (face centered cubic). In metals with hexagonal and tetragonal crystal structures, the thermal expansion varies with the direction in the crystal; consequently, temperature changes give rise to stresses at the grain boundary where two crystals of different orientations adjoin and free expansion is hindered. Furthermore, it was found that the rate of heating and cooling of the specimens during the cyclic temperature treatment had no effect on the deformation in that particular test. This study concluded that deformation can occur in noncubic metals under thermal cycling due to their intrinsic anisotropy of thermal expansion. The order of magnitude of the stresses introduced during the cyclic temperature treatment of noncubic metals was also estimated and expressed as

$$\sigma = (\alpha_1 - \alpha_2) \Delta_T \frac{E_1 E_2}{E_1 + E_2},$$

where $\alpha_1$ and $\alpha_2$ are the thermal expansion coefficients of the two adjoined crystals, $\Delta_T$ is the difference of upper and lower temperature, and $\varepsilon_1$ and $\varepsilon_2$ are elastic moduli of crystal 1 and crystal 2. As can be seen, the stress is directly proportional to the difference in thermal expansion between the two crystals, and the temperature range.

In alloys in which two phases are present, deformation due to different thermal expansion of the two phases is superimposed. For tin alloys—as an example, in the tin-rich tin-antimony alloys[22]—its solid solution matrix has the tetragonal structure of tin and the second phase has the intermetallic compound SnSb which possesses a cubic crystal structure. This second phase showed a stiffening effect on the matrix. The alloy had very little distortion in the region of grain boundaries in contrast to the behavior of similar alloys without the presence of the hard second phase after the same cyclic temperature treatment. Such alloys without the presence of the hard second phase showed marked deformation. The tin alloys consisting of the tin-rich matrix in which particles of a hard second phase of cubic crystal structure were embedded demonstrated considerably smaller deformation in the region of the boundaries between the crystals of the two phases than in the region of the crystal boundaries of the anisotropic matrix without a hard second phase.

## 9.7 CORROSION-ENHANCED FATIGUE

The fatigue strength of many metals is often reduced by the presence of a corrosive environment. Corrosion is normally electrochemical in nature. Thermodynamically, a pure metallic surface which is in a high energy state has a tendency to convert to a lower energy state such as oxides, sulfides, halides, or other compounds. Although thermodynamics does not tell the rate at which the reactions take place, types of reaction can be anticipated, in principle. In addition to knowing what reactions are thermodynamically favorable, the factors that may affect the reaction rate should be studied. These include surface condition, composition of the metallic surface, chemical components in the media, physical state of the media, and temperature and state of the surrounding environment.

The mechanism for corrosion-enhanced fatigue is not well understood. Corrosion fatigue is, however, known to be both a frequency and a time-dependent process.[23, 24] Since the reaction is always temperature-dependent, it is expected that time, temperature, and frequency are three major factors to be considered. The effects of "corrosion" environment on fatigue life are illustrated in Figures 9.4, 9.5, and 9.6. As can be seen, fatigue limit in the presence of corrosive

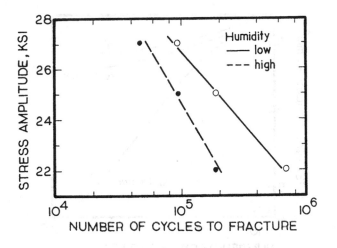

**Figure 9.4**   S/N curves of a magnesium alloy in high and low humidity.[26]

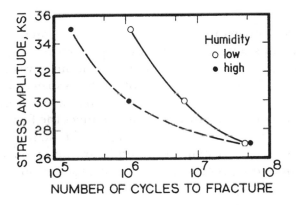

**Figure 9.5**   S/N curves of an aluminum alloy in high and low humidity.[25]

atmosphere has drastically dropped. The crack rate in an aggressive environment is always higher than that in an inert environment under otherwise identical conditions. The lowest crack rate is normally observed in a completely inert environment as vacuum, or dry inert gas. Liu and Corten indicated that the mean fatigue life of an aluminum alloy wire could be changed by a factor of two to three by the moisture content of air.[26]

A corrosive environment was found to affect both the nucleation and the propagation stages by shortening nucleation period and

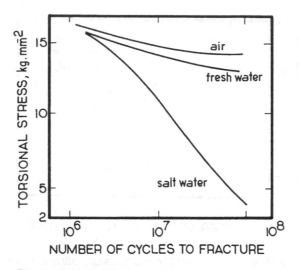

**Figure 9.6** S/N curves for a marine propeller shaft steel.[25]

accelerating propagation rate. Generally, corrosion media can be in either a gaseous state or liquid state. A gaseous environment is less aggressive than a liquid environment. In a liquid environment, local "etching" could likely produce pits which act as stress raisers at either selective places of higher slip activity or at nonselective places on the surface. The corrosion environment enhances the formation of slip stress-raisers. It is believed that in higher slip activity areas, slight differences in electrochemical potential inside or outside the slip band promote the process of stress-raisers formation.

## 9.8  COATING CONSIDERATION

With the corrosion principle in mind, and in view of the surface vulnerability, it is natural to consider a surface coating as potential protection in order to expand fatigue life. A lot of efforts have been made in developing protective coatings for various metallic surfaces.[27, 28, 29] Figures 9.7 and 9.8 indicate the beneficial effect of an organic coating on fatigue for smooth and notched surfaces, respectively.

In the literature, many coating techniques for metallic surfaces have been established, including hot-dipping, vapor deposition, metal-cladding, electroplating, spraying, brushing, and immersion.

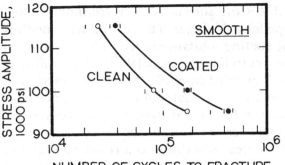
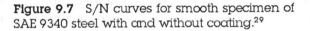

Figure 9.7   S/N curves for smooth specimen of
SAE 9340 steel with and without coating.[29]

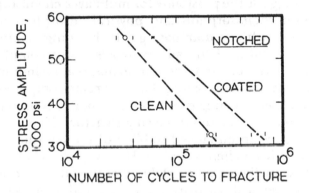

**Figure 9.8**   S/N curves for notched specimen of
SAE 4340 steel with and without coating.[29]

In category, two types of coating materials are commonly used: metallic and organic. It is perceived that organic coating would be more relevant to soft solder alloys, therefore a few more words are appropriate here. In general, surface protection can be viewed as operating through three basic types of function: (1) to serve as an "inert surface", (2) to act as sacrificial medium, and (3) to provide formation of "stable" compounds on the surface.

The first type can be well illustrated in the hydrophobic liquid-film–forming coating. In such liquid-film–forming coatings, the active ingredient must contain a molecular structure having both hydrophilic and hydrophobic functional groups. The metallic surface adsorbs the film at the polar end and leaves the nonpolar end to cover and wet the surface with hydrophobic film to achieve a moisture repellent character. The second type involves the mechanism that the coating substance reacts with the active constituents in the environment to lower the availability of harmful environmental components. The third type relies on the interaction of the coating with the to-be-protected material to form a more inactive surface. Regardless of the functional type, one key prerequisite is proper adsorption by either chemical or physical linkage.

The ideal physical adsorption is a result of weak intermolecular forces between the adsorbate and adsorbent, and the binding energy is considered to be related to the heat of vaporization. In chemisorption, the binding forces are those which are covalent in nature. The occurrence of the covalent bond suggests preferred adsorption sites and in turn suggests the possibility for multilayer chemisorption. In practice, both chemisorption and physical adsorption may exist simultaneously with one dominating. The bonding or binding energy and the stability of the adsorbate-adsorbent complex which depends on its structure and orientation may play an important role in the effectiveness of the coating. The adsorbate composed of more than one electron donating group (multidentate) coupled with the potential formation of a desirable ring structure (five- or six-membered ring) gives an enhanced stability of the coordination compound formed from chemisorption. However, this chelate effect is not the sole contributor to the stability of the adsorbate-adsorbent complex. Its chemical reactivity and configuration are also factors.

In order to achieve protection, properties such as adhesion, film integrity, permeability, and chemical characteristics require attention. For solder joints, organic coating appears to be a rewarding area to be researched. Specifically, for high reliability joints serving adverse conditions, the development of a compatible coating in conjunction with a practical technique would make a significant contribution to the reduction of corrosion enhanced fatigue.

## 9.9 INTERMETALLIC COMPOUNDS

The alloys, such as binary alloy, where two metals have a limited mutual solubility, may form new phases at certain ratios of the two components (stochiometric) when the solid solution solidifies. These new phases possess different crystal structures from those of either component. These new phases are called intermetallic compounds. (Basic alloy structure and alloy formation are covered in Sections 4.1 and 4.2.)

The properties of resulting intermetallic compounds generally differ from the component metal exhibiting less metallic characteristics such as reduced density, ductility, and conductivity. The formation of intermetallic compounds between solder and the substrate is the metallurgical reaction as just defined. The existence of this type of intermetallic compound has been observed at or near the interface of solder alloy and some substrates in the solder joined assemblies. By using phase diagrams, the potential formation of the intermetallic compounds and its extent between the known composition of solder and the substrates can be anticipated, although some intermetallic compounds that do not appear in the equilibrium phase diagrams have been identified.

When solder is in contact with the common substrate as listed in Table 9.2 for long enough time at high enough temperature, the potential formation of intermetallic compounds or incompatibility may exist. Table 9.2 also lists the compounds which have been established in the equilibrium phase diagrams and identified in the assemblies. Both tin-based and indium-based solder have potential to form intermetallic compounds with common solderable substrates.

At the temperature below the solder liquidus temperature, the formation is primarily a solid state diffusion process and thus highly depends on temperature and time. As solder is in molten state, the solubility of the element of substrate in molten solder is expected to accelerate the rate of formation.

The adverse effect of intermetallic compounds on the solder joint integrity is believed to be attributed to the brittle nature of such compounds and to the thermal expansion property, which may differ from the bulk solder. This differential thermal expansion could contribute to the internal stress development. In addition, the excessive intermetallic compound formation may impair the solderability of some systems where intermetallic compound depletes one element of the contact surface. One example is the tin (Sn) depletion from Sn/Pb coating on copper (Cu) leads due to Sn/Cu intermetallic compound formation, resulting in poor solderability of the component leads. In this case, the interface area is expected to have $Cu_3Sn$ phase next to copper substrate, followed by $Cu_6Sn_5$ phase

**TABLE 9.2**

**Potential Intermetallic Compounds Formation and Incompatibility Between Solder and Common Substrates**

| Solder | Substrate | Intermetallic Compounds or Incompatibility |
|--------|-----------|--------------------------------------------|
| Sn-containing | Cu | $Cu_6Sn_5$, $Cu_3Sn$ |
| Sn-containing | Au-based | $AuSn$, $AuSn_2$, $AuSn_4$ |
| Sn-containing | Ag-based | (Intermediate phases) |
| Sn-containing | Pd-based | $PdSn_4$ |
| Sn-containing | Ni | $Ni_3Sn_4$ |
| In-containing | Cu | $Cu_6In_5$ (Intermediate phases) |
| In-containing | Au-based | $AuIn$, $AuIn_2$ |
| In-containing | Ag-based | $Ag_3In$, $Ag_2In$, $AgIn_2$ |
| In-containing | Ni | $Ni_3In_7$, $Ni_4In$, $NiIn$, $Ni_2In$, $Ni_3In$ |
| Sb-containing | Brass or Zn-containing | $ZnSb$, $Zn_3Sb_2$ |
| Sn/Bi | Pb-containing (Sn/Pb Solder) | (Low-melting ternary phase) |
| High In-containing | Sn-containing (Sn/Pb solder) | (Low-melting Sn/In alloy) |

and lead-rich region. The excess intermetallic compound formation may also alter the solder joint appearance.

The excessive intermetallic compound formation and its mobility in the assembly of 63-Sn/37-Pb solder on Cu substrate were found where the assembly was held at reflow peak temperature for a prolonged time. The examination of the cross section of the resulting solder joint indicates Cu/Su intermetallic phase has moved away from the interface area. Under SEM examination, Figure 9.9(a) shows the presence of the needle-like dark phase in the solder, which was identified as Cu/Sn composition by semiquantitative x-ray energy dispersion analysis, as shown in Figure 9.9(b).

The formation of intermetallic compounds has been identified as one of the main sources of solder joint failure. When an appreciable thickness of intermetallic compound is developed along the solder–substrate interface, cracks are often initiated around the interfacial area under stressful conditions. Figure 9.10 indicates the formation of intermetallic compound(s) where the Sn/Ag eutectic

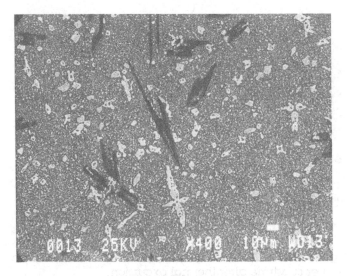

**Figure 9.9(a)** SEM micrograph showing the presence of needle-like Cu/Sn intermetallic compound.

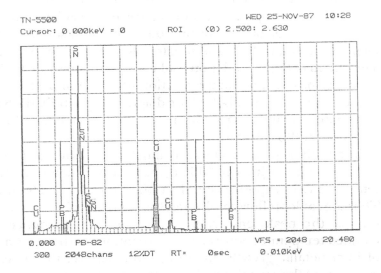

**Figure 9.9(b)** EDAX showing Cu/Sn composition.

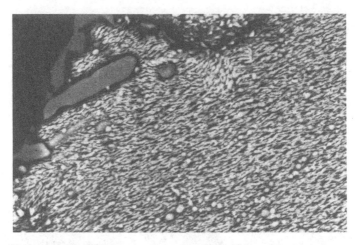

**Figure 9.10**  SEM micrograph of Sn/Ag solder joint on copper substrate after thermal excursion.

solder is used on the copper substrate, and then subjected to a long-time exposure of temperature cycling. In Figure 9.11, as can be seen, the crack propagates along the interface of Sn/Ag solder and the intermetallic compound containing Cu and Sn is identified by X-ray energy dispersion analysis.

The combination system of Sn/Pb solder on silver, gold, or silver- or gold-containing substrates is not considered to be solely an intermetallic compound problem, since a leaching solubilizing problem of silver or gold in tin due to the high solubility of these elements in tin, is the main initiator, as discussed in Section 10.12. The system of Sn/Bi solder on lead-containing substrate such as Sn/Pb solder has the potential formation of a low-temperature ternary phase, as shown in Sn-Pb-Bi phase diagram (Appendix II). Similarly, high indium-containing solder may form low-melting alloy (SnIn) on Sn or Sn/Pb substrate.

In addition to the material compatibility, external factors also affect the rate of intermetallic compound formation, namely, the temperature of exposure and the time at the elevated temperature. Thus the solder reflow conditions, such as the peak temperature, the time at the peak temperature, and the total residence time at elevated temperatures are expected to influence the rate and extent of intermetallic growth. In storage and service, the temperature exposure of the assembly is obviously another external factor for the intermetallic growth in the "vulnerable" combination systems.

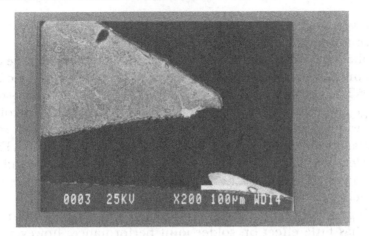

**Figure 9.11**  SEM micrograph of solder joint crack along the intermetallic and solder interface.

Therefore, it can be summarized that the formation of intermetallic compound between solder and the substrate material in contact depends upon the following factors:

• compatibility of solder composition with substrate
• time of soldering
• temperature of soldering
• paste-solder storage conditions
• service conditions

The formation of appreciable intermetallic compounds has been shown to adversely affect solder joint strength and integrity. This metallurgical reaction is dependent on time and temperature. With an understanding of the metallurgical reactions, in conjunction with the utilization of phase diagrams, control of the time and temperature is the way to minimize the intermetallic compound growth. (More studies are discussed in Section 9.11.)

## 9.10  SOLDER JOINT VOIDS

Voiding in solder joint is known as one of the adverse phenomena to solder joint integrity and reliability. Voids in the joint may potentially cause weakening in joint strength, reduction in electrical and thermal conductivity, and induce crack initation.

Mahalingam and others analyzed the effect of soft solder die bond voids on thermal performance of ceramic, metal, and plastic packages.[30] The thermal transient, measured as change in drain-source voltage before and after applying the heat pulse as a function of heat pulse width, was obtained for TO-3 metal packages with soft solder die attach over a range of void contents. The data as shown in Figure 9.12 clearly indicate that void affects thermal dissipation, and high void content jeopardizes the useful lifetime of a power device due to the increased function temperature.

It is generally expected that a low volume of small well-dispersed voids has little effect on solder joint performance; however, high volume and/or large size voids may degrade the joint in electrical, thermal, or mechanical properties. As an example, the large void in solder joint as shown in SEM micrograph Figure 9.13 indicates crack initiation and propagation at the void after being exposed to temperature cycling.

With respect to solder paste, factors affecting voids in the resulting joint include flow characteristics of flux/vehicle system, thermal and physical properties of flux/vehicle system, and metal load. The void development varies with different solder paste products. In addition,

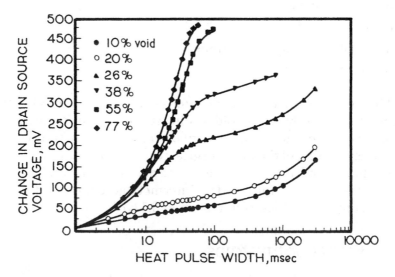

**Figure 9.12**   Effect of voids in solder joint die attach.[30]

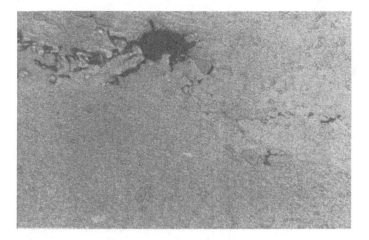

**Figure 9.13**   SEM micrograph exhibiting crack initiation at void.

processing parameters and solder joint design are equally important factors.

In order to minimize voiding, processing parameters and joint design should also be optimized. One paste reflowed under two different heating conditions shows different void development, as depicted in Figure 9.14.

In addition to the selection of solder paste, the paste deposit dosage, deposit thickness, joint configuration, including the interface area, reflow time, cooling rate, and wettability all affect the void development.

## 9.11   STATE-OF-THE-ART STUDIES

With the complexity of alloy behavior, and versatility of assembly design, detailed failure mechanisms for surface mount solder joint design are not well understood. This section is to summarize the major findings so far obtained and the state-of-the-art conclusions with respect to failure mechanism about the solder joint integrity in surface mount assemblies.

In the literature, studies have been largely devoted to the effects of temperature and power cycling on solder joints and the mismatch of thermal expansion coefficients among the materials in an assembly.

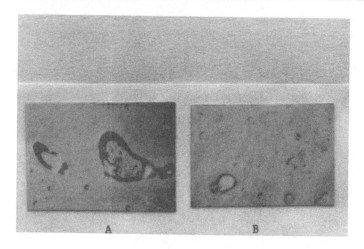

**Figure 9.14** Void development of a solder paste reflowed under two conditions.

Hagge estimated the thermal cycling fatigue life of solder joints for leadless ceramic chip carriers based on Mason-Coffin fatigues equation covering both plastic and elastic strain range.[31] The work predicted that the total strain range applied for the 63-Sn/37-Pb solder joint to exceed 100 cycles of fatigue life must not exceed 4.5%, and to exceed 1000 cycles, the strain range applied must be kept below 1.5%. The study suggested that the typical failure mode was repeated plastic deformation, and that fatigue ductility is considered to be the most important property. The contamination in solder joint by precious metals such as gold significantly reduced the fatigue life, which was interpreted as a decrease in fatigue ductility. Hagge also discussed the effect of grain structure of solder joint. The interparticle spacing in the grain structure influences the fatigue life, and the greater fatigue life is attributed to the smaller interparticle spacing. The interparticle spacing was believed to depend on the cooling rate during solidification, with the formation of finer particles, as solder is quickly cooled down. A relationship describing the effective strain, solder joint, and leadless component in thermal cycling condition is expressed in an equation:

$$\text{effective strain} = b \frac{\Delta\alpha\Delta_T L}{2t}$$

where $b$ is geometry factor, $\Delta\alpha$ is the difference in thermal expansion coefficient, $\Delta_T$ is the temperature range, $L$ is the length of leadless component, and $t$ is the thickness of solder joint. The equation indicates that a lower strain is produced when temperature range and thermal expansion coefficient difference are low, leadless component is smaller, and solder joint is thicker.

The effects of temperature and power cycling and cooling in leadless chip carrier on four-layer polyimide printed wiring board assemblies were studied by Heller.[32] The effectiveness of solid metal core (Cu-Mo-Cu) and heat pipe core was also experimented. Under the conditions of having a temperature cycle with temperature extreme of +71° to –34°C and 60 minutes at each extreme for a total of three hours per cycle, and having a power cycle with 1.5 hours for on and off power, the assembly of leadless ceramic chip carrier/four-layer polyimide glass board containing a polyimide glass insulating layer and Cu-Mo-Cu core withstood over 1000 cycles without solder joint crack. In contrast, the same assembly, differing only in the insulating layer being Kapton in place of polyimide glass, showed solder joint crack at 679 cycles. As expected, two solder alloys, 63-Sn/37-Pb and 62-Sn/76-Pb/2-Ag were found to perform equally well under the above described testing conditions. Furthermore, thermal pads and heat pipe were found to facilitate heat dissipation.

Engelmaeir derived an analytical method to predict the power cycle life expectancy of the solder joints between a leadless chip carrier and the substrate.[33] The solder joint performance can be optimized by designing the thermal expansion coefficient of substrate to satisfy the equation

$$\alpha_s = \alpha_c \frac{T_c - T_o}{T_s - T_o},$$

where $\alpha_s$ is the thermal expansion coefficient of substrate, $\alpha_c$ is the thermal expansion coefficient of leadless chip carrier, $T_o$ is the steady-state temperature, and $T_s$ and $T_c$ are the temperatures of substrate and chip carrier, respectively.

The reliability of solder joints between clip leaded 100 I/O ceramic chip carrier and FR-4 epoxy glass board was investigated.[34] The results indicate that the compliance of clip leads provides drastically improved fatigue life of solder joints, even for systems containing thermal expansion mismatch and large components. It is also indicated that leaded components may require a tailored thermal expansion coefficient of the substrate, but to a less exacting degree than is necessary for leadless components, to provide high reliability. The effect of leaded surface mounted devices on solder joint with respect to thermal stresses was also analyzed by Riemer, showing that the leaded devices impose significantly lower thermal stresses onto solder joints than do the leadless devices.[35]

Riemer, Hall, Howard, and others have studied the solder joint integrity between leadless ceramic chip carrier and alumina substrate.[35, 36, 37] In this type of assemblies, the thermal expansion coefficients of components and substrate are essentially in match.

Solder joint cracks during temperature cycling in the leadless ceramic chip carrier/alumina assembly were found, suggesting that mismatch in thermal expansion coefficient is not the only cause for solder joint cracks.[35]

Chalco developed a mathematical model to predict the solder joint strength and showed that the joint strength was improved by increasing solder fillet length and solder thickness, and by decreasing the amount of voids in the joint.[38] The advantage of increased solder joint thickness was found to improve the fatigue life under temperature cycling by Hwang, Lau and Engelmaier.[33, 39, 40] Figure 9.15 illustrates the effect of solder joint thickness on the shear strength across a range of temperature cycling. Under test condition

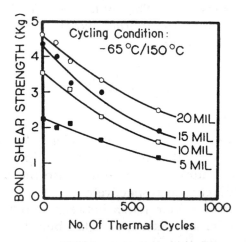

**Figure 9.15**  Shear strength after temperature cycling for a range of solder joint thickness.[39]

with temperature extremes of –65°–150°C and 10 minutes at each extreme, the solder joints made from wet paste thickness of 10 mils or less displayed cracks at a significant percentage of total joints tested after 660 cycles, and solder joints made from wet paste thickness of 20 mils showed no joint cracks after the same number of temperature cycles.

The influence of solder joint shape on its reliability was analyzed by Sherry et al.[41] The solder joint with designed shape demonstrated substantially better fatigue life than the conventional, unshaped joint in the leadless ceramic chip carrier and printed wiring board assembly.

Riemer and Saulsberry studied the power cycling of ceramic chip carrier and ceramic substrate, and indicated that a temperature gradient existed inside the chip carrier between the center area and the outside region with solder connected to the substrate, but little temperature differential across solder joint was detected.[42] The lack of temperature gradient in solder joint, and therefore, low stresses in the joint, is considered to be the explanation for the absence of solder joint crack during power cycling. The test results are not consistent with the thermal stress analysis of the system. This is attributed to the efficient cooling of the chip carrier package being tested. This analysis also leads to the conclusion that void-free solder joint to facilitate thermal conductivity is required in order to keep temperature drop across solder joint to a minimum.

Hall has measured the strains applied in leadless ceramic chip carrier and the FR-4 printed wiring board during temperature cycling to infer the resultant strain in the solder joint.[43] It was found that the shear angle to which solder joint is subjected is 2.4 for the modified in-line solder post and 1.6 for the solder of staggered pillars with FR-4 board. However, with copperclad-Invar FR-4 board, the shear angle for the solder of staggered pillars is 0.4. The large shear strain applied in solder joint between FR-4 board and leadless ceramic chip carrier is expected to develop plastic flow in the solder. Hall indicated that below room temperature, most expansion mismatch is accommodated by "bi-metallic strip" type bending of LCC and PWB; above room temperature, mismatch is accommodated by shear in solder, however, upon repeated temperature cycling resulting in repeated stress-strain hysteresis, plastic deformation in solder may occur.

FR-4 printed wiring board having thermal expansion coefficient matched to leadless chip carrier such as incorporating with copperclad Invar core has been studied by Smeby.[44] The 63-Sn/37-Pb solder in both systems of thermally matched and mismatched FR-4 board/ leadless chip carrier were examined. The strain levels were found to be a function of chip carrier size and the composition of the printed wiring board. Printed wiring board with copperclad-Invar core showed less strain level as compared with FR-4 material. Thermal cycling results indicated that leadless chip carrier mounted on FR-4 printed wiring board had solder joint failure at much fewer thermal cycles than that on FR-4 board that is incorporated with copperclad-Invar core. As expected, the use of copperclad-Invar core in FR-4 to lower its thermal expansion coefficient improves solder joint life under thermal cycling. This is consistent with Taylor's findings.[45] Further examination reveals that failed solder joints associated with a thermally matched system (FR-4 with copperclad-Invar) do exhibit intergranular cracking and voids formation within the solder structure, and that cracks propagate in many directions, following the boundaries between tin-rich and lead-rich phases. The cracks are also found to originate at the surface of the fillet where it is not the high strain region according to the stress distribution analysis.[44] That cracks originate at surface in a thermally matched system is further supported by other studies,[39] as shown in Figure 9.16(a). Solder joint volume in this case affects the solder integrity. Under identical test conditions, the solder joint made from paste wet print of 20 mil thickness did not show joint failure, as shown in Figure 9.16(b). In contrast, solder joint which was made from 10 mil wet print failed after 660 temperature cycling (–65°C/150°C), as indicated in Figure 9.16(a). Microscopic examination confirmed that the solder joint had undergone significant microstructure change during temperature cycling in the thermally matched system—namely, particle growth and spheroidization.

Other investigations suggest that the lifetime of solder joints undergoing thermal/isothermal fatigue correlates well with creep phenomenon.[46, 47] Weertman has measured creep rate of tin-lead solders and indium solders and found that, as expected, the creep rate of indium solder is significantly higher than that of tin-lead solder.[48] For example, 7.8 At % Sn/Pb alloy displayed a creep rate of $0.32 \times 10^{-3}$ min$^{-1}$ at 423°K and $27 \times 10^6$ dynes cm$^{-2}$ in comparison with 7.9 At % Sn/In alloy having a creep rate of $62 \times 10^{-3}$ min$^{-1}$ at 413°K and $7.2 \times 10^6$ dynes cm$^{-2}$.

The integrity of solder joints connecting chip resistors to metallized substrate was investigated.[49] During temperature excursion a

**CROSS SECTION OF SOLDER JOINTS**
**10 mil print**

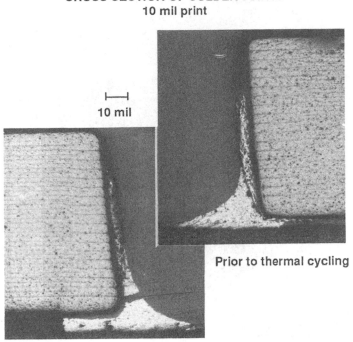

10 mil

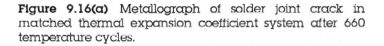

Prior to thermal cycling

After 660 thermal cycles

**Figure 9.16(a)** Metallograph of solder joint crack in matched thermal expansion coefficient system after 660 temperature cycles.

change in composition of termination coating on resistor chip in conjunction with the change in the vicinity of interface was found. This compositional change evidences a metallurgical reaction between solder and termination coating with the formation of Sn/Au, Sn/Pd, Sn/Pt intermetallics. The intermetallic formation leads to degradation in termination adhesion strength as well as in solder joint integrity. A nickel coating is found to provide a barrier for such reactions.

As discussed in Section 9.9, intermetallic compounds can be readily formed between tin-lead solder and constituents of substrate such as Au, Pd, or Cu. Compounds $AuSn_4$, $AuPd_4$, $AuPd_2$, $PdSn_4$, $Cu_3Sn$, and $Cu_6Sn_5$ were identified by x-ray diffraction and electron microprobe analysis.[50, 51, 52] It was found that there is a progressive transition of $Cu_3Sn$ formed during soldering to $Cu_6Sn_5$ upon aging.

## CROSS SECTION OF SOLDER JOINTS
### 20 mil print

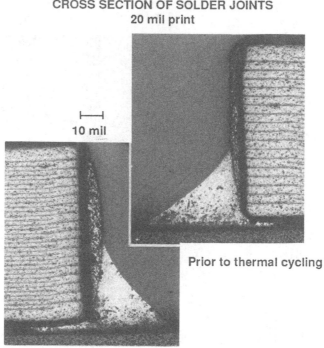

10 mil

Prior to thermal cycling

After 660 thermal cycles

**Figure 9.16(b)**  Metallograph of solder joint crack in matched thermal expansion coefficient system after 660 temperature cycles.

The aging effect on 60-Sn/40-Pb solder joint, as shown in Figure 9.17, presents the mean peel strength versus aging time at the temperature range of 75°–180°C. Microscopically, the growth of intermetallic compounds is identified by microstructure.

After being isothermally aged at 125°–170°C, the solder cladded leads connected to Pt-Pd-Ag thin film integrated circuit have shown different levels of degradation in strength, depending on the solder alloy.[53] The strength of Sn/Ag solder (96.5-Sn/3.5-Ag) fell at a faster rate than did Sn/Pb (60-Sn/40-Pb) or Sn/Pb/Ag (62-Sn/36-Pb/2-Ag) solders. This result is believed to be due to the rejection of Pb at the solder intermetallic interface, where solder composition changes continually during aging. The addition of lead can significantly change the growth kinetics of the intermetallic layer. The

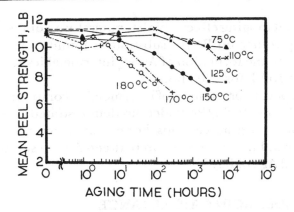

**Figure 9.17** Aging effect on solder peel strength at temperature range of 78°–180°C.[53]

strength is related to the growth of intermetallic layer as expressed by

$$\sigma = \sigma_0 - C\sqrt{Dt} ,$$

where $\sigma_a$ is initial strength, $C$ a constant, $\sqrt{Dt}$ is related to the thickness of the intermetallic layer, $D$ is effective diffusion coefficient for compound growth, and $t$ is the aging time.

One long-term solder joint failure was well illustrated in the core memory module after having served in the field for about 10 years.[54] The module consisted of a stack of PWBs (printed wiring boards) with copper wire being soldered between adjacent boards as board-to-board interconnection. The module design was reported to impart temperature fluctuation around solder joints and to operate at approximately 60°C. It was observed that solder exhibited cracking and exfoliation at the copper wire-solder interface. Microscopic examination suggested that the joint failed as a result of fracture within the tin-depleted region of the solder, which is adjacent to the intermetallic layers formed on the copper wire. The examination confirmed, by using x-ray energy dispersion analysis, the existence of a three-layer structure—($Cu_3Sn$) next to copper, ($Cu_6Sn_5$), and nearly pure lead—formed at the joints. The eutectic solder structure also changed to larger grains of alpha and beta terminal solutions as shown in the tin-lead phase diagram.

Equivalent to aging effect is high temperature storage. The 62-Sn/ 36-Pb/2-Ag solder on Pt/Au substrate is found to be severely degraded due to the formation of $AuSn_4$ intermetallics as the assembly is stored at 150°F for 1000 hours.[55]

With respect to prevention of intermetallic compound growth, an alloying addition to Sn/Pb solder has demonstrated retarding effect on such metallurgical reactions. For example, antimony is found to be effective in hindering reaction between Ag/Pd substrate and 10-Sn/88-Pb/2-Ag solder.[56, 57, 58]

## 9.12 SOLDER JOINT APPEARANCE

As outlined in Section 9.13, recently, x-ray technique, laser thermal technique and other examinations of solder joint quality have been extensively developed in addition to automated optical inspection. Nevertheless, visual inspection has been commonly adopted to determine the solder joint quality. Visual inspection is still a criterion called for by the military specification at the present time. Thus, a few words are in order about the appearance of solder joints. The factors which affect solder joint appearance in terms of luster, texture, and intactness are

- inherent alloy luster
- inherent alloy texture
- residue characteristics after paste reflow
- degree of surface oxidation
- completeness of solder powder coalescence
- microstructure
- mechanical disturbance during solidification
- foreign impurities in solder
- intermetallic compounds
- phase segregation
- cooling rate during solidification
- subsequent heat excursion, including aging, temperature cycling, power cycling, and high temperature storage

It is known that the process of solidification from melt is crucial to the microstructure development of an alloy. Microstructure in turn affects alloy strength and surface condition. It is observed that an

ideal Sn/Pb eutectic structure as shown in Figure 9.18 imparts a
bright and smooth solder surface. Significant deviation from the
eutectic composition as shown in Figure 9.19 usually results in a

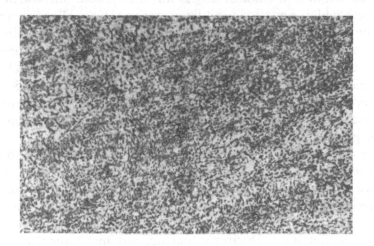

**Figure 9.18**   Metallograph of 63-Sn/37-Pb.

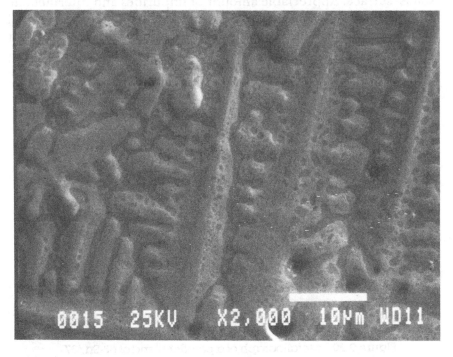

**Figure 9.19**   SEM micrograph of 25-Sn/75-Pb.

visually duller joint, primarily due to the presence of a lead-rich phase as inherited in alloy composition and/or larger grain size, as a result of reflow process. For eutectic Sn/Pb solder, solder joint appears bright and smooth unless an undesirable microstructure develops, which can be caused by excessive slow cooling, excessive heating at elevated temperature, or incomplete alloying in the composition. Figure 9.20 shows the 63-Sn/37-Pb microstructure of a two-metal powder mix instead of pre-alloyed powder, which was reflowed under the same condition as that in Figure 9.18. The difference is attributed to insufficient reflow time to achieve the eutectic alloy structure. Its microstructure also explains a slightly duller joint.

The residue interference on the molten surface, and during its solidification as a result of sluggish flow, can contribute to the rough texture on the solder surface. Inadequate heating time and/or temperature will cause incomplete coalescence of solder particles, which also contributes to unsmooth surface and possibly to an inferior solder joint. During soldering, any mechanical agitation which disrupts the solidification process, may lead to an uneven solder surface. Appreciable amount of impurities and intermetallic compounds formation due to foreign elements may result in dull appearance. After the completion of reflow and solidification, heat excursion is expected to have a significant impact on solder joint appearance, whether it is only a surface reaction, or internal struc-

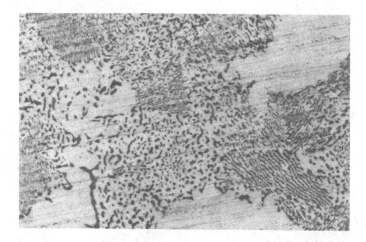

**Figure 9.20**  Metallograph of a powder blend of 63-Sn/37-Pb.

ture change. Heat excursion can come from different sources, such as high temperature storage, aging, temperature fluctuation during service and power cycling during functioning. The effects of heat excursion have been discussed in the other sections.

In many cases, a duller and/or unsmooth joint may not be necessarily a "defective" joint in a functional sense. Tests may also confirm that the duller and unsmooth joints have equivalent mechanical strength. However, one must be cautious in drawing the conclusion regarding the joint reliability in relation to joint surface appearance. It should be noted that surface condition of metal is considered to be one of the variables that affect failure mechanisms of the alloy. Thoroughly designed tests, considering all variables relevant to the service condition of solder joint, will provide significant information to predict the solder joint reliability, although a test to precisely simulate the conditions is quite difficult.

In the author's view, the smoothness and brightness to the level that the alloy should exhibit, reflect a "proper" joint having been produced from a "proper" process.

## 9.13  SOLDER JOINT INSPECTION

Inspections on solder joints currently can be categorized by following four techniques:

1.  Visual inspection with or without optical aid.
2.  Automated optical inspection.
3.  Real-time x-ray inspection.
4.  Laser/infrared inspection.

Visual inspection of the quality of solder joints as a criterion to reject or accept has been adopted as the most prevalent technique, until recent demands posed by surface mount technology and high density circuits. With increased number of solder joints per board, increased density of circuitry and miniaturized size of solder joints, visual inspection becomes more humanly strenuous and in some cases very difficult. For the last couple of years, new techniques using x-ray and laser/infrared technology, as well as advanced automated optical systems, have been developed and are under evaluation in the production sites.

Solder joint defects can be external and/or internal structure-related. Visual inspection detects missing joints, solder bridging, and wetting problems, but not for internal structure defect without destructive testing. On the average, the accuracy of visual inspection is believed to be below 75–85%, with the limit of throughput about 10 joints per second. Visual inspection varies with operator, as shown in Figure 9.21. By using x-ray technology, voids, solder balling, insufficient solder, lead-related defects, device-related defects, and solder bridging are identified.[59, 60] The laser/infrared inspection system using 30-watt continuous wave Nd:YAG laser as a heating source and a visible-band (HeNe) laser for the target is shown in Figure 9.22. The system is composed of an optical fiber and special optics to align the laser beam, and the infrared detector field of view to receive the thermal emission from the solder joint. Any defect which causes an anomaly in the heating rate and/or in the cooling rate from the "reference signature" can be detected since the heating and/or cooling rate of the targeted joint is a function of the surface properties and internal structure.[61, 62] It is obvious that the determination of the standard of "reference" joint is crucial, which is dependent on component or device, design, and board. The plot of surface tem-

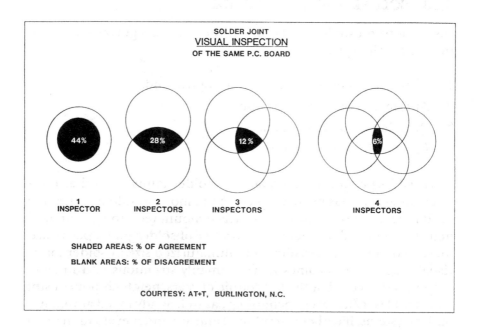

**Figure 9.21** Operator dependency of solder joint visual inspection. *(Courtesy of AT&T–Federal Systems Division, Edward Barnes.)*

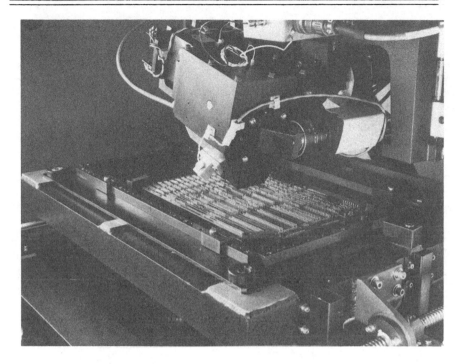

**Figure 9.22** Laser/infrared inspection system. *(Courtesy of Vanzetti Systems, Incorporated.)*

perature versus time during the heating and cooling, called "thermal signature," is an identification tool. Figure 9.23 provides an example of the thermal signature of a void defect of a PLCC solder joint in comparison with the standard. As can be seen, the joint containing air voids beneath the lead gets "excessively hot" resulting in a greater surface temperature, due to the fact that heat flow is hindered by the presence of voids. The signatures of defects are therefore expected to fall either above or below the "reference signature." Defects such as lifted leads of IC package, rough solder surface, flux residue or contaminated solder, deformed or off-centered leads, insufficient solder, and excessive intermetallics show higher peak temperature in their thermal signatures, while excess solder and solder bridging exhibit lower signatures. Solder balls that are lodged underneath the device are reported not to be detected; however, clusters of solder balls on the surface of the joint usually can produce higher temperature readings.[61] In general, the thermal signatures of various solder joint defects are represented in Figure 9.24.

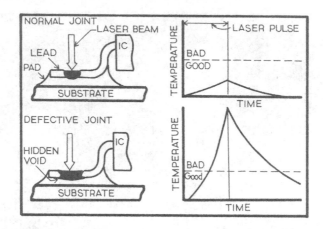

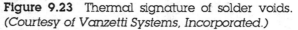

**Figure 9.23** Thermal signature of solder voids. *(Courtesy of Vanzetti Systems, Incorporated.)*

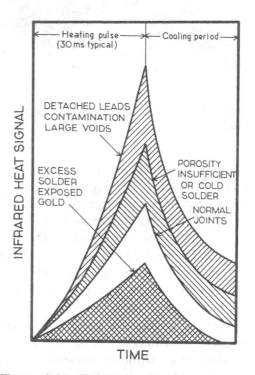

**Figure 9.24** Thermal signatures of defects of lap solder joints. *(Courtesy of Vanzetti Systems, Incorporated.)*

Success in laser/infrared inspection has been measured.[63, 64] Streeter has investigated the defects as specified by military specifications MIL-P-46843 and MIL-S-45748. Among the 19 types of defects specified, the defect detection confidence in the range of 95–100% include inadequate solder plug, cold solder joints, rosin joints, cracks, holes, voids, gaps, blow holes, scars, insufficient solder-component side, disturbed solder, unclean solder, and unsoldered connection. Solder bridging, dewetting-solder side, dewetting-component side, insufficient solder, lead obscured by solder (excessive solder) on solder side, and exposed copper-lead ends not covered with solder are listed with low defect detection confidence.

In addition to solder joint inspection, to assure the quality and to eliminate unnecessary cost as defects go further down the production line, an inspection system on printed solder paste prior to reflow is also recommended. An automated optical inspection system has been developed to focus on the inspection of solder paste patterns.[65]

To assure solder joint integrity of surface mounted boards, techniques which are performance- and cost-effective demand continuing efforts. The discussion in this chapter on basic failure processes that are believed to be related to solder joint failure and on the state of the art in solder joint reliability studies, though oversimplified, is hoped to shed some light on the complexity of failure causes and failure mechanisms in solder joints. It is also hoped that the obvious need to understand solder joint properties in relation to reliability will stimulate researchers' further efforts in this area.

# REFERENCES

1. J. E. Dorn and L. A. Shepard, "What We Need to Know About Creep," *ASTM Spec. Tech. Publ.* 165 (1954): 3–30.

2. R. E. Frenkel, O. D. Sherby, and J. E. Dorn, *ACTA Metalurgica* 3 (1955): 470.

3. O. D. Sherby, R. J. Orr, and J. E. Dorn, *Trans. A.I.M.E.* 200 (1954): 71.

4. F. A. Mohamed, K. L. Murty, and J. W. Morris, *Metallurgical Trans.* 4 (1973): 935.

5. W. Lange, A. Hassnet, and I. Berthold, *Phys. Status Solids* 1 (1961): 50.

6. J. E. Breen and J. Weertman, *Trans. A. I. M.E.* 207 (1955): 1230.

7. R. W. Evans and B. Wilshire, "Creep of Metals and Alloys," *The Institute of Metals* (1985).

8. C. Herring, *J. Appl. Phys.* 21 (1950): 437.

9. N. J. Mott, *Proc. Phys. Soc.* 64B (1951): 729.

10. W. R. Cannon and O. D. Sherby, *Metallurgical Trans.* 1 (1970): 1030.

11. J. Weertman, *Trans. TM-AIME* 218 (1960): 207.

12. C. R. Barrett and O. D. Sherby, *Trans. TMS-AIME* 233 (1965): 1116.

13. Derek Hull, *Introduction to Dislocations* (Pergamon Press, 1965).

14. H. J. Grover, S. A. Gordon and L. R. Jackson, "The Fatigue of Metals and Structures," U.S. Government Printing Office (1954).

15. Joseph Leibowitz, William Winters, and Jack Kelkin, "Graphite Layers in SMT Boards Control Thermal Mismatch," *Electronic Packaging and Production* (Jun. 1985).

16. L. F. Coffin, "Symposium on Internal Stress and Fatigue of Metals" (Elsevier Publishing Co., 1959).

17. S. S. Manson, "Behavior of Materials Under Conditions of Thermal Stress," *NASA Tech. Note 2933* (1954).

18. E. Glenny, J. E. Northwood, S. K. W. Shaw and T. A. Taylor, *J. Inst. Metals* 85 (1958–59): 294.

19. E. Glenny and T. A. Taylor, *J. Inst. Metals* 88 (1960): 449.

20. W. K. Rey, *NASA Tech. Note 4284* (1948).

21. W. Boas and R. W. K. Honeycombe, *Proc. Roy. Soc. A.* 186 (1945): 57.

22. W. Boas and R. W. K. Honeycombe, *Proc. Roy. Soc. A.* 188 (1946): 28.

23. O. J. Horger, "Metals Engineering Design," *Am. Soc. Mech. Eng. Handbook* (New York: McGraw Hill, 1953).

24. W. J. Harris, *Metallic Fatigue* (London: Pergamon Press, 1961).

25. H. Saburo, 1956 International Conference on Fatigue of Metals, Inst. Mech. Engrs.: 348.

26. W. H. Liu and H. T. Corten, "Theoretical and Applied Mechanics Dept. Report No. 566," Univ. of Illinois (Dec. 1958).

27. L. Ferguson and G. M. Bouton, "The Spirit of Coating of Polybutene on the Fatigue Properties of Lead Alloys," 1944 Symposium on Stress Corrosion Cracking of Metals, Am Soc Test Spec. Tech. Pub. No. 64: 743.

28. W. Gilde, *British Welding Journal* (Mar. 1960): 208.

29. B. P. Haigh and B. Jones, *J. Inst. Met.* 43 (1930): 271.

30. Mali Mahalingam, Madhukar Nagarkar, Lynn Loftran, James Andrews, Dennis R. Olsen, and Howard M. Berg, "Thermal Effect of Die Bond Voids," *Semiconductor International* (Sept. 1984).

31. J. K. Hagge, SMT, International Electronics Packaging Society: 547.

32. P. Heller, SMT:2, International Electronics Packaging Society: 457.

33. W. Engelmaier, SMT, International Electronics Packaging Society: 539.

34. W. Engelmaier, "Surface Mount Attachment Reliability of Clip-leaded Ceramic Chip Carriers on FR-4 Circuit Boards," proceedings, 1987 International Electronics Packaging Soc.: 104.

35. D. E. Reimer and J. D. Russel, proceedings, 1983 International Hybrid Microelectronics Symposium: 217.

36. P. M. Hall, *Solid State Technology* (Mar. 1983).

37. R. T. Howard, S. W. Sobeck, and C. Sanetia, *Solid State Technology* (Feb. 1983).

38. P. A. Chalco, proceedings, 1983 International Hybrid Microelectronics Symposium: 223.

39. Jennie S. Hwang and N. C. Lee, proceedings, 1985 International Symposium on Microelectronics: 23.

40. John H. Lau and Donald W. Rice, "Effects of Standoff Height on Solder Joint Fatigue," SMT:3, International Electronics Packaging Society: 685.

41. W. M. Sherry, J. S. Erich, M. K. Bartschat, and F. B. Prinz, "Analytical and Experimental Analysis of LCCC Solder Joint Fatigue Life," proceedings, 1985 Electronic Components Conference: 81. © 1985–IEEE.

42. D. E. Reimer and C. W. Saulsberry, proceedings, 1984 International Hybrid Microelectronics Symposium: 480.

43. P. M. Hall, proceedings, 1984 Electronic Components Conference: 107. © 1984–IEEE.

44. J. M. Smeby, proceedings, 1984 Electronic Components Conference: 117. © 1984–IEEE.

45. J. R. Taylor and D. J. Pedder, "Joint Strength and Thermal Fatigue in Chip Carrier Assembly," proceedings, 1982 International Hybrid Microelectronics Symposium: 209.

46. D. O. Ross, SMT: 2, International Electronics Packaging Soc.: 468.

47. M. C. Shine, L. R. Fox, J. W. Sofia, SMT: 2, International Electronics Packaging Soc.: 478.

48. J. Weertman, Trans. of the Metallurgical Soc. of AIME. 218 (1960): 207.

49. D. W. Hamer and G. R. Sellers, proceedings, 1985 International Hybrid Microelectronics Symposium: 11.

50. K. N. Tu, Acta Metallurgica, 21 (1973): 347.

51. P. L. Blum, J. Polissier, and G. Silvestre, *Solid State Technology* 16 (1973): 55.

52. T. F. Marinis and R. C. Reinert, proceedings, 1985 Electronic Components Conference: 73. © 1985–IEEE.

53. Harry N. Keller, "Reliability of External Connections Condensation Soldered to Ti-Pd-Au Thin Film," proceedings, 1978 International Hybrid Microelectronics Symposium: 297.

54. R. E. Moore, *Connection Technology* (May 1985).

55. G. A. Walker, P. W. DeHaven, and C. C. Goldsmith, proceedings, 1984 International Hybrid Microelectronics Symposium: 125.

56. A. Adrari and W. L. Green, proceedings, 1983 International Hybrid Microelectronics Symposium: 229.

57. B. Chalmers, *Physical Metallurgy* (New York: Wiley, 1959).

58. C. Wagner, *Thermodynamics of Alloys* (Addison Wesley Publications, 1952).

59. Charles D. Goodwin, "Real-Time X-Ray Process for Solder Joint Integrity," proceedings, 1987 National Electronic Packaging and Production Conference, West: 533.

60. Terry A. Deane and William A. Gruver, "Application of X-Ray Vision for Automating the Solder Quality Inspection of Printed Circuit Boards," 1987 Smart III Conference, EIA & IPC.

61. Dr. Alan Traug and Richard Alpen, "Laser/Infrared Technology for Inspecting SMD Solder Joint Quality," proceedings, 1987 National Electronic Packaging and Production Conference, West: 669.

62. Dr. Riccardo Vanzetti, "Laser-Infrared Solder Joint Inspection, The Last Step to Full Automation," The Fifth International Conference of the Israel Society for Quality Assurance, 1984.

63. Douglas H. Ensign, "Laser Inspection Signature Analysis of Electrical Solder Connections, proceedings, 1987 National Electronic Packaging and Production Conference, West: 571.

64. J. P. Streeter, "Laser Inspection of Solder Joints: Recent Results and Defect Detection Capabilities," proceedings, 1987 National Electronic Packaging and Production Conference, East: 511.

65. David Trail and David Smith, "Solder Paste Inspection System Focuses on High Production Yields," *Hybrid Circuit Technology* (Feb. 1986).

# CHAPTER 10

# Special Topics in Surface Mount Soldering Problems and Other Soldering-Related Problems

The previous chapters have discussed the technology and scientific principles behind the major aspects of solder paste technology and applications. This chapter is to condense the common concerns and problems encountered by using solder paste one by one into a check list. It is hoped this chapter can be used as a troubleshooting guide with the objective that the users be able to establish a "trouble-free" process by intelligently learning from others' experiences.

## 10.1 SURFACE MOUNT J-LEAD WICKING

Surface mount J-lead wicking is a phenomenon in which molten solder climbs up the J leads of a PLCC or SOIC component during solder reflow, consequently resulting in starved or open joints, as shown in Figure 10.1. This mainly is driven by relative metallurgical wetting rate.[1] Preferential wetting between the lead of the component and the solder pads on the printed circuit board may occur when there is a significant transient temperature differential. The quantitative temperature difference required to initiate this problem is not easy to measure, since the phenomenon is a transient kinetics-related problem. Nonetheless, the transient temperature difference as the principal driving force is evidenced.

It has been observed that the wicking problem is more often linked

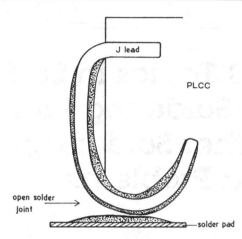

**Figure 10.1**  Surface mount J-lead wicking phenomenon.

to the process in populating the large size of printed circuit boards in conjunction with the use of single fluid vapor phase reflow. The single fluid vapor suitable to reflow the solder paste heats the leads of the component faster than the solder pads on the large printed circuit boards due primarily to the thermal mass difference and low thermal conductivity of printed boards. The large printed circuit board can also be viewed as a large heat sink. The phenomenon is not observed with hybrid board assembly or even with small size printed circuit board. This substantiates that the cause is the temperature differential between the leads and the solder pads. The problem may be aggravated by the lack of coplanarity of leads. Therefore the potential causes and remedies are summarized as follows:

## Potential Causes

1.  Large size of printed circuit board in conjunction with very fast heating as provided by single fluid vapor phase reflow.
2.  Transient temperature differential between the lead and solder pad.
3.  Lack of coplanarity of leads.
4.  Large curvature of leads.

## Potential Remedies

1.  Selecting and adjusting the reflow process with emphasis on heating rate, to minimize temperature gradient—for example,

adding a preheating step or secondary vapor in vapor phase reflow process or utilizing infrared reflow with controlled profile to compensate for the slower heating rate of the large board. With individually controlled top and bottom heating zones in most infrared equipment, a proper profile can be achieved to alleviate the heating-related cause of the problem, as well as to utilize the inherent slower heating of the infrared furnace.

2.  Slowing down the wetting process through flux/vehicle formulation. The flux/vehicle composition affects the wetting rate, yet it is not effective enough to offset the metallurgical driving force.

3.  Slowing down the soldering kinetics through a metallurgical approach, for example—utilizing *in situ* alloying by using blended powders as initiated by Texas Instruments. The time expended to alloy the two powders retards the melting of solder, which in turn retards the molten solder wetting.

4.  Lowering surface tension of molten solder by selecting solder alloys, if feasible.

5.  Improving the coplanarity of leads.

6.  Reducing the curvature of leads.

## 10.2  INSERTED LEAD WICKING

The solder wicking occurrence is not limited to surface mount assemblies. When the assemblies and the reflow process meet the criteria of wicking at an instant moment, molten solder will move to the preferred substrate. Figure 10.2 shows the section of a lead which was inserted into a package and then soldered with a 63-Sn/37-Pb solder paste. The lead is made of copper with tin plate. After reflow the surface of the lead is obviously altered to a porous and rough appearance near the solder joint, and the other end of the lead remains smooth (identical to original finish of the lead). The x-ray energy dispersion analysis identifies that the smooth portion contains essentially tin, and the rough portion contains both tin and lead, as shown in Figure 10.3 (a) and (b). It was suspected that the flux system in the paste may contribute to the rough appearance. The appearance and composition of the lead after being fluxed is then investigated. Under identical heating conditions, except the deletion of solder alloy, the lead does not show the porous and rough appearance and the near-surface composition of the fluxed lead as

**Figure 10.2**  SEM micrograph of rough, porous portion of a component lead.

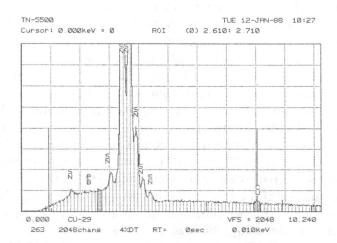

**Figure 10.3(a)**  X-ray energy dispersion analysis of smooth portion of a component lead.

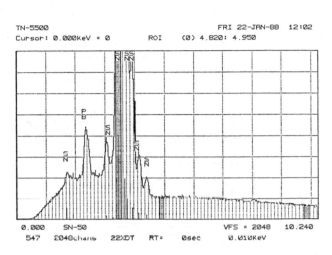

ZAF CORRECTION     25.00 KV       40.00 Degs

No. of Iterations  2
----    K     [Z]     [A]     [F]   [ZAF]  ATOM.%  WT.%
PB-M  0.057  1.080  1.110  0.998  1.198   3.88   6.57
SN-L  0.942  0.995  1.038  1.000  1.033  96.12  93.43
 * - High Absorbance

SSQ:

TN-5500                          FRI 22-JAN-88  12:02
Cursor: 0.000keV = 0        ROI    (0) 4.820: 4.950

0.000    SN-50                        VFS = 2048   10.240
 547    2048chans    22%DT    RT=    0sec    0.010keV

**Figure 10.3(b)** X ray energy dispersion analysis of rough, porous portion of a component lead.

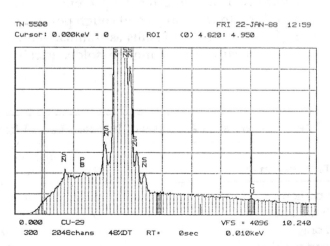

ZAF CORRECTION     25.00 KV       40.00 Degs

No. of Iterations  1
----    K     [Z]     [A]     [F]   [ZAF]  ATOM.%  WT.%
PB-M  0.001  1.085  1.118  0.998  1.212   0.10   0.17
SN-L  0.997  1.000  1.001  0.999  1.001  99.61  99.67
CU-K  0.001  0.902  1.118  1.000  1.009   0.29   0.16
 * - High Absorbance

SSQ:

TN-5500                          FRI 22-JAN-88  12:59
Cursor: 0.000keV = 0        ROI    (0) 4.820: 4.950

0.000    CU-29                        VFS = 4096   10.240
 300    2048chans    48%DT    RT=    0sec    0.010keV

**Figure 10.3(c)** X-ray energy dispersion analysis of a fluxed lead.

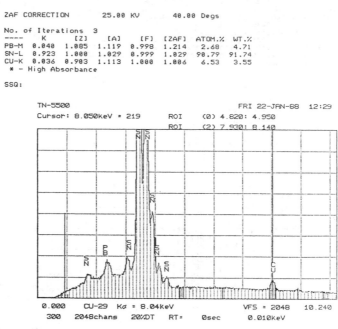

ZAF CORRECTION        25.00 KV        40.00 Degs

No. of Iterations  3
----    K      [Z]      [A]     [F]    [ZAF]  ATOM.%   WT.%
PB-M  0.040  1.085   1.119   0.998  1.214   2.68   4.71
SN-L  0.923  1.000   1.029   0.999  1.029  90.79  91.74
CU-K  0.036  0.903   1.113   1.000  1.006   6.53   3.55
 * - High Absorbance

SSQ:

TN-5500                                        FRI 22-JAN-88  12:29
Cursor: 8.050keV = 219        ROI   (0) 4.820: 4.950
                              ROI   (2) 7.930: 8.140

0.000     CU-29   Kα = 8.04keV           VFS = 2048   10.240
300     2048chans   20%DT    RT=    0sec    0.010keV

**Figure 10.3(d)**  X-ray energy dispersion analysis of a lead with simulated wicking.

identified by x-ray dispersion analysis consists of essentially tin, as shown in Figure 10.3 (c). A simulated wicking test provides visual evidence by witnessing the molten solder wicking up the lead under conduction reflow. X-ray analysis of the simulated specimen as shown in Figure 10.3 (d) indicates the presence of lead (Pb) with similar composition to the porous and rough portion of the lead as indicated in Figure 10.2. The results as summarized in Table 10.1 substantiate the occurrence of molten solder wicking along the

## TABLE 10.1

### X-Ray Energy Dispersion Analysis of Inserted Lead Wicking

| Lead Test Specimen | Composition (Wt %) | | |
|---|---|---|---|
|  | Pb | Sn | Cu |
| Rough portion | 6.57 | 93.43 | — |
| Smooth portion | 0 | 99.06 | — |
| Fluxed only | 0.17 | 99.67 | 0.16 |
| Simulated wicking | 4.71 | 91.74 | 3.55 |

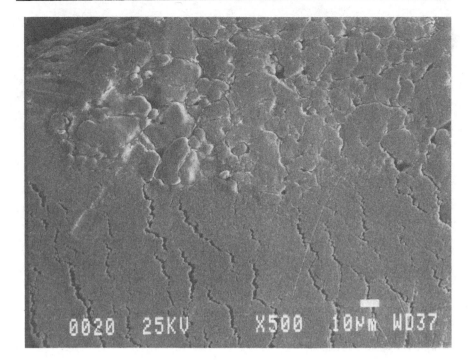

**Figure 10.4**  SEM micrograph comparing wicked and nonwicked portion of inserted lead surface.

inserted lead. The SEM micrograph as shown in Figure 10.4 also indicates the surface morphology difference between the wicked area and area without solder wicking. It is noted that the coating on the nonwicked area after reflow process has crazes on the surface.

Although the inserted lead is structured in a different way from that for surface mount J lead, the solder wicking can occur, and the potential causes are fundamentally the same. Thus the potential remedies, as discussed in Section 10.1, with deletion of the items related to lead coplanarity and curvature, are applicable here.

## 10.3  CLIP-ON LEAD WICKING

The two wicking phenomena as outlined in Sections 10.1 and 10.2 are primarily attributed to molten solder. Another problem area which is also called wicking occurs in the lead attachment assemblies by using clip-on type of leads or lead frames with one or two prongs on one side of substrate surface and one or two prongs on the other

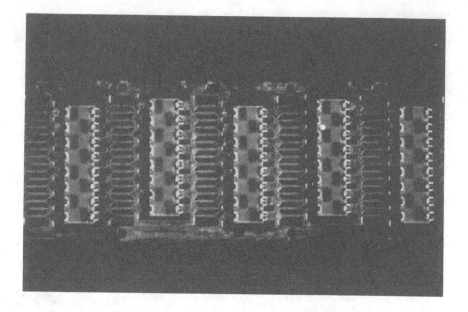

**Figure 10.5** An example of clip-on lead assembly. *(Courtesy of Bourns, Incorporated.)*

side of substrate, as shown in the photograph of Figure 10.5. The solder paste is normally deposited on the substrate after the lead frame is positioned. The solder is observed to flow down to the lead instead of localizing in the solder joint area, and consequently, not only resulting in the starved joint as shown in Figure 10.6 (arrowed area) but also causing the subsequent molding and sealing problem. This phenomenon should be distinguished from the other two wicking phenomena. The main contribution to this problem is the rheology of solder paste in response to rising temperature and gravitational force. The molten solder wetting down the lead in this case plays a minor role unless the capillary conditions are met. The solution to this type of wicking problem when assembling lead frame to components or modules therefore includes

  selecting a solder paste possessing flow control or flow restrictivity in response to rising temperature,

  positioning the paste deposits strategically to avoid meeting capillary conditions and being overwhelmed by the gravitational force, and

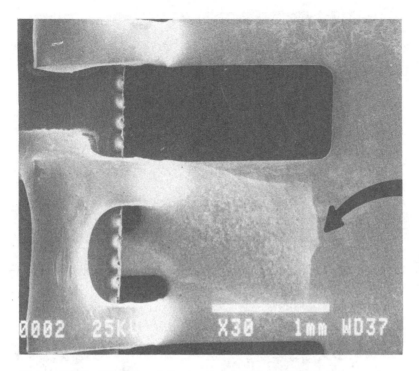

**Figure 10.6**  Wicking phenomenon of clip-on leads.

assuring the lead frame is crimped in the right angle and position to achieve a structure without the risk of capillary drawing.

With the proper selection of solder paste and control of process, good solder joints can be obtained in lead frame attachment, as shown in Figures 10.7 and 10.8.

## 10.4 TOMBSTONING

Tombstoning is a phenomenon encountered during surface mount assembly with boards containing small two-termination chips. In the presence of moving forces, the small chips tend to dislocate during solder reflow, resulting in misalignment; and in the extreme case, they stand up on one end of metal terminations, resembling tomb-stones, as illustrated in Figure 10.9. The moving force(s) is a result of an unbalanced force between the two-chip termination and solder pad interfaces.

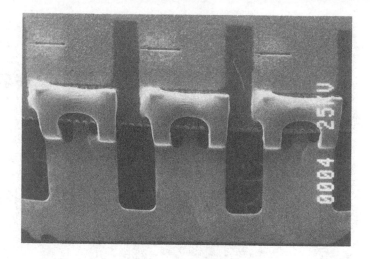

**Figure 10.7** Lead frame attachment solder joints without wicking phenomenon. *(Courtesy of Bourns, Incorporated.)*

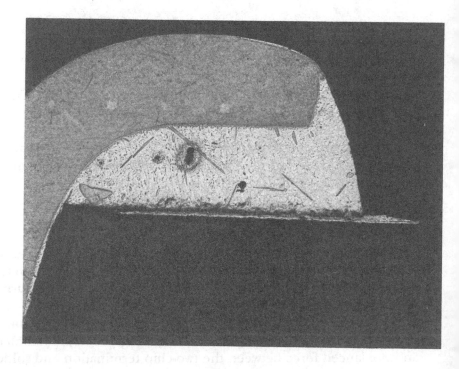

**Figure 10.8** Solder joint cross section of lead frame attachment. *(Courtesy of Bourns, Incorporated.)*

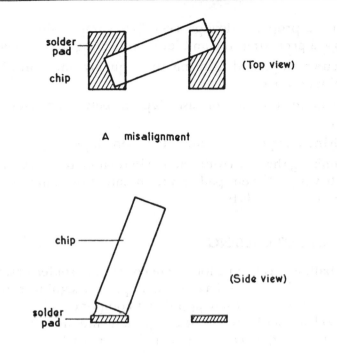

A   misalignment

B   tombstoning

**Figure 10.9**   Component misalignment and tombstoning phenomena.

The factors that contribute to the moving forces are

- incompatible reflow process
- violent fluxing
- excessive volatiles
- uneven solderability between two terminations of a chip
- uneven solderability between two solder pads
- uneven paste deposition on two solder pads
- inadequate placement force to make intimate contact between the paste and the termination of the chip
- improper pad design

Therefore, the solution to eliminate the problem may be one or a combination of the steps that follow:

- Selecting a compatible solder pad.

- Setting a proper reflow process, for example—in vapor phase reflow, a predrying step and/or a preheating step is needed.
- Selecting chips and assuring their quality and consistency in termination pads.
- Assuring the evenness of paste deposits between two termination pads.
- Applying adequate placement force on chips.
- Minimizing the extra width of pad in relation to the width of chip and having sufficient pad and termination overlap area, as illustrated in Figure 10.10.

## 10.5 SOLDER BALLING

Solder balling has been a long-time concern in solder paste reflow. Recently, solder pastes have been improved significantly in this regard. The more densely populated components in conjunction with low clearance between the component and the board in surface mount technology makes solder-ball–free, or nearly solder-ball–free, one of the performance requirements for solder paste.

The solder balling phenomenon can be defined as the situation that results when small spherical particles with various diameters are

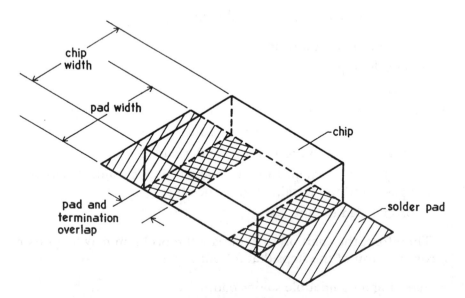

**Figure 10.10**  Relative dimensions of chip termination and solder pad.

formed away from the main solder pool during reflow, and do not coalesce with the solder pool after solidification. Solder balling appears to be a simple phenomenon, yet its mechanism can be complex. Solder balls can be formed from different origins, although the results appear the same. Six likely sources are as follows:

1.  Solder paste with inefficient fluxing with respect to solder powder and/or substrate, resulting in discrete particles which do not coalesce, due either to paste design or to subsequent paste degradation.

2.  Incompatible heating with respect to paste prior to solder melt (preheating or predrying), which degrades fluxing activity.

3.  Paste spattering due to too fast heating, forming discrete solder particles or aggregates outside the main solder pool.

4.  Solder paste contaminated with moisture or other high "energy" chemicals that promote spattering.

5.  Solder paste containing extra fine solder particles that are carried away from the main solder pool by organic portion (flux/vehicle portion) during heating, resulting in small solder balls.

6.  Interaction between solder paste and solder mask.

(The measurement method for solder balls as an indication of the performance of a solder paste is covered in Chapter 11.)

To achieve solder-ball–free joints on surface mounted board, one deals with several issues:

*   selection of a solder paste with ultimate performance
*   understanding of the characteristics of solder paste selected
*   set-up of a reflow process which best fits the solder paste selected
*   assurance that materials, such as solder mask, are compatible with solder paste
*   assurance of consistency and quality of solder substrates such as leads of the components, terminations of chips, and solder pads on the board
*   development of a cleaning process with capability of removing small solder balls, if present (Section 8.11)

From the solder paste end, researchers and developers have been striving to maximize performance latitude in the paste to accommodate less than ideal conditions. It requires a delicate balancing of all ingredients in the flux/vehicle system and the compatibility of the solder powder being used with the flux/vehicle system, as well as the compatibility of paste with various substrates and reflow conditions.

## 10.6  RESIDUE CLEANING

The cleaning issues covered in Chapter 8 are not reiterated here; however, the importance of a thorough testing for a whole system including paste, equipment, and solvent system prior to making any final set-up cannot be overemphasized.

## 10.7  PAD BRIDGING

As discussed in Section 3.13, potential causes for pad bridging problem are

- paste spreading as temperature rises
- molten solder flowing
- paste contact between pads during component placement
- paste contact between pads due to poor printing or dispensing

  Potential remedies include

- selecting solder paste with restrictive flow in response to temperature rising in paste state
- selecting solder paste with restrictive flow in molten solder
- assuring paste deposition in good resolution and quality without slump and smear
- using proper volume of solder paste
- using proper pressure for component placement

## 10.8  PASTE TACK TIME

Section 3.6 discusses the need of tack time and its measurement. A proper selection of solder paste is essential.

## 10.9  PASTE OPEN TIME

Section 3.14 outlines this characteristic. Again, a proper selection of solder paste is the solution.

## 10.10  LOT-TO-LOT CONSISTENCY

If a solder paste product can deliver perfect performance only occasionally, it is not much better than other products without good performance. Lot-to-lot consistency is the key to making quality solder joints, which in turn is essential to the yield and quality in producing the final parts. Achieving lot-to-lot consistency begins with an understanding of the product formulation and process, followed by proper process control and quality control. A control specification should be designed collaboratively between the paste vendor and the user so that a working and realistic specification can be derived and a consistent product can be supplied.

## 10.11  ELECTROMIGRATION

The electromigration of metals in electronics has been recognized for over 30 years. The phenomenon causes electrical short as the two adjacent conductor paths are bridged together through the growth of metal dendrites. In principle, a metal which can readily form ions in the presence of moisture, and have reasonable mobility in a medium under electrical voltage potential, would exhibit electromigration. Thus the requirements of electromigration of metals are electrolyte, electrical voltage, time, and temperature. To monitor the level of electromigration in a circuit, the time needed to reach a certain electric current between two conductors consisting of the metal to be tested and separated at a specified distance in an electrolytic environment under an applied voltage is normally measured. The electrolytic environment is typically moisture with or without ionic species.

Under favorable conditions, the metal ions can migrate from cathode (−) to anode (+) and form a dendrite-like structure between two electrodes. Table 10.2 lists the results of electromigration among different metals. The metals are grouped into three classes in terms of susceptibility to electromigration. Electromigration of the first group can occur with the presence of only moisture and voltage. The second group has the additional requirement of halogen-containing contaminants. The third group is quite passive in a normal environment. Among the metals, silver is most prone to electromigration, due to its high activity in the presence of moisture. Except silver, the metals related to soldering such as lead, tin, copper, and bismuth all have electromigration potential when the conditions are met.

## TABLE 10.2

**Electromigration Conditions* for Different Metals[2]**

| Metals That Migrate with Distilled Water | Metals That Migrate with Distilled Water and Halogen Contaminant | Metals That May Need Other Conditions to Migrate |
|---|---|---|
| Bismuth | Gold | Aluminum |
| Cadmium | Indium | Antimony |
| Copper | Paladium | Chromium |
| Lead | Platinum | Iron |
| Silver | | Nickel |
| Tin | | Rhodium |
| Zinc | | Tantalum |
| | | Titanium |
| | | Vanadium |

*Bias 1–45 volts

For thick film conductors, on hybrid circuits, the silver-based systems are normally built with platinum and/or palladium. If the particle size of the metal powders is small enough, silver and the second metal can readily achieve alloying during the firing process, as shown in the phase diagrams in Figures 10.11 (Ag/Pd) and 10.12 (Au/Pd) resulting in a composition which possesses more electromi-

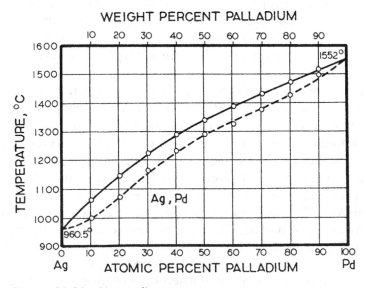

**Figure 10.11** Phase diagram of Ag/Pd.

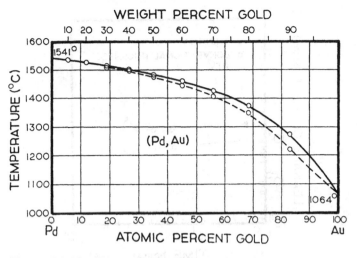

**Figure 10.12**   Phase diagram of Au/Pd.

gration resistance. The resistance can be further enhanced by using coprecipitated silver/palladium or silver/platinum systems. The copper-based thick film is considered to deliver better migration resistance. However, the study indicates that soldered copper thick film systems are not totally immune from electromigration and lead (Pb) migration is found in the soldered copper thick film.[3] The same study on dendritic growth on bulk solder and soldered thick film indicates that dendrites originate from the solder/thick film interface, and therefore the migration can be reduced by completely covering the thick film by solder. Among the bulk solder alloys, the study shows that migration is retarded by high tin-containing solders due to the formation of a nonconductive tin oxide film. Table 10.3

## TABLE 10.3

### Electromigration Time* of Some Solder Alloys[3]

| Solder Alloy | Typical Migration Time | Dendrites Identity |
| --- | --- | --- |
| 94-Sn/4-Ag | >720 sec | tin oxide filaments |
| 30-Sn/70-Pb | 200 sec | lead |
| 10-Sn/88-Pb/2-Ag | 150 sec | lead |
| 10-Sn/90-Pb | 125 sec | lead |
| 93-Pb/5-In/2-Ag | 40 sec | lead |

*Tests were conducted by applying 10 volts DC to the conductors separated by 20 mils in deionized water. The migration time was defined as the time to obtain 100 microamps of current.

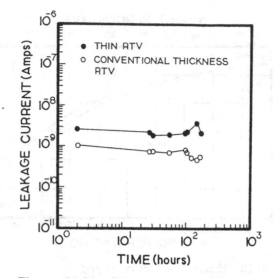

**Figure 10.13** Effect of encapsulant on high voltage circuits.[4]

contains the results of migration tests among solder alloys. It is indicated that the key element which is necessary for the occurrence of electromigration in circuits is moisture. Figure 10.13 illustrates the effect of encapsulant on the high voltage circuits. The observed difference in leakage currents under the accelerated aging and bias (85°C, 85% relative humidity, 1000 volts) is attributed to the thickness of RTV encapsulant which serves as a sealing material for circuits. The thinner encapsulant provides less protection from moisture permeation, and therefore higher electric leakage occurs in the thinner encapsulant package.

## 10.12 LEACHING

Leaching is recognized as a situation where noble metals, specifically silver (Ag) and gold (Au), are deprived from the thick film conductor compositions, coatings, and other compositions during soldering, as a result of the ready solubility of silver and gold in molten tin. The phase diagrams in Figures 4.11 and 10.14 show that the solubility of silver and gold in tin is appreciable and increases with increasing temperature.

The solubility of silver (Ag) in molten tin (Sn) is as follows: approximately up to 7 wt % at 250°C; up to 10 wt % at 300°C; and up to 16 wt % at 350°C. The solubility of gold (Au) in molten tin (Sn)

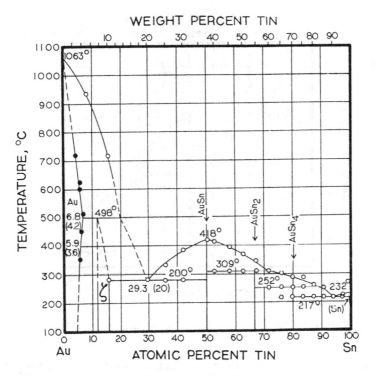

**Figure 10.14**  Phase diagram of Au/Sn.

are even higher. At 250°C, 300°C, and 350°C, solubilities are approximately up to 17 wt %, 35 wt %, and 47 wt %, respectively. In the same temperature range, metals such as nickel (Ni) and platinum (Pt) have negligible solubility in molten tin, in reference to the phase diagrams in Figures 10.15 and 10.16 . Figure 10.17 illustrates the relationship of the maximum solubility of Au, Ag, Pt, and Ni in molten tin with temperature. Solubility of Au and Ag in tin increases rapidly with increasing temperature, and solubility of Pt and Ni is negligible over the specified temperature range.

Therefore, in tin-lead solder alloys, the solubilities of silver, gold, platinum, and nickel are expected to be close to the magnitude as shown in Figure 10.17 and to vary drastically with increasing temperatures.

The leaching phenomenon can cause problems in both solder and in the alloy composition being leached. For example, the leached thick film conductor composition may lose adhesion to the substrate and even be totally peeled off, and in the meantime, the solder may incorporate with undesirable intermetallic compounds

WEIGHT PER CENT TIN

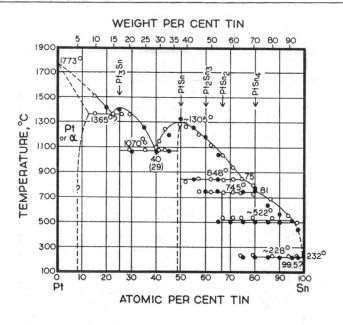

**Figure 10.15** Phase diagram of Pt/Sn.

that consequently cause adverse effects in solder joint integrity, as discussed in Section 9.9. The common assembling process which may encounter leaching problems include the component attachment on thick film solder pads of hybrid circuit boards, lead attachment on hybrid circuits, lead frame attachment on thick film solder pads in making passive components, and soldering capacitors and resistors with thick film terminations.

The leach resistance of the solder pads can be readily tested by immersing the substrate with solder pads in a molten solder pot containing specified solder alloy (e.g., 60-Sn/40-Pb, 63-Sn/37-Pb, or 62-Sn/36-Pb/2-Ag) at specified temperature (e.g., 210°—250°C) for a specified time. The solder pads are normally fluxed prior to immersion. The number of immersions for the test varies with the performance requirements. After the completion of the immersion test, the leach resistance is evaluated by examining the existence and the extent of erosion on the composition of solder pads with visual aid (e.g., optical microscopy, 20X). The leach usually starts at the corners of pads. The criteria of leach resistance and the test for it should match the soldering process. For example, 1–3 immersions and 5 sec of each immersion in molten solder (60-Sn/40-Pb) at 235°C are common test conditions, and the erosion exceeding 10%

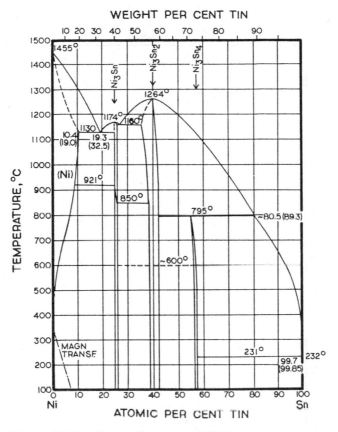

**Figure 10.16** Phase diagram of Ni/Sn.

of the pads is normally considered to be unacceptable. Among the reflow processes, the demand on the leaching resistance in increasing order is: laser reflow, hot air reflow, vapor phase reflow, conduction reflow, infrared reflow, and convection reflow.

The leaching problem is a kinetics-controlled metallurgical process. The problem should therefore be tackled from both the material and the process end. In the material aspects, the steps are

1.  Using silver-containing solder to saturate the solubility level.

2.  Using leach-resistant thick film conductor compositions, such as compositions containing appreciable amount of palladium and/or platinum in appropriate particle sizes. Selection guidelines are readily available from thick film materials manufacturers. Both palladium and platinum have little solubility in molten

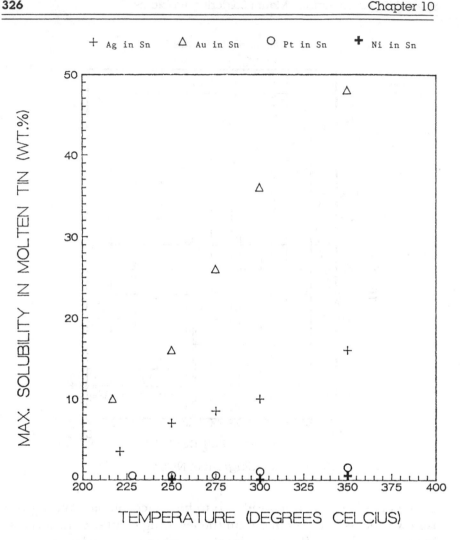

**Figure 10.17** Maximum solubility of Au, Ag, Pt, and Ni in molten tin versus temperature.

tin. Their alloy ability with silver or gold, as shown in Figures 10.11 and 10.12, will retard the dissolution of silver or gold in solder during reflow process.

3. For the terminations of capacitors and resistors, use of a nickel barrier as illustrated in Figure 10.18 is found to be effective to prevent the leaching from the silver-containing thick film terminations during soldering.

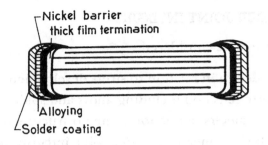

**Figure 10.18** Chip capacitor termination with barrier layer.

With respect to process, the temperature used in the soldering process and the residence time at peak temperature are equally important as, if not more important than, materials to leaching control. Particularly, while working on paste/lead frame attachment assembling using furnace-type reflow, the high temperature solder alloys required for the solder joint and the longer time of heat exposure as inherent in furnace reflow process make the process control vital to the leaching problem. The reflow process should be adopted to minimize heat exposure, particularly the time at which the solder is in the molten state. The reflow profile set-up is discussed in Chapter 7. The reflow method comparison indicates that reflow methods such as hot gas and laser provide fast heating with minimal heat exposure.

The leaching rate of a silver- or gold-containing composition in molten solder is strongly dependent on temperature and time. Selecting a thick film conductor with good leach resistance and maintaining good control of reflow temperature profile will prevent the leach problem.

## 10.13 SOLDERABILITY

One of the common concerns in reflowing solder paste is solderability (Sections 3.10 and 3.11). After proper selection of solder paste and the establishment of a reflow process, the key factors in achieving good solderability are how to maintain the quality of solder paste, the quality of substrates to be soldered (i.e., the leads of components and the solder pads), and the control of reflow process.

## 10.14  SOLDER JOINT INTEGRITY

Four issues are to be considered:

1.  Assembly design in relation to stress development.
2.  Solder paste selection including alloys and solderability.
3.  Reflow parameters in relation to microstructure development.
4.  Test method to measure solder joint integrity in relation to service conditions.

The criteria of ultimate solder joint integrity depends on a specific set of service conditions. The major fundamental solder joint failure processes and state-of-the-art studies are discussed in Chapter 9.

## REFERENCES

1.  Jennie S. Hwang, "Surface Mounting—Its Challenges to Solder Pastes," proceedings, 1987 Surface Mount and Reflow Technology Conference, IPC/EIA.
2.  A. DerMarderosian, "The Electrochemical Migration of Metals," proceedings, 1987 International Hybrid Microelectronics Symposium: 134
3.  Richard G. Gehman, "Dendritic Growth Evaluation of Soldered Thick Film," proceedings, 1983 International Hybrid Microelectronics Symposium: 239.
4.  S. D. Schlough and E. F. O'Connell, "Cleaning Processes for High Voltage Circuits with Soldered Chip Carriers," proceedings, 1981 International Hybrid Microelectronics Symposium: 28.

# Quality Assurance and Tests

## 11.1 QUALITY

One way to define quality is that quality is the total conformance to the specified requirements for a specific use. It is for today and in the future a universal target, but quality does not come without effort. The well-known Genichi Taguchi's quality ideas are effectively interpreted and extended by R. N. Kackar.[1] To name two of many points,

1. In a competitive economy, continuous quality improvement and cost reduction are necessary for staying in business.
2. The final quality and cost of a manufactured product are determined to a large extent by the engineering designs of the product and its manufacturing process.

These two points appear to be clear, and they are the foundations of a productive and therefore competitive process.

Looking at the engineering processes, design, material, equipment, and process control are four fundamental elements. Neglecting any of the elements will not make a successful engineering process. The processes in general can be characterized by four types:

1. An extremely well-planned process covering all four elements perfectly, with no surprises, and able to deliver the targeted yield. In other words, all possible occurrences and problems have been anticipated and therefore the process is built to combat all problems.

2.  A reasonably well-planned process having considered all four
    elements with the objective to achieve the designated yield; but
    as the process is being operated in the real world, some ailments
    emerge due either to unanticipated aspects of any of the four
    elements or to occurrences of undesirable interaction among
    the four elements. Under a normal production environment, the
    process team, under a tight time constraint, has to struggle to
    solve the unanticipated problems. During the course of debug-
    ging, not only are the problems resolved (with some sweat), but
    also some insights into the subtlety of the process are gained.
    After a debugging period, the process runs smoothly and even-
    tually achieves the targeted yield.
3.  A copy of another process, which is easily established with
    minimal effort and operated with high efficiency at the begin-
    ning. In this situation, a caution should be noted that when
    copying a process, an adequate understanding of the engineer-
    ing aspects of the process is necessary.
4.  A free-style process without adequate planning, understanding,
    and control.

In view of the competitiveness in the electronics world in terms of
product performance, cost, and quality, it is not hard to conclude
that the fourth type is not going to survive. The chance to fall into the
first type—perfection—most readers would probably agree, is very
slim, though not impossible, considering the complexity of most
processes and the precision that is demanded. The operation of the
copy process is quite risky if not handled properly. When copying a
process, a fundamental understanding of the process, the equipment, and the materials involved need also be "copied." By doing so,
one would not push every button and turn every dial as a problem
occurs, to create a "mystery." Engineering process does not consist of
any mystery, and the perceived mystery is attributed to the lack of
sufficient knowledge and understanding. The second type is perhaps
the most common one; therefore, the objective is to minimize the
time and effort of the debugging period through thorough planning
coupled with good understanding of technologies involved, so that
high yield and quality product are achieved.

With the established and controlled process, the quality and
consistency of raw materials are mandatory to achieve a do-it-right-
the-first-time environment. Solder paste is a major raw material to

form solder interconnections for electronics assemblies. Its quality is crucial to overall productivity of a process. **Efforts on quality always pay off.**

This chapter is thus dedicated to the issue of quality assurance and tests for the specific area—solder paste application—and is tuned to fit the solder paste users' needs. With the consensus that quality of solder paste is vital, the next question is how to assure the quality, which leads to other questions—what parameters are to be tested and how to test the parameters.

With the emphasis on performance characteristics the following sections outline the parameters and corresponding test procedures that are deemed appropriate for assuring the suitability and quality of a solder paste. In addition to the author's views, the recommended tests also incorporate the existing test procedures issued by Military and Federal Specification Agencies, American Testing and Material Society (ASTM), and the Institute of Interconnecting and Packaging for Electronic Circuits (IPC). The tests are grouped into five parts: paste, vehicle, powder, reflow, and post-reflow. Table 11.1 summarizes the tests in each of the five parts.

## 11.2  APPEARANCE

Appearance is considered to be subjective, but to experienced eyes appearance is the quickest way to distinguish differences. Right appearance is a necessary condition for a right product though not a sufficient condition. In other words, the right product is always associated with the right appearance, but right appearance does not guarantee a right product. Therefore appearance is the first sign of quality, and is complementary to the subsequent tests and results. Appearance of solder paste is to be examined in two forms—namely, the paste packed in the jar, and the paste packed in the syringe. In either case, the paste should be homogeneous, without crust, liquid separation, and lumps. In addition, the paste packed in syringe, which normally is made for dot-dispensing through a fine needle, is to be examined for any air voids and stratifications. The presence of air voids or any indication of stratification is a sign of poor quality and may lead to poor dispensing performance.

It should be noted that the two mesh sizes of solder powder commonly employed in the paste possess inherent difference in appearance of the paste, regardless of the systems or vendors. To the human

## TABLE 11.1

### Summary of Quality Assurance Tests for Solder Paste

| Part | Tests |
|------|-------|
| Paste | 1. Appearance |
| | 2. Metal content and flux/vehicle percentage |
| | 3. Density |
| | 4. Viscosity |
| | 5. Viscosity versus shear rate |
| | 6. Cold slump |
| | 7. Hot slump |
| | 8. Molten flow |
| | 9. Tack time |
| | 10. Dryability |
| | 11. Dispensability |
| | 12. Printability |
| | 13. Shelf stability |
| | 14. Storage, handling, and safety |
| Flux/Vehicle | 15. Water extract resistivity |
| | 16. Copper mirror corrosion |
| | 17. Chloride and bromide |
| | 18. Acid number |
| | 19. Infrared spectrum fingerprint and other spectroscopies |
| Solder Powder | 20. Alloy composition |
| | 21. Particle size—sieve |
| | 22. Particle size distribution—SediGraph |
| | 23. Particle shape |
| | 24. Particle surface condition |
| | 25. Dross |
| | 26. Melting range |
| Reflow | 27. Solder ball |
| | 28. Solderability |
| | 29. Exposure time |
| | 30. Soldering dynamics |
| Post-reflow | 31. Cleanliness—resistivity of solvent extract |
| | 32. Surface insulation resistance, before and after cleaning |
| | 33. Solder joint appearance |

Table 11.1 *(continued)*

| Part | Tests |
|------|-------|
| Post-reflow | 34. Solder voids |
| | 35. Joint strength |
| | 36. Power cycling |
| | 37. Temperature cycling |
| | 38. Vibration test |
| | 39. Simulated aging |
| | 40. Thermal shock |

eye, the paste containing the coarser powder (nominal −200 /+325 mesh) does not have the same degree of smoothness as compared with the paste containing finer powder (nominal −325 mesh).

## 11.3  METAL CONTENT

The metal content, indicating the amount of solder powder in the paste, is normally measured as weight percentage. Its volume equivalent can be calculated based on a specific solder alloy and a specific vehicle/flux system, as shown in Table 3.3.

## A.  Apparatus

crucible

hot plate

balance

solvent

clean lint-free towel

## B.  Procedure

1.  Weigh a crucible and record weight, *A,* to the third decimal place.
2.  Sample approximately 20 g of paste into the crucible and weigh the crucible again with sample to the nearest 0.001 g. Record the weight, *B.*
3.  Heat the crucible until all solder melts at the bottom of the mass.
4.  Remove the heat and allow the molten solder to solidify. Carefully pour off the organic liquid on the top.

5. Add 10 ml of suitable solvent (isopropyl alcohol, M-Pyrol) and reheat the crucible to melt the solder.

6. Remove the heat and allow the solder to solidify. Pour off the solvent and allow the solder button to cool to ambient temperature.

7. Remove the solder button and wash off any traces of residue, and dry with a lint-free towel.

8. Weigh the solder button to the nearest 0.001 g and record weight, *C.*

9. Calculate the % metal content of the paste:

$$\% \text{ metal} = \frac{C}{B-A} \times 100\%$$

10. Compare the measured % metal content with the theoretical % metal. The actual metal content should be well within ± 1%.

11. The flux/vehicle percentage can be calculated by

$$\text{flux/vehicle percentage} = \frac{B-A-C}{B-A} = 100\%$$

## 11.4  DENSITY

### A.  Apparatus

weight-per-gallon cup (Gardner Laboratory, regular 83.2 ml)
balance
thermometer
spatulas

### B.  Procedure

1. Weigh the empty weight-per-gallon cup.

2. Fill cup with paste, and avoid air bubble entrapment by tapping cup on tabletop.

3. Carefully place cover on cup and push down to settle cover completely in place.

4. Clean off excess paste which has been forced out of opening in cover.

5.  Carefully clean outside of entire cup.

6.  Weigh cup and paste.

7.  Calculate weight-per-gallon.

$$\frac{\text{net weight in grams}}{10} = \text{pounds per gallon}$$

$$\frac{\text{net weight in grams}}{10 \times 8.3} = \text{grams per milliliter}$$

## 11.5  VISCOSITY

The single point viscosity measured by a specific technique as discussed in Chapter 5 is a convenient indicator for a known paste system. It is valid only when the paste system is given and the measuring technique is specified. In addition to the type of spindle and speed of revolution, the container size and spindle position in relation to the level of the paste, the paste handling prior to the spindle descending in (i.e., how the paste is mixed) affect the results. The following is a procedure consisting of a specified container and specific mixing steps by using helipath TF spindle/5 rpm which is in congruence with federal specifications and IPC specifications.

When using cylindrical spindles, the spindle should be immersed in the paste with surface leveled to the middle of the indentation in the spindle shaft. This may sound like a small point, but failure to do so may result in erratic results.

## A.  Apparatus

Brookfield viscometer RTV model

helipath TF spindle

$1^{1}/_{4}$"-diameter, 3-blade propeller

electric motor–driven stirrer

tachometer kit (50–10,000 rpm)

4-fl oz jar ($3^{1}/_{4}$" height, $2^{1}/_{4}$" diameter)

timer

temperature bath

## B.  Procedure

1. Calibrate the viscometer
2. Prepare 4-oz jar and fill the jar with paste to be measured up to about ¼" from the rim.
3. Set up viscometer with helipath TF spindle 5 rpm.
4. Set up stirrer with ¼"-diameter shaft and 1¼"-diameter, 3-blade propeller.
5. Set stirrer at $170 \pm 50$ rpm; check with tachometer at about 4–6" distance between the probe and the reflective tape on the shaft, and at a perpendicular angle. (Read operating instructions.)
6. Mix 4-oz jar of paste for 3 min with stirrer. During mixing, the operator needs to move the jar up and down to assure all of the paste is being mixed.
7. Measure the temperature of paste. If it exceeds $78° \pm 1°F$ ($25.5°$ $\pm 0.5°C$) allow paste to reach $78° \pm 1°F$ by using temperature bath.
8. Place the jar under viscometer with spindle tip just above the surface of paste.
9. Start descending the spindle, and start the time.
10. At 2 min after descent press the dial lever and read the dial.
11. Multiply the dial reading by 20,000 to obtain viscosity in cps (centipoises).

## 11.6   VISCOSITY VERSUS SHEAR RATE

Depending on the availability of equipment, the Brookfield viscometer or a rheometer can be used to obtain the response of viscosity to shear rate. The viscosity ($\eta$) versus shear rate ($\dot{r}$) curve generated by a rheometer can be used as a fingerprint reference as shown in the following:

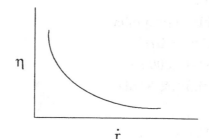

A viscosity ratio can be readily obtained by using a Brookfield viscometer, which indicates the characteristics of a paste in response to shear rate. For example

$$\text{viscosity ratio} = \frac{\eta_1}{\eta_{10}} = \frac{1,500,000 \text{ cps}}{200,000 \text{ cps}} = 7.5$$

where $\eta$ is the viscosity measured at 1 rpm, and $\eta_{10}$ is the viscosity measured with the same spindle TF at 10 rpm.

## 11.7  COLD SLUMP

Regardless of the application, the paste deposit, either through fine-gauge needle dispensing or a screen/stencil printing, should not change its shape and dimension at ambient temperature.

### A.  Apparatus

   slump test apparatus (volume of well: 0.30 cubic inch)

   thermometer

   timer

### B.  Procedure

1.  Adjust paste temperature to 78° ± 1°F (25.5° ± 0.5°C).
2.  Fill opening in slump test apparatus. Be sure plunger is at lowest position.
3.  Level off paste with a straight-edged spatula.
4.  Set plate in holding fixture at 30° angle.
5.  Raise plunger to upper position in one smooth operation.
6.  Take scale reading of amount of slump after 30 sec.
7.  Take final scale reading of amount of slump after 5 min.

## 11.8  HOT SLUMP

In contrast to cold slump of paste at ambient temperature, the hot slump test is to measure the paste slump at elevated temperature which reflects the flow characteristics of paste in the paste state during reflow.

## A. Apparatus

clean copper coupon ($^1/_{32}$" thickness, $^3/_4$" diameter)

hot plate or conveyored-belt hot plates.

## B. Procedure

1. Set hot plate temperature at 150°C.
2. Deposit the paste with 0.25" (0.64 cm) diameter and 0.020" (0.05 cm) thickness on copper coupon.
3. Place paste coupon on hot plate for 30 sec.
4. Measure diameter of paste spread.
5. Compare the diameter (in inches) after heat and before heat.

$$\frac{0.25"}{\text{diameter after heat}} \times 100\% = \begin{array}{l}\text{paste slump resistance \%}\\ \text{or}\\ \text{paste flow restrictivity \%}\end{array}$$

Thus, the result of 100% indicates zero hot slump and complete shape retention.

## 11.9  MOLTEN FLOW

This is a continuation of hot slump test to measure the spread of molten solder, not including the paste slump. The coupon of Section 11.8 is heated at the reflow temperature. The diameter of solidified solder is thus measured. A comparison of the diameter of solidified solder with the diameter of paste after heat in Section 11.8 gives the further spread contributed by molten solder, and in reference to the diameter of paste before heat, it gives the total spread of a paste.

$$\frac{0.25"}{\text{diameter of solidified solder}} \times 100\% = \begin{array}{l}\text{total slump resistance \%}\\ \text{or}\\ \text{total restrictivity \%}\end{array}$$

Federal Specification QQ-S-571E calls for a spread factor to be 80 minimum when solder paste is tested as specified in 4.7.7.2. By examining the procedure and the following formula by which the spread factor is calculated,

$$\text{spread factor (percent)} = \frac{D - H}{D} \times 100,$$

where $H$ is an averaged value of solder height of five solder spot samples, and $D$ is the diameter of solder sphere made of solder paste with equal weight of solder paste of five samples, the following should be noted:

The value of $D$ can be considered as a "constant" for a given solder alloy and solder content, and therefore, the spread factor is essentially determined by the value of $H$ as actually measured. The value of $H$ reflects the height of solder spot resulting from the slump and spreading of solder paste in both paste and molten solder states under the specified heating condition. It is observed that decreasing height, $H$, primarily relates to the collapse of paste during heating prior to solder melting. In other words, the lower the value of $H$, the higher the value of spread factor, is not necessarily an indication of better or worse wetting and fluxing action. As the paste is built in with a zero slump rheology, the $H$ value will be at a maximum (no collapse), and the spread factor is at a minimum. The lower spread factor in such cases does not mean a poor fluxing or wetting action.

## 11.10 TACK TIME

The measuring technique is summarized in Section 3.6.

## 11.11 DRYABILITY

The solder paste can be constituted as dryable and nondryable in the sense of readiness of solvent evaporation as well as the drying effect on solderability. The extent of physical dryability can be tested by the steps similar to Section 11.8.

1. Set hot plate at a specified temperature.
2. Deposit paste with 0.25" (0.64 cm) diameter and 0.020" (0.05 cm) thickness on copper coupon.
3. Weigh paste coupon and record weight, $A$, to nearest 0.0001 g.
4. Place paste coupon on test plate for a specified time.
5. Weigh dryed paste coupon and record weight, $B$, to nearest 0.0001 g. The extent of physical dryability can be estimated by

$$\frac{A - B}{A} \times 100\% = \text{weight loss } \%$$

A higher weight loss indicates a more dryable paste. The drying temperature and time to be employed depend on a specific paste. The proper conditions can be obtained from the paste vendor, and they should correlate with the reflow process. Usually, the range of drying temperature is 80°–150°C and the time duration is 1–30 min. It should be noted that some pastes do not need a drying step and some pastes may deteriorate under drastic drying conditions, thus adversely affecting solderability. The solderability check should always accompany the drying test.

## 11.12  DISPENSABILITY

### A.  Materials and apparatus

dispenser and accessories
packaged syringe of paste
balance

### B.  Procedure

1.  Set up dispenser according to manufacturer's instructions at a proper pressure and pulse length.
2.  Start test (discard the first several dots); dispense on a suitable substrate.
3.  Weigh every five dots on substrate.
4.  Average the weight of each dot.

The interval of weighing depends on the extent of test desired. For an established paste, the samplings at the beginning, middle, and the end of the whole syringe dispensing are the minimal to assure consistent dispensing rate. The qualitative free flowing without clogging, needless to say, is a must. In addition, clean break-off to produce a good-shaped dot and a free-of-dripping condition is included in the total quality of paste dispensability. For a new paste or for an extensive test, every dispensing dot can be monitored in shape and weight by using an automatic X-Y table.

## 11.13  PRINTABILITY

Major parameters involved in printing are discussed in Chapter 6. The paste should be tested with the printer to be used for production at a given set of parameters. The print pattern is to be examined under the optical microscope or automated inspection equipment (if available) for resolution, definition, and printing defects such as bridging, slump, and bleeding. In addition, devices are available to measure print thickness and the uniformity of thickness (Section 3.12).

## 11.14  SHELF STABILITY

The paste shelf stability varies with the grades, vendor, and shelf conditions. Under a mutually agreed shelf time and condition between vendor and user, the paste should maintain its characteristics without degradation. The paste packed in jars sometimes exhibits a slight liquid layer on the surface which may not create problem as long as it can be reconstituted. But paste packed in syringes demands absolute absence of separation. The paste shelf stability depends on the conditions in which it is stored, especially the temperature. In general, the lower the temperature, the longer the life is, with the exception of when the paste vehicle is an emulsion in nature. In such cases, freezing and thawing should be avoided. Always follow the vendor's recommendation for storage conditions.

## 11.15  HANDLING AND SAFETY

Solder pastes contain a mixture of chemicals and lead-containing alloys. Avoid contact with skin and eyes. Follow the material safety data sheet of each product to be used and observe safe working practices and good industrial hygiene.

## 11.16  WATER EXTRACT RESISTIVITY

This test is to measure the conductivity of the flux/vehicle system, which reflects the content of ionic species of the paste. The flux/vehicle system has to be extracted from the paste. This step is obviously important to the outcome of the results. The extracted solution is also to be used in the copper mirror and chloride and bromide tests. The key steps for extraction specified in Federal Specification QQ-S-571E 4.7.3.1.2 are as follows:

1.  Place 200 ml reagent grade, 99% isopropyl alcohol in a clean Erlenmeyer flask.

2.  Add 40 ± 2 g of solder paste to the flask, cover with a watch glass, and boil for 10–15 min using medium heat.
3.  Allow the powder to settle for 2–3 min. Decant the hot solution into a funnel with filter paper, and collect the extract in a clean vessel (save the solder powder portion).
4.  Concentrate the extract to approximately 35% by weight through evaporation of excess solvent.
5.  Determine the exact solids content of the extract.

Key steps to measure water extract resistivity (QQ-S-571E 4.7.3.2) are as follows:

1.  Fill five 200-ml beakers with 50 ml distilled water each (the temperature of beakers should be maintained at 23 ± 2 C).
2.  Measure the resistivity of the distilled water in each beaker with a conductivity bridge using a conductivity cell.
3.  Add 0.100 ± 0.0054 cm$^3$ of the concentrated extract solution into each of the three beakers (two beakers are retained as controls) by using a calibrated dropper or microliter syringe.
4.  Heat all five beakers simultaneously.
5.  Let the beakers boil for 1 min, and quickly cool them down to-the-touch in ice water.
6.  Maintain the beakers at 23° ± 2°C, and measure the conductivity of five beakers with a conductivity bridge.

During the course of measurement, good wet chemistry analytical technique should be practiced, such as using watch glasses to cover beakers throughout the whole test and preventing glassware from being contaminated. The resistivity of each of the controls should not be less than 500,000 ohm-cm. The resistivity number is taken as the reciprocal of the conductivity reading.

## 11.17  COPPER MIRROR

The purpose of the test is to examine the corrosivity of flux/vehicle in the paste. The copper mirror specimens used for the test should conform to the conditions as specified in Federal Specification QQ-S-571E 4.7.9.2.

## Test Procedure:

1. Place approximately 0.05 ml of the concentrated extract solutions and 0.05 ml of the control standard flux (35 wt % Class A type II, Grade WW rosin of LLL-R-626 in reagent grade 99% isopropyl/alcohol) on the specified copper mirror.

2. Place mirror in humidity cabinet at $23° \pm 2°C$, $50 \pm 5\%$ relative humidity for 24 h.

3. Remove the solution and control standard flux by immersing the copper mirror in clean isopropyl alcohol.

4. Examine the mirror.

The test result is judged by examining any complete removal of the copper film. Any complete removal of the copper film is construed as failure, but discoloration of the copper is not considered as failure.

## 11.18 CHLORIDE AND BROMIDE

U.S. Federal Specification QQ-S-571E calls for the silver chromate paper (brownish red) test. The color change of the spot at which the extract solution is placed to off-white (yellow) indicates the presence of chloride and bromide and the area of the off-white (yellow) spot reflects the approximate concentration of chloride (bromide), as shown in Figure 11.1.

The color changes follow this reaction:

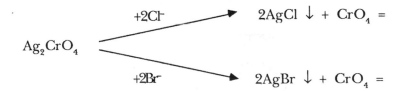

The silver nitrate wet chemical titration can also be used to detect chloride and bromide by means of the following reaction:

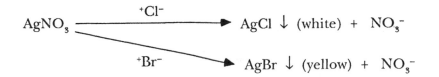

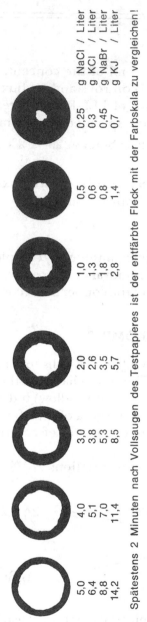

| | | | | | |
|---|---|---|---|---|---|
| 5,0 | 4,0 | 3,0 | 2,0 | 1,0 | 0,5 | 0,25 | g NaCl / Liter |
| 6,4 | 5,1 | 3,8 | 2,6 | 1,3 | 0,6 | 0,3 | g KCl / Liter |
| 8,8 | 7,0 | 5,3 | 3,5 | 1,8 | 0,8 | 0,45 | g NaBr / Liter |
| 14,2 | 11,4 | 8,5 | 5,7 | 2,8 | 1,4 | 0,7 | g KJ / Liter |

Spätestens 2 Minuten nach Vollsaugen des Testpapieres ist der entfärbte Fleck mit der Farbskala zu vergleichen!

**Figure 11.1** Chloride and bromide spot test. (*Courtesy of Macherey-Nagel.*)

The white (yellow) precipitates indicate the presence of chloride ions (bromide ions). It should be noted that the extracted solution should be adjusted to acidic by adding dilute (0.1M) nitric acid. It also should be noted that, due to the nature of flux/vehicle system in solder pastes, a reference test by adding distilled water (without silver nitrate) is necessary to compare the reaction of flux/vehicle extract to pure water. In some cases, the extracted solution by itself may turn turbid or produce white flocculants when dispersed in water, which should not be interpreted as a positive result. The turbidity or white flocculants may be caused by the agglomerates of water insolubles present in the system which are "salted-out" as the aqueous silver nitrate solution is added. Only the presence of a heavy amount of white (yellow) precipitates with high density is an indication of the presence of chlorides (bromides).

For detecting low concentration, chromatography and atomic absorption spectroscopy in conjunction with ASTM method D-808 for chlorine determination are viable techniques.

## 11.19   ACID NUMBER

The acid number of the extracted solution can be measured in accordance with ASTM method D-465.

## 11.20   INFRARED SPECTRUM FINGERPRINT AND OTHER SPECTROSCOPIES

The extract solution as discussed in Section 11.16 can be used to obtain an infrared spectrum of flux/vehicle solution, with isopropyl alcohol in the reference cell. The spectrum provides the fingerprint of the flux/vehicle which is compared with a standard for indication of lot-to-lot consistency. Other chemical instrumentations can be utilized, such as nuclear magnetic resonance (NMR), ultraviolet spectroscopy (UV), and mass spectroscopy.

A brief introduction about these four primary instrumentations for organic chemical identification will follow. Considering the energy levels of a molecule, its total energy is the sum of electronic energy ($E_e$), vibrational energy ($E_v$), and rotational energy ($E_r$), with decreasing magnitude in the order of $E_e$, $E_v$, and $E_r$.

Energy absorbed in the ultraviolet (UV) region of electromagnetic radiation spectrum, as shown in Figure 7.3, produces changes in the electronic energy of the molecule through the transitions of valence electrons between the molecular orbitals. The specific absorption wavelengths depend on the energy level of the molecule. As the ultraviolet radiation reaches the molecule, the energy is absorbed according to the energy level of the molecule that is the characteristic of the molecule. This results in a UV spectrum relating the specific absorption wavelengths to their intensity. Although the energy level of a molecule is quantized, the discrete UV absorption line is not obtained due to its superimposition upon vibrational and rotational sublevels.

The energy level for vibrational and rotational transitions constitutes infrared (IR) absorption. As the radiation in the infrared region is absorbed by the molecule, transitions between vibrational and rotational energy level result. These energy transitions are also quantized. However, vibrational transitions often appear as bands rather than discrete lines because a single vibrational energy is composed of a number of rotational energy changes. The resulting IR spectrum, wavelength versus intensity, reflects the characteristics of the molecule. Within the radio-frequency region, the molecule can absorb the radiation by the transitions between energy levels created by the nuclei spin orientation under the applied magnetic field. Each molecular structure possesses distinct transitions governed by the magnitude of nuclear magnetic moment. A plot of peak absorption frequency versus peak intensity constitutes an NMR spectrum.

Mass spectrometry provides information about the fragmentation as the molecule is bombarded by an electron beam. The spectrum of mass/charge fragments versus peak height which is proportional to the number of ions of each mass reflects the molecular structure. Combining IR, UV, NMR, and mass spectroscopies, chemical identification can be fulfilled.

## 11.21  ALLOY COMPOSITION

The major metal elements of solder powder as separated from the extract of flux/vehicle composition by isopropyl alcohol can be determined by either wet-chemistry or spectroscopy. ASTM Method E-46 specifies chemical analysis of lead- and tin-based solder metal; ASTM Method E-51 specifies spectrochemical analysis of tin alloys; and ASTM Photometric Method E-87 specifies chemical analysis of lead, tin, antimony, and their alloys. The alloy composition is confirmed by the test as in Section 11.26 for melting range.

## 11.22   PARTICLE SIZE—SIEVE

The power portion as separated from Section 11.16 is to be thoroughly cleaned by warm isopropyl alcohol through settling and decanting until the decanted liquid is clear, and then the powder is to be dried at 230°F (110°C) until constant weight is obtained. The particle size sieve analysis is carried out in accordance with ASTM B-214. The results are to be compared with the values as specified.

## 11.23   PARTICLE SIZE DISTRIBUTION—SEDIGRAPH

The particle size distribution curve as shown in Figure 4.36 can be obtained by the sedimentation technique discussed in Section 4.9. For control purpose, the curve is to be compared with the standard.

## 11.24   PARTICLE SHAPE

The micrograph of scanning electron microscopy (SEM) provides good indication of particle shape, as shown in Figures 4.31 and 4.32.

## 11.25   PARTICLE SURFACE CONDITION

The micrograph obtained from scanning electron microscopy (SEM), as indicated in Figure 4.32, shows the morphology and surface condition, including metallurgical phases and microstructure. The particle surface composition can be analyzed by Auger electron spectroscopy (AES) and electron spectroscopy for chemical analysis (ESCA). With AES technique, a specimen in high vacuum is bombarded with a focused beam of electrons. The incident electrons create electron vacancies and leave atoms in an ionized state. When an ionized atom relaxes to a lower energy state, an x-ray photon or Auger electron is emitted. Surface composition can thus be determined by the energy distribution of the emitted electrons, which represent constituent elements. Surface analysis by ESCA is accomplished by irradiating a specimen with monoenergetic soft x-ray. The energy of the emitted electrons is analyzed, which characterizes constituent elements.

## 11.26  DROSS

The test is to determine the oxide content plus other foreign impurities contained on or in the solder powder. The solder powder separated from isopropyl alcohol extraction may not be exactly identical to the virgin solder powder. However, if the technique is maintained the same, the comparative results indicate the relative quality and consistency of solder powder used in the paste.

### A.  Apparatus

analytical balance

hot plate

surface thermometer

30-ml beaker

20-ml graduated cylinder

solvent (GAF M-Pyrol)

acetone

lint-free towel

timer

rosin solution (50 wt % M-Pyrol and 50 wt % rosin)

### B.  Procedure

1. Set hot plate at 25°C above liquidus of the solder alloy temperature.
2. Weigh 30-ml beaker to 0.0001 g and record as weight $A$.
3. Place $10 \pm 0.5$ g of the powder in beaker. Weigh to 0.0001 g and record as weight $B$.
4. Add approximately 20 g of rosin solution to beaker.
5. Heat beaker on hot plate and stir the mixture.
6. Heat at the melting point for 5 min.
7. Remove heat and let solder solidify.
8. Add 10 ml of M-Pyrol with mild agitation.
9. Pour off the diluted rosin solution and rinse the metal button with fresh M-Pyrol.
10. Add 10 ml warm M-Pyrol to free the button of rosin residue.

11. Rinse the metal button with hot water followed by acetone.

12. Dry the button with lint-free towel.

13. Weigh the button to 0.0001 g and record as weight *C.*

14. Calculate dross percentage by

$$\text{dross \%} = \frac{B - A - C}{B - A} \times 100\%$$

## 11.27 MELTING RANGE

Melting range can be measured by using a Fisher-Johns melt point apparatus. For a more precise measure, the thermogram of differential thermal analyzer (DTA) or differential scanning calorimeter (DSC) indicates the exact solidus and liquidus temperature for noneutectic alloys, as shown in Figure 4.30. The peaks of the DTA or DSC curve, which correspond to phase transitions, should be in reference to the corresponding phase diagram.

## 11.28 SOLDER BALL

The extent of solder balling under a specified reflow condition is an indication of the quality of solder paste. A paste reflects good quality when tested on nonwettable substrate ($Al_2O_3$), as outlined below, exhibiting one discrete solder pool without small solder balls, debris, and halo around solder pool, as shown in Figure 11.2 (a). In contrast, the result as shown in Figure 11.2 (b) indicates the poor quality of paste that produced severe solder balling.

### A. Apparatus

clean alumina substrate (2" x 2" x 0.025", 96% $Al_2O_3$)

electric hot plate or conveyored hot plate

surface thermometer

hand-held squeegee (or scraper)

hand-held 10-mil (0.025 cm) thick mask (metal or plastic) with 0.125" (0.313 cm) diameter opening

forceps with flat and bent tip

optical microscope

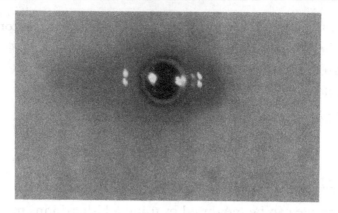

**Figure 11.2(a)**   Paste without solder balling.

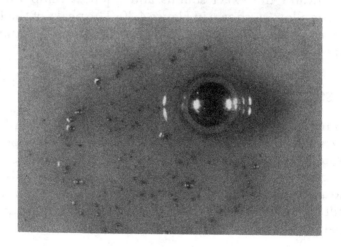

**Figure 11.2(b)**   Paste producing severe solder balling.

## B.   Procedure

1.   Set the hot plate at 25°C above the liquidus temperature of the tested alloy.

2.   Print three specimens of the paste to be tested by hand-held squeegee and mask on the clean alumina substrate. The deposit

should be a well-contoured paste disc. If the deposit is smeared, reprint the paste.

3.  Place the alumina substrate on the hot plate and let the paste heat at the specified temperature for $35 \pm 5$ sec to obtain complete reflow.

4.  Carefully remove the substrate and examine under the optical microscope at 20-X magnification.

5.  Count the discrete solder balls away from the main solder pool, and observe the presence of any gray debris or halo around the solder pool. Record the results.

The solder pool should be a single sphere without any gray debris or halo, and should have less than the specified number of discrete solder balls (on average).

## 11.29 SOLDERABILITY

The test is to determine the wettability of solder paste on substrate. The procedure involves depositing a given amount of paste to be tested on a copper coupon or thick film conductor substrate, reflowing the paste at a specified condition, and then examining the quality of wetting. The phenomena of dewetting, nonwetting, unsmooth boundary, or larger than 75° wetting angle are considered to be indicators of poor solderability.

As discussed in Section 3.10, the solderability of solder pads on substrate can be tested by the Edge Dip Test in reference to ANSI/IPC-S-804A. The test comprises the following steps:

1.  Flux the test specimen in an isopropyl alcohol solution containing 25 wt % of WW-grade gum rosin by dipping to the full depth to be soldered for 5-10 sec.

2.  Drain the specimen for about 60 sec, and conduct solder dip test immediately.

3.  Immerse the specimen into clean molten solder edgewise to a depth of $25 \pm 6$ mm ($1.0 \pm 0.25$ in.) for $3.0 \pm 0.5$ sec.

4.  Withdraw the specimen and maintain in vertical position.

5.  Allow the solder to solidify by air cool.

6.  Evaluate the wetting quality and area in reference to the required standard.

For testing the solderability of the leads and terminations of components, three methods are designed in reference to MIL-STD-202: dip test, wetting balance, and globule. Dip test is similar to the test designed for solder pads as just described. The wetting quality and the area are to be examined; the wetting balance method is to measure the test time at zero balance point and the positive 300 dynes/cm point, respectively; and the globule test is to compare wetting time.

The test primarily comprises the following steps:

1. Age the specimen by exposing it to the saturated steam in a container made of borosilicate glass or stainless steel for 4–8 h.
2. Apply flux by immersing the surface to be tested in isopropyl alcohol solution containing 25 wt % of WW-grade rosin for a minimum of 1 sec.
3. Drain the specimen and then immerse in the molten solder at $245° \pm 5°C$ by one of the three test methods.
4. Evaluate the solderability according to the criteria of each method or the specified standard.

## 11.30  EXPOSURE TIME

In addition to tack time, the exposure duration is the time that the paste can endure atmospheric exposure without any degradation of its quality. A simple test is to repeat the solder ball test except letting the printed paste disc be exposed to the atmosphere (monitoring humidity and temperature) for a specified time, and to compare the results between the fresh sample and exposed sample.

## 11.31  SOLDERING DYNAMICS

This test is to observe the activity of the paste during heating. The printed paste on a copper coupon subjected to a specified heat rate should not exhibit violent activity such as popping and splattering.

## 11.32  CLEANLINESS

The resistivity of solvent extract method and alternate methods are specified in Military Specification MIL-P-28809 4.8.2 and 6.6.1(Appendix III). The methods are basically to measure the concentration of ionic species which are left behind after a cleaning process and able to be extracted by a test solution consisting of 75 wt % ACS reagent grade isopropyl alcohol and 25 wt % distilled/deionized water by means of a "rinsing" technique as specified in MIL-P-28809A 4.8.2.3. Three commercial equipment limits are accepted as equivalent to the resistivity of the solvent extract method. The equivalence factors for these methods are listed in Table 11.2 in terms of microgram equivalents of sodium chloride per unit area.

For example, the use of the Ionograph system, the ionic contamination level in terms of µg NaCl per square centimeter is calculated as

$$\text{ionic contamination level} = f \times \frac{\text{sample count}}{\text{sample area, cm}^2,}$$

where $f$ is the calibration factor.

$$f = \frac{(\mu\text{L of std solution}) \times (\mu\text{g NaCl/L})}{\text{final integrator count average}}$$

The test method is designed for conventional printed wiring assemblies. When testing the densely packed surface mount circuit board, it is questionable whether the hard-to-reach areas can be rinsed by the test solution in order to completely collect the ionic contaminants. The question is not without basis since the residual contaminants which cannot be cleaned by a regular cleaning process may not be reached and rinsed out by the test solution used in this

**TABLE 11.2**

**Equivalence Factors (MIL-P-28809A)**

| Method | Equivalence Factor | Instrument "Acceptance Limit" | |
|---|---|---|---|
| | | µg NaCL/cm² | µg NaCl/in.² |
| MIL-P-28809 Beckman | 1 | 1.56 | 10.06 |
| MIL-P-28809 Markson | 1 | 1.56 | 10.06 |
| Omega Meter | 1.39 | 2.2 | 14 |
| Ionograph | 2.01 | 3.1 | 20 |
| Ion Chaser | 3.25 | 5.1 | 32 |

test. More vigorous rinsing procedures are in development. None-theless, this test in conjunction with surface insulation resistance test provides an indication of "cleanliness" with respect to ionic species.

For nonionic contaminants, IPC-TM-650 Test Methods 2.3.38 and 2.3.39 provide nondestructive methods to determine the presence or absence of contaminants. Test Method 2.3.38 is a rinsing proce-dure to collect the contaminants, and Test Method 2.3.39 is infrared spectroscopic analysis of the collected contaminants.

## 11.33 SURFACE INSULATION RESISTANCE

Military Specification MIL-P-55110D specifies that insulation resis-tance for printed wiring assemblies shall be performed in accor-dance with IPC-TM-650 Method 2.6.3 and shall have a minimum of 500 megohms of resistance between conductors at $100 \pm 10$ volts dc polarizing voltage during chamber exposure. IPC-TM-650 Method 2.6.3 specifies two test patterns: Y patterns and comb patterns, as shown in Figure 11.3, and three test conditions, as listed in Table 11.3. The criterion in the magnitude of resistance for acceptance should be in reference to the virgin board. It should be specified in accordance with the reliability requirement of the assembly. Further-more, the test on the specific circuit design in question always adds more assurance in reliability.

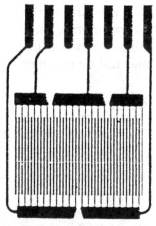

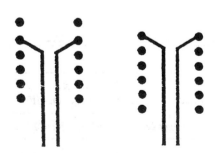

**Figure 11.3(a)** Insulation resistance test pattern: Y pattern. *(Courtesy of IPC.)*

**Figure 11.3(b)** Insula-tion resistance test pattern: comb pattern. *(Courtesy of IPC.)*

**TABLE 11.3**

**Insulation Resistance (IPC-TM-650 Method 2.6.3) and Corresponding Classes**

| Class | Test Conditions | As Received | After Exposure to Moisture |
|-------|-----------------|-------------|-----------------------------|
| Class 1 | 35° ± 5°C 90–98% RH 4 days (static) | maintain electrical function | maintain electrical function |
| Class 2 | 50° + 5°C 90–98% RH 7 days (static) | 500 megohms | 100 megohms |
| Class 3 | 25° ± 2°C to 65° ± 2°C 90–98% RH 6²/₃ days (cycling) | 500 megohms | 500 megohms |

## 11.34  OTHER POST-REFLOW EXAMINATIONS AND TESTS

In addition to cleanliness and electrical insulation resistance tests, other examinations and tests include the solder joint appearance, solder joint voids, joint strength, power cycling effect, temperature cycling effect, vibration effect, simulated aging effect, and thermal shock effect. The solder joint appearance and joint voids are discussed in Sections 9.12 and 9.10, respectively. The solder joint voids can be inspected by the nondestructive Vanzetti laser/ infrared technique or x-ray technique. For qualification tests and quality audit purposes, metallographic cross-section examination and electron microscopic examination are viable techniques. Military Specification MIL-STD-883 method 1010.5 specifies different test conditions for the thermal cycling tests. Vibration test and thermal shock test are specified in MIL-STD-810 Method 514 and MIL-STD-883C Method 1011.4, respectively.

## 11.35 REQUIRED TESTS

Table 11.1 lists 40 tests or characteristics to be examined, which may not even be exhaustive for some cases. The natural questions are: Is every test mandatory? If not, which tests are required? With respect to quality, there are two aspects to be covered for each of the raw materials used on actual production floor: One is to determine the suitability of the raw material, which is fulfilled by a qualification process. The other is to assure the quality and consistency of the already-qualified raw materials.

For qualification process, which is a one-time evaluation, the tests should be conducted as thoroughly and exhaustively as practical. After having conducted a qualification process, the quality assurance program for incoming raw materials can be set up on a collaborative basis with vendors. Incoming raw material tests can be significantly simplified through good understanding and agreement between users and venders and effective utilization of vendors' SPC programs.

Another important aspect of quality is to set up the acceptance criteria for each of the characteristics and performance parameters. Too-tight criteria would limit the sources as well as the latitude of raw materials, and not-tight-enough would jeopardize the quality. The ultimate acceptance criteria are those which are in conformance with the performance requirements in the real world.

## REFERENCE

1.   R. N. Kackar, "Taguchi Quality Philosophy: Analysis and Commentary," *Quality Progress* (Dec. 1986).

# PART V

# Future Tasks and Emerging Trends

# CHAPTER 12

# Future Developments

## 12.1 QUALITY-DRIVEN

The future of electronics is quality in addition to continued innovations. The emergence of more complex and densely integrated circuits demand materials, technologies, and processes capable of relating them to the outside world. This creates motivation to the development of subsequent packaging and interconnections. Surface mount technology for which Chapter 1 provides a brief introduction has been, and will continue to play, a major role for the second level interconnections. The utilization of surface mount technology has been growing, and its benefits have been evidenced in a variety of manufacturing facilities. Figure 12.1 depicts the manufacturing process at the Lexington, Kentucky facility of IBM Corporation using surface mount technology to manufacture printed circuit boards for its printers and typewriters, and Figure 12-2 is an eight-Megabyte memory board which is a double-sided surface mount board manufactured by Xetel Corporation in Austin, Texas.

In the electronics packaging area, the objective of soldering as a whole is obviously to achieve quality and yield. As the number of solder joints increases on each assembly, the quality as indicated by the defect rate becomes vital to the overall yield of the process. Based on probability theory, the relationship between first-time yield and defect rate for the hypothetical conventional through-hole board and surface mount board with equivalent size is well illustrated.[1] In this illustration, the surface mount board contains approximately 7500 joints, and the through-hole board contains 1500 joints. The following lists the defect rates for these two types of board and corresponding yields.

**Figure 12.1** Automated surface mount assembling. *(Courtesy of IBM Corporation.)*

| DEFECT RATE (PPM) | | YIELD (%) |
|---|---|---|
| *Through-hole* | *SMT* | |
| 319 | 66 | 60 |
| 139 | 29 | 80 |
| 66 | 14 | 90 |
| 32 | 7 | 95 |
| 6 | 1 | 99 |

In other words, the presence of solder joint defects will reduce the yield much more drastically for surface mount board than that for through-hole board, and the effect will be accelerated as the number of solder joints increases. The quality of solder joints depends upon the quality of the processes and materials involved, of which solder paste plays on important role.

**Figure 12.2** Surface mounted eight-megabyte memory board. *(Courtesy of Xetel Corporation.)*

## 12.2  MIRROR AND MARRIAGE

The evolution of electronics has been demonstrated by two empirical "laws" proposed by Moore and Loeb for integrated circuits and printed circuit boards.[2] According to W. Loeb, the complexity of printed circuit boards doubles every two and one-half years, as indicated by the evolution of single-sided one-layer board in the early 1950s, to two-layer 125-mil grid printed through-hole board in 1960, and multilayer 100-mil grid printed circuit board in the 1970s, then to more dense 100-mil grid board and surface mount 50-mil pitch or finer pitch board in the 1980s. The integrated circuitry based on Moore's analysis has doubled in its complexity (overall function)

every year and in its density (function per area) every two years from
25 μm line width in 1960 to submicron line width today. By combining both analyses, the development of complexity for integrated
circuits and printed circuits reveals interesting mirror images.

With the success in development and implementation of surface
mount technology, another obvious trend is the merging of printed
circuits and hybrid circuits. This natural marriage provides more
versatility and flexibility in electronic packaging and interconnections.

## 12.3  CONDUCTIVE ADHESIVE

Among the new and renewed developments such as tape automated
bonding (TAB), chip on board (COB), and polymer-based materials, the conductive adhesive has been considered to be intimately
related to solder paste in terms of function. Theoretically, they both
provide interconnections for electronics packaging. However, in
contrast to solder paste as a material to form metallurgical bond, the
conductive adhesive makes an organic bond. The conductive adhesive is composed of a polymer system and conductive metal particles.
The polymer system acts as a binder to hold conductive particles and
as an adhesive to make a permanent bond through one or more of
five operating mechanisms: adsorption, mechanical interlocking,
electrostatic attraction, diffusion, and deformation/rheology. The
conductive particles dispersed in the polymeric system provide
electrical and thermal conductivity.

Basically, both conductive adhesives and solder paste can offer
mechanical strength and electrical and thermal conductivity in
joining electronic devices on substrates. Further evaluation of their
relative attributes falls logically into three areas: (1) conductivity and
mechanical property, (2) processing efficiency/cost, and (3) reliability. The electrical conductivity of a solder joint is far superior to
what commercially available conductive adhesive can offer. Nevertheless, in some applications such as silicon die attachment, the
performance level of conductive adhesive, in this regard has proven
to be adequate. While the use of conductive adhesive may simplify
the processing steps, the materials of conductive adhesives are more
costly than solder. The resilient nature of polymer-based systems is
considered to be able to minimize the failure as a result of thermal
stress. The reliability is, however, a much more complex issue than

a single property parameter, and an in-depth discussion of such is beyond the scope of this book.

The conductive adhesive therefore is viewed as a complementary material rather than a competitive one. The choice between conductive adhesive and solder paste is dictated by the specific application for which the material and the performance properties are required to obtain optimum processing and reliability.

## 12.4 TASKS AND ISSUES

Lately, there has been significant advancement in technology and product development in solder paste, yet numerous areas need to be further researched. With respect to chemical aspects, broadly speaking, development of synthetic resins and systems, and in-depth studies of specific fluxing activity and mechanisms, of quantitative temperature dependency of flux activity, of the relationship between chemical structure and fluxing, as well as of correlation between key rheological parameters and printing/dispensing performance particularly for large-scale automated production of fine pitch patterns, are fertile areas for research. In addition, development of residue-free solder paste and process system, and no-need-to clean solder paste that provide reliable performance are desirable product development tasks.

In metal-related applications, extensive efforts have been devoted to the modification of basic alloys to provide the desired characteristics for a specific application. The results have been well ultilized in numerous applications related to different industries. The impact of modifications ranges from hardness of base alloy to conductivity and to fatigue or creep resistance. However, information on the modification in intrinsic properties of common solder compositions by alloying, such as the well-known antimony strengthening effect, and about relationship between the modification and the performance is scarce. The feasibility studies are obviously warranted.

Studies on fundamental mechanical properties of solder joint (particularly surface mount solder joint) and its failure mechanisms are other warranted areas. Testing techniques and evaluation criteria of specific solder joint designs calls for dedicated efforts.

Although solder compositions are relatively low performance alloys in metal field, most compositions are heat treatable; that is, the characteristics of the composition can be altered by heating and cooling. A study on the effect of heating and cooling from both theoretical and practical points of view, to derive the parameters of temperature effect on the characteristics of solder alloys, will be informative. The characteristics include the extent of residual stress and how it relates to solder performance in an adverse environment.

After a solder joint is made, feasibility studies on materials and techniques to protect and enhance the solder joint integrity is another broad area where research would contribute to the electronics packaging and interconnections. It is indicated that the length of a task list can continue and the tasks should be considered from both chemical and metallurgical aspects.

Recently, the concerns about the stratospheric ozone depletion has led to U.S. federal regulatory response and global response (Montreal Protocol) for control of chlorofluorocarbon (CFC) use, which is the major constituent in most cleaning solvent for the electronics industry. The terms of agreement of the Montreal Protocol are to freeze the CFC production at the 1986 level seven months after entry into force (entry into force is scheduled for January 1989, providing 11 ratifications, representing two-thirds of 1986 global consumption), and to have 20% reduction of CFC production and consumption at the 1986 level starting July 1993, and an additional 30% reduction starting July 1998.

Chemically, the contribution of CFC to potential ozone depletion is by the weight of chlorine in the compound and the overall stability of the molecule in the atmosphere. It is considered that the compounds containing no chlorine atoms or the compounds containing at least one hydrogen atom are some of the potential alternatives to reduce the ozone depletion problem. With the current proposed regulation and continued effort to control the ozone depletion problem, opportunities to develop an aqueous or organic solvent system suitable for electronics are for both existing and new solvent manufacturers. It is also an opportunity to enhance the performance of current solvent systems.

## 12.5  CONCLUSION

Section 12.4 lists only some examples for further efforts in the subject area and is by no means exhaustive. It has been said that the best science/technology is produced by a combination of four elements: an overriding commitment to (1) scientific excellence, (2) vision,

(3) intuition, and (4) initiative. Solder paste technology is no exception. Therefore, the ultimate goal is to meet versatile demands on solder paste material and continue to add to the pool of technology by applying and utilizing fundamental sciences and technologies to serve the needs in the ever-changing and growing electronics industry.

## REFERENCES

1.  A. N. Arslancan and D. K. Flattery. Radiant Technology Corporation, Cerritas, California.
2.  William E. Loeb. W. E. Loeb & Associates, Soquel, California.

# PART VI

# Appendix

# Federal Specification
# QQ-S-571E
# and Amendment 4

QQ-S-571E
May 5, 1972
SUPERSEDING
Fed. Spec. QQ-S-571d
July 10, 1963

FEDERAL SPECIFICATION

SOLDER, TIN ALLOY: TIN-LEAD ALLOY; AND LEAD ALLOY

This specification was approved by the Commissioner, Federal
Supply Service, General Services Administration, for the use
of all Federal agencies.

1. SCOPE AND CLASSIFICATION

1.1 Scope. This specification covers tin-silver, tin-antimony, tin-lead, lead-antimony, and
lead-silver solders in the form of solid bars, ingots, powder, and special forms, and in the form
of solid and flux-cored ribbon and wire and solder paste (see 6.1).

1.2 Classification.

1.2.1 Type designation. The type designation shall be in one of the following forms, and as
specified (see 6.2):

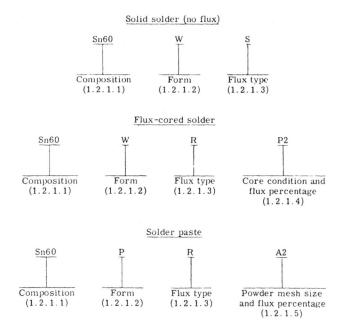

FSC 3439

QQ-S-571E

1.2.1.1 <u>Composition</u>. The composition is identified by a two-letter symbol and a number. The letters indicate the chemical symbol for the critical metallic element in the solder and the number indicates that nominal percentage, by weight, of the critical element in the solder (see 3.2.1).

1.2.1.2 <u>Form</u>. The form is indicated by a single letter, in accordance with table I.

TABLE I.  Form

| Symbol | Form 1/ |
|---|---|
| B - - - - - - - - - - - - | Bar |
| I - - - - - - - - - - - - | Ingot |
| P - - - - - - - - - - - - | Powder |
| R - - - - - - - - - - - - | Ribbon |
| S - - - - - - - - - - - - | Special 2/ |
| W - - - - - - - - - - - - | Wire |

1/ See 6.3 to 6.3.2 inclusive, as applicable.
2/ Includes pellets, preforms, etc. (see 6.2).

1.2.1.3 <u>Flux type.</u> The flux type is indicated by a letter or combination of letters, in accordance with table II.

TABLE II.  Flux type

| Symbol | Flux type |
|---|---|
| S | Solid metal (no flux) |
| R | Rosin flux |
| RMA | Mildly activated rosin flux |
| RA | Activated rosin or resin flux |
| AC | Nonrosin or nonresin flux 1/ |

1/ Includes acid, organic chloride, inorganic chloride, etc.

1.2.1.4 <u>Core condition and flux percentage (applicable only to flux cored solder)</u>. The core condition and flux percentage is identified by a single letter and a number, in accordance with table III.

TABLE III.  Core condition and flux percentage

| Condition symbol | Condition | | |
|---|---|---|---|
| D - - - - - - - - - - | Dry powder | | |
| P - - - - - - - - - - | Plastic | | |
| **Percentage symbol** | **Flux percentage** | | |
| | Nominal | Minimum | Maximum |
| 1 - - - - - - - - - - | 1.1 | 0.8 | 1.5 |
| 2 - - - - - - - - - - | 2.2 | 1.6 | 2.6 |
| 3 - - - - - - - - - - | 3.3 | 2.7 | 3.9 |
| 4 - - - - - - - - - - | 4.5 | 4.0 | 5.0 |
| 6 1/ - - - - - - - - | 6.0 | 5.1 | 7.0 |

1/ Not applicable to flux types R, RMA, and RA.

2

1.2.1.5 <u>Powder mesh size and flux percentage (applicable only to solder paste).</u> The powder mesh size and flux percentage is identified by a single letter and a number, in accordance with table IV.

TABLE IV. Powder mesh size and flux percentage

| Size symbol | Powder mesh size | |
|---|---|---|
| A - - - - - - - - - - | 325 | |
| B - - - - - - - - - - | 200 | |
| C - - - - - - - - - - | 100 | |

| Percentage symbol | Flux percentage | |
|---|---|---|
| | Minimum | Maximum |
| 1 - - - - - - - - - - | 1 | 5 |
| 2 - - - - - - - - - - | 6 | 10 |
| 3 - - - - - - - - - - | 11 | 15 |
| 4 - - - - - - - - - - | 16 | 20 |
| 5 - - - - - - - - - - | 21 | 25 |
| 6 - - - - - - - - - - | 26 | 30 |
| 7 - - - - - - - - - - | over 30 | |

## 2. APPLICABLE DOCUMENTS

2.1 The following documents, of the issue in effect on date of invitation for bids or request for proposal, form a part of this specification to the extent specified herein.

Federal Specifications:

| | | |
|---|---|---|
| NN-P-71 | - | Pallets, Material Handling, Wood, Double Faced, Stringer Construction. |
| QQ-C-576 | - | Copper Flat Products with Slit, Slit and Edge-Rolled, Sheared, Sawed, or Machined Edges, (Plate, Bar, Sheet, and Strip). |
| QQ-S-698 | - | Steel, Sheet and Strip, Low-Carbon. |
| QQ-S-781 | - | Strapping, Steel, Flat and Seals. |
| LLL-R-626 | - | Rosin: Gum, Wood, and Tall Oil. |
| PPP-B-585 | - | Boxes, Wood, Wirebound. |
| PPP-B-601 | - | Boxes, Wood, Cleated-Plywood. |
| PPP-B-621 | - | Boxes, Wood, Nailed and Lock-Corner. |
| PPP-B-636 | - | Boxes, Shipping Fiberboard. |
| PPP-C-96 | - | Cans, Metal, 28 Gage and Lighter. |
| PPP-T-60 | - | Tape: Packaging, Waterproof. |
| PPP-T-76 | - | Tape, Pressure-Sensitive Adhesive Paper, (For Carton Sealing). |

Federal Standards:

| | | |
|---|---|---|
| Fed. Std. No. 123 | - | Marking for Domestic Shipment (Civil Agencies). |
| Fed. Test Method Std. No. 151 | - | Metals; Test Methods. |

(Activities outside the Federal Government may obtain copies of Federal Specifications, Standards, and Handbooks as outlined under General Information in the Index of Specifications and Standards and at the prices indicated in the Index. The Index, which includes cumulative monthly supplements as issued, is for sale on a subscription basis by the Superintendent of Documents, U.S. Government Printing Office, Washington, DC 20402.

(Single copies of this specification and other Federal Specifications required by activities outside the Federal Government for bidding purposes are available without charge from Business Service Centers at the General Services Administration Regional Offices in Boston, New York, Washington, DC, Atlanta, Chicago, Kansas City, MO, Fort Worth, Denver, San Francisco, Los Angeles, and Seattle, WA.

3

QQ-S-571E

(Federal Government activities may obtain copies of Federal Specifications, Standards, and Handbooks and the Index of Specifications and Standards from established distribution points in their agencies.)

Military Specifications:

| | | |
|---|---|---|
| MIL-P-116 | - | Preservation, Methods of. |
| MIL-C-45662 | - | Calibration System Requirements. |
| MIL-F-55561 | - | Foil, Copper, Cladding for Printed Wiring Boards. |

Military Standards:

| | | |
|---|---|---|
| MIL-STD-105 | - | Sampling Procedures and Tables for Inspection by Attributes. |
| MIL-STD-129 | - | Marking for Shipment and Storage. |
| MIL-STD-130 | - | Identification Marking of US Military Property. |
| MIL-STD-147 | - | Palletized and Containerized Unit Loads 40" x 48" Pallets, Skids, Runners, or Pallet-Type Base. |
| MIL-STD-202 | - | Test Methods for Electronic and Electrical Component Parts. |

(Copies of Military Specifications and Standards required by contractors in connection with specific procurement functions should be obtained from the procuring activity or as directed by the contracting officer.)

3. REQUIREMENTS

3.1 Qualification. Sn60 type S solder and Sn60 flux-cored wire solder and solder paste types R, RMA, and RA, furnished under this specification, shall be a product which has been tested, and has passed the qualification tests specified in 4.5, and has been listed on or approved for listing on the applicable qualified products list (see 6.4). Any change in the formulation of a qualified product will necessitate its requalification. The material supplied under contract shall be identical, within manufacturing tolerance, to the product receiving qualification.

3.2 Material.

3.2.1 Alloy composition. When solder is tested as specified in 4.7.2 to 4.7.2.2 inclusive, as applicable, the solder-alloy composition shall be as specified in table V, as applicable.

3.2.2 Flux.

3.2.2.1 Types.

3.2.2.1.1 Type R. Type R flux shall consist of rosin conforming to class A, type I, grade WW or WG of LLL-R-626. When solvents or plasticizers are added, they shall be nonchlorinated (see 4.3).

3.2.2.1.2 Type RMA. Type RMA flux shall consist of rosin conforming to class A, type I, grade WW or WG of LLL-R-626. When solvents or plasticizers are added, they shall be nonchlorinated. Incorporated additives shall be such as to provide a material meeting type RMA (see 4.3).

3.2.2.1.3 Type RA. Type RA flux shall be composed of rosin or resin conforming to class A, type I, grade WW or WG of LLL-R-626. When solvents or plasticizers are added, they shall be nonchlorinated. Incorporated additives shall be such as to provide a material meeting type RA requirements (see 4.3).

3.2.2.1.4 Type AC. Type AC flux shall be composed of one or more acids or salts with or without an organic binder and solvents (see 4.3).

3.2.2.2 Resistivity of water extract (applicable only to flux types R, RMA, and RA). When the extracted flux is tested as specified in 4.7.3.2, the mean of the specific resistances of the water extracts shall be at least 100,000 ohm-centimeters (ohm-cm) for flux types R and RMA, and 45,000 ohm-cm for flux type RA.

4

QQ-S-571E

3.2.2.3 <u>Chlorides and bromides (applicable only to flux types R and RMA)</u>.    When the extracted flux is tested as specified in 4.7.3.3, the test paper shall show no chlorides or bromides present by a color change of the paper to off-white or yellow white (see figure 1).

PASS                                                              FAIL

FIGURE 1.  <u>Chlorides and bromides test results</u>,

3.2.2.4 <u>Core</u>.  The core shall be of any construction, provided it is symmetrically disposed in the solder, continuous, homogeneous, and uniform in cross section.  The core shall be sealed within the solder at both ends by crimping or other means (see 4.7.1.1.1).  Core condition D shall be a dry powder core; core condition P shall be a plastic core (see 4.7.1.1.1).

3.2.3 <u>Solder paste</u>.  Solder paste shall consist of prealloyed solder powder suspended in a flux media such that it has a smooth texture and shows no caking or separation of flux and solder (see 4.7.1.2).

3.2.3.1 <u>Solder powder mesh size</u>.  When the extracted solder powder is tested as specified in 4.7.4, the mesh size shall be as specified (see 1.2.1 and 1.2.1.5).

3.2.3.2 <u>Viscosity</u>.  When solder paste is tested as specified in 4.7.5, the viscosity shall be 400,000 to 600,000 centipoises.

3.3 <u>Flux percentage (applicable only to flux types R, RMA, RA, and AC)</u>.  When flux-cored solder or solder paste is tested as specified in 4.7.6, the flux percentage by weight shall be as specified (see 1.2.1, 1.2.1.4, and 1.2.1.5, as applicable).

3.4 <u>Fluxing action</u>.

3.4.1 <u>Solder pool (applicable only to composition Sn60, flux types R, RMA, RA, and AC)</u>.  When flux-cored solder or solder paste is tested as specified in 4.7.7.1, the flux shall promote the spreading of the molten solder over the coupon to form integrally thereon a coat of solder which shall feather out to a thin edge.  The complete edge of the solder pool shall be clearly visible through the flux residue, and there shall be no evidence of spattering, as indicated by the presence of flux particles outside the main pool of residue.

3.4.2 <u>Spread factor (applicable only to composition Sn60, flux types R, RMA, and RA)</u>.  When flux-cored solder or solder paste is tested as specified in 4.7.7.2, the spread factor shall be 80, minimum.

3.5 <u>Dryness (applicable only to flux types R, RMA, and RA)</u>.  When flux-cored solder or solder paste is tested as specified in 4.7.8, the surface of the residue shall be free from tackiness, permitting easy removal of applied powdered chalk.

5

QQ-S-571E

TABLE V. Solder-alloy compositions

| Composition | Tin (%) | Lead (%) | Antimony (%) | Bismuth, max (%) | Silver (%) | Copper, max (%) | Iron, max (%) | Zinc, max (%) | Aluminum, max (%) | Arsenic, max (%) | Cadmium, max (%) | Total of all others, max (%) | Solidus °C | Liquidus °C |
|---|---|---|---|---|---|---|---|---|---|---|---|---|---|---|
| Sn96 | Remainder | 0.10, max | --- | --- | 3.6 to 4.4 | 0.20 | --- | 0.005 | --- | 0.05 | 0.005 | --- | 221 | 221 |
| Sn70 | 69.5 to 71.5 | Remainder | 0.20 to 0.50 | 0.25 | --- | 0.08 | 0.02 | 0.005 | 0.005 | 0.03 | --- | 0.08 | 183 | 193 |
| Sn63 | 62.5 to 63.5 | Remainder | 0.20 to 0.50 | 0.25 | --- | 0.08 | 0.02 | 0.005 | 0.005 | 0.03 | --- | 0.08 | 183 | 183 |
| Sn62 | 61.5 to 62.5 | Remainder | 0.20 to 0.50 | 0.25 | 1.75 to 2.25 | 0.08 | 0.02 | 0.005 | 0.005 | 0.03 | --- | 0.08 | 179 | 179 |
| Sn60 | 59.5 to 61.5 | Remainder | 0.20 to 0.50 | 0.25 | --- | 0.08 | 0.02 | 0.005 | 0.005 | 0.03 | --- | 0.08 | 183 | 191 |
| Sn50 | 49.5 to 51.5 | Remainder | 0.20 to 0.50 | 0.25 | --- | 0.08 | 0.02 | 0.005 | 0.005 | 0.025 | --- | 0.08 | 183 | 216 |
| Sn40 | 39.5 to 41.5 | Remainder | 0.20 to 0.50 | 0.25 | --- | 0.08 | 0.02 | 0.005 | 0.005 | 0.02 | --- | 0.08 | 183 | 238 |
| Sn35 | 34.5 to 36.5 | Remainder | 1.6 to 2.0 | 0.25 | --- | 0.08 | 0.02 | 0.005 | 0.005 | 0.02 | --- | 0.08 | 185 | 243 |
| Sn30 | 29.5 to 31.5 | Remainder | 1.4 to 1.8 | 0.25 | --- | 0.08 | 0.02 | 0.005 | 0.005 | 0.02 | --- | 0.08 | 185 | 250 |
| Sn20 | 19.5 to 21.5 | Remainder | 0.80 to 1.2 | 0.25 | --- | 0.08 | 0.02 | 0.005 | 0.005 | 0.02 | --- | 0.08 | 184 | 270 |
| Sn10 | 9.0 to 11.0 | Remainder | 0.20, max | 0.03 | 1.7 to 2.4 | 0.08 | --- | 0.005 | 0.005 | 0.02 | --- | 0.10 | 268 | 290 |

Approximate melting range 1/

1 See footnote at end of table.

6

QQ-S-571E

TABLE V. Solder-alloy compositions (Continued)

| Composition | Tin | Lead | Antimony | Bismuth, max | Silver | Copper, max | Iron, max | Zinc, max | Aluminum, max | Arsenic, max | Cadmium, max | Total of all others, max | Solidus °C | Liquidus °C |
|---|---|---|---|---|---|---|---|---|---|---|---|---|---|---|
| | | | | | | | | | | | | | Approximate melting range [1] | |
| Sn5 | 4.5 to 5.5 | Remainder | 0.50, max | 0.25 | --- | 0.08 | 0.02 | 0.005 | 0.005 | 0.02 | --- | 0.08 | 308 | 312 |
| Sb5 | 94.0, min | 0.20, max | 4.0 to 6.0 | --- | --- | 0.08 | 0.08 | 0.03 | 0.03 | 0.05 | 0.03 | 0.03 | 235 | 240 |
| Pb80 | Remainder | 78.5 to 80.5 | 0.20 to 0.50 | 0.25 | --- | 0.08 | 0.02 | 0.005 | 0.005 | 0.02 | --- | 0.08 | 183 | 277 |
| Pb70 | Remainder | 68.5 to 70.5 | 0.20 to 0.50 | 0.25 | --- | 0.08 | 0.02 | 0.005 | 0.005 | 0.02 | --- | 0.08 | 183 | 254 |
| Pb65 | Remainder | 63.5 to 65.5 | 0.20 to 0.50 | 0.25 | --- | 0.08 | 0.02 | 0.005 | 0.005 | 0.02 | --- | 0.08 | 183 | 246 |
| Ag1.5 | 0.75 to 1.25 | Remainder | 0.40, max | 0.25 | 1.3 to 1.7 | 0.30 | 0.02 | 0.005 | 0.005 | 0.02 | --- | 0.08 | 309 | 309 |
| Ag2.5 | 0.25, max | Remainder | 0.40, max | 0.25 | 2.3 to 2.7 | 0.30 | 0.02 | 0.005 | 0.005 | 0.02 | --- | 0.03 | 304 | 304 |
| Ag5.5 | 0.25, max | Remainder | 0.40, max | 0.25 | 5.0 to 6.0 | 0.30 | 0.02 | 0.005 | 0.005 | 0.02 | --- | 0.03 | 304 | 380 |

[1] For information only.

7

QQ-S-571E

3.6 Effect on copper mirror (applicable only to flux types **R** and **RMA**). When tested as specified in 4.7.9, the flux-cored solder or solder paste shall have failed the test if there is any complete removal of the copper film, as evidenced by the white background showing through (see figure 2). Discoloration of the copper due to a superficial reaction or only a partial reduction of the thickness of the copper film shall not be cause for failure.

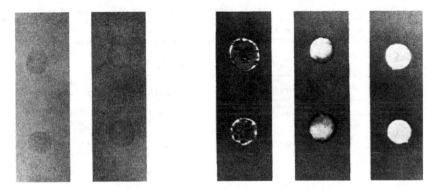

PASS                                                                    FAIL

FIGURE 2. Copper mirror test results

3.7 Dimensions (applicable only to ribbon and wire solder). The dimensions shall be as specified (see 6.2 and 6.3.1). For wire solder, the tolerance on the specified outside diameter (see 6.2) shall be ±5 percent or ±0.002 inch, whichever is greater.

3.8 Unit weight. The unit weight of all forms of solder, as applicable, shall be as specified (see 6.2 and 6.3.2).

3.9 Marking. Bar and ingot forms shall be marked in accordance with MIL-STD-130, with the type designation, the manufacturer's name or code symbol, and the unit weight.

3.10 Workmanship. Solder shall be processed in such a manner as to be uniform in quality and shall be free from defects that will affect life, serviceability, or appearance.

4. QUALITY ASSURANCE PROVISIONS

4.1 Responsibility for inspection. Unless otherwise specified in the contract or purchase order, the supplier is responsible for the performance of all inspection requirements as specified herein. Except as otherwise specified in the contract or order, the supplier may use his own or any other facilities suitable for the performance of the inspection requirements specified herein, unless disapproved by the Government. The Government reserves the right to perform any of the inspections set forth in the specification where such inspections are deemed necessary to assure supplies and services conform to prescribed requirements.

4.1.1 Test equipment and inspection facilities. Test and measuring equipment and inspection facilities of sufficient accuracy, quality and quantity to permit performance of the required inspection shall be established and maintained by the supplier. The establishment and maintenance of a calibration system to control the accuracy of the measuring and test equipment shall be in accordance with MIL-C-45662.

4.2 <u>Classification of inspections</u>. The inspections specified herein are classified as follows:

    (a) Materials inspection (see 4.3).
    (b) Qualification inspection (see 4.5).
    (c) Quality conformance inspection (see 4.6).

4.3 <u>Materials inspection</u>. Materials inspection shall consist of certification supported by verifying data that the materials listed in table VI, used in compounding the solder, are in accordance with the applicable referenced specifications prior to such compounding. The certification and verifying data applicable to a qualification test sample shall be made a part of the qualification test report.

TABLE VI. Materials inspection

| Material | Requirement paragraph | Applicable specification |
|---|---|---|
| Rosin or resin | 3.2.2.1.1 | LLL-R-626 |
| | 3.2.2.1.2 | LLL-R-626 |
| | 3.2.2.1.3 | LLL-R-626 |
| Solvents or plasticizers 1/ | 3.2.2.1.1 | --- |
| | 3.2.2.1.2 | --- |
| | 3.2.2.1.3 | --- |
| Additives 2/ | 3.2.2.1.1 | --- |
| | 3.2.2.1.2 | --- |
| | 3.2.2.1.3 | --- |
| Acids or salts Organic binder 2/ Solvents 2/ | 3.2.2.1.4 | --- |

1/ Verification of solvents or plasticizers as nonchlorinated.
2/ Verification of presence of or absence of additives, organic binders, or solvents.

4.4 <u>Inspection conditions</u>. Unless otherwise specified herein, all inspections shall be performed in accordance with the test conditions specified in "GENERAL REQUIREMENTS" of MIL-STD-202.

4.5 Qualification inspection (applicable only to Sn60 type S solder and to Sn60 flux-cored wire solder and solder paste, flux types R, RMA, and RA). Qualification inspection shall be performed at a laboratory acceptable to the Government (see 6.4) on samples produced with equipment and procedures normally used in production.

4.5.1 <u>Sample size</u>. Two pounds (minimum) of Sn60 of either solder paste or form W (0.062 inch diameter) solder shall be submitted for qualification inspection. Solder paste shall be flux percentage symbol 2; flux-cored solder shall be flux percentage symbol 3.

4.5.2 <u>Inspection routine</u>. The sample shall be subjected to the inspections specified in table VII.

4.5.3 <u>Failures</u>. One or more failures shall be cause for refusal to grant qualification approval.

**QQ-S-571E**

TABLE VII. Qualification inspection

| Examination or test | Requirement paragraph | Method paragraph |
|---|---|---|
| Visual and dimensional examination: | | |
| Core | 3.2.2.4 | 4.7.1.1.1 |
| Solder paste | 3.2.3 | 4.7.1.2 |
| Dimensions 1/ | 3.7 | 4.7.1.1 |
| Unit weight | 3.8 | 4.7.1.1, 4.7.1.2, and 4.7.1.3 |
| Marking2/ | 3.9 | 4.7.1.3 |
| Workmanship | 3.10 | 4.7.1.1, 4.7.1.2, and 4.7.1.3 |
| Material: | | |
| Alloy composition | 3.2.1 | 4.7.2 to 4.7.2.2 incl, as applicable |
| Resistivity of water extract 3/ | 3.2.2.2 | 4.7.3.2 |
| Chlorides and bromides 4/ | 3.2.2.3 | 4.7.3.3 |
| Solder powder mesh size 5/ | 3.2.3.1 | 4.7.4 |
| Viscosity 5/ | 3.2.3.2 | 4.7.5 |
| Flux percentage 3/ | 3.3 | 4.7.6 |
| Fluxing action: | 3.4 | 4.7.7 |
| Solder pool 3/6/ | 3.4.1 | 4.7.7.1 |
| Spread factor 3/6/ | 3.4.2 | 4.7.7.2 |
| Dryness 3/ | 3.5 | 4.7.8 |
| Effect on copper mirror 4/ | 3.6 | 4.7.9 |

1/ Applicable only to wire solder.
2/ Applicable only to bar and ingot solder.
3/ Applicable only to flux types R, RMA, and RA.
4/ Applicable only to flux types R and RMA.
5/ Applicable only to solder paste.
6/ Applicable only to composition Sn60.

4.5.4 Retention of qualification. To retain qualification, the supplier shall forward a report at 12-month intervals to the qualifying activity. The qualifying activity shall establish the initial reporting date. The report shall consist of:

(a) A summary of the results of the tests performed for inspection of product for delivery, groups A and B, indicating as a minimum the number of lots that have passed and the number that have failed. The results of tests of all reworked lots shall be identified and accounted for.
(b) A summary of the results of tests performed for qualification verification inspection, groups C and D, including the number and mode of failures. The summary shall include results of all qualification verification inspection tests performed and completed during the 12-month period. If the summary of the test results indicates nonconformance with specification requirements, and corrective action acceptable to the qualifying activity has not been taken, action may be taken to remove the failing product from the qualified products list.

Failure to submit the report within 30 days after the end of each 12-month period may result in loss of qualification for the product. In addition to the periodic submission of inspection data, the supplier shall immediately notify the qualifying activity at any time during the 12-month period that the inspection data indicates failure to the qualified product to meet the requirements of this specification.

10

QQ-S-571E

In the event that no production occurred during the reporting period, a report shall be submitted certifying that the company still has the capabilities and facilities necessary to produce the item. If during two consecutive reporting periods there has been no production, the manufacturer may be required, at the discretion of the qualifying activity, to submit representative solder to testing in accordance with the qualification inspection requirements.

4.5.5 Extent of qualification. Qualification of Sn60 flux-cored solder of a specific flux type and percentage will be the basis for qualification of any solder composition, form, dimensions, and flux percentage for the flux type. Qualification of Sn60 solder paste of a specific flux type and percentage will be the basis for qualification of any solder composition, powder mesh size, and flux percentage for the flux type. Each flux type shall be qualified separately. Qualification of Sn60, type S solder will be the basis for qualification of any type S solder composition , form, and dimensions; qualification of a flux-cored solder will be extended to cover qualification of any type S solder composition, form, and dimensions.

4.6 Quality conformance inspection.

4.6.1 Inspection of product for delivery. Inspection of product for delivery shall consist of groups A and B inspections.

4.6.1.1 Unit of production. The unit of product shall consist of either spools, coils, cans of solder, bars, or ingots having the same type designation and produced from the same batch of raw materials under essentially the same conditions.

4.6.1.2 Inspection lot.

4.6.1.2.1 Groups A, B, and C inspections. An inspection lot, as far as practicable, shall consist of all the solder of the same type designation, produced from the same batch of raw materials under essentially the same conditions, and offered for inspection at one time.

4.6.1.2.2 Group D inspection. An inspection lot shall consist of all the solder of the same core mixture, produced from the same batch of raw materials under essentially the same conditions, and offered for inspection at one time.

4.6.1.3 Group A inspection. Group A inspection shall consist of the examination specified in table VIII.

4.6.1.3.1 Sampling plan. Statistical sampling and inspection shall be in accordance with MIL-STD-105 for general inspection level II. The acceptable quality level (AQL) shall be as specified in table VIII. Major and minor defects shall be as defined in MIL-STD-105.

4.6.1.3.2 Rejected lots. If an inspection lot is rejected, the supplier may rework it to correct the defects, or screen out the defective units, and resubmit for reinspection. Resubmitted lots shall be inspected using tightened inspection. Such lots shall be separate from new lots, and shall be clearly identified as reinspected lots.

TABLE VIII.  Group A inspection

| Examination | Requirement paragraph | Method paragraph | AQL (percent defective) | |
|---|---|---|---|---|
| | | | Major | Minor |
| Visual and dimensional examination: | | | | |
| Core | 3.2.2.4 | | | |
| Solder paste | 3.2.3 | | | |
| Dimensions 1/ | 3.7 | 4.7.1 | 2.5 | 6.5 |
| Unit weight | 3.8 | | | |
| Marking 2/ | 3.9 | | | |
| Workmanship | 3.10 | | | |

1/ Applicable only to ribbon and wire solder.
2/ Applicable only to bar and ingot solder.

QQ-S-571E

4.6.1.4 Group B inspection. Group B inspection shall consist of the tests, as applicable, specified in table IX and the sample shall be selected from inspection lots which have passed group A
inspection.

4.6.1.4.1 Sampling plan. The sampling plan shall be in accordance with MIL-STD-105 for
special inspection level S-4. The sample size shall be based on the inspection lot size from which
the sample was selected for group A inspection. The AQL shall be 6.5 percent defective.

4.6.1.4.2 Rejected lots. If an inspection lot is rejected, the supplier may rework it to correct
the defects, or screen out the defective units, and resubmit for reinspection. Resubmitted lots
shall be inspected using tightened inspection. Such lots shall be separate from new lots, and shall
be clearly identified as reinspected lots.

TABLE IX. Group B inspection

| Test | Requirement paragraph | Method paragraph |
|---|---|---|
| Flux percentage 1/ | 3.3 | 4.7.6 |
| Fluxing action: | 3.4 | 4.7.7 |
| Solder pool 1/ 2/ | 3.4.1 | 4.7.7.1 |
| Chlorides and bromides 3/ | 3.2.2.3 | 4.7.3.3 |

1/ Applicable only to flux types R, RMA, RA, and AC.
2/ Applicable only to composition Sn60.
3/ Applicable only to flux types R and RMA.

4.6.1.4.3 Disposition of samples. Samples which have been subjected to group B inspection
shall not be delivered on the contract or purchase order.

4.6.2 Qualification verification inspection. Qualification verification inspection shall consist of
groups C and D. Except where the results of these inspections show noncompliance with the applicable requirements (see 4.6.2.3), delivery of products which have passed groups A and B shall not
be delayed pending the results of these qualification verification inspections.

4.6.2.1 Group C inspection. Group C inspection shall consist of the test specified in table X.
Group C inspection shall be made on samples selected from inspection lots which have passed the
groups A and B inspections.

TABLE X. Group C inspection

| Test | Requirement paragraph | Method paragraph |
|---|---|---|
| Alloy composition | 3.2.1 | 4.7.2 to 4.7.2.2 incl, as applicable |

4.6.2.1.1 Sampling plan. Samples shall be selected from the first lot and thence from one lot
in every 25 or once each year, whichever is more frequent, in accordance with table XI. For
spools and coils, the second 6-foot piece of solder shall be tested.

TABLE XI. Sampling plan for group C inspection

| Size of lot (pounds) | Number of samples (spools, coils, containers, or pieces) |
|---|---|
| Up to 1,000, inclusive | 3 |
| Over 1,000 to 10,000 inclusive | 5 |
| Over 10,000 | 10 |

12

4.6.2.1.2 <u>Failures</u>. If one or more samples fail to pass group C inspection, the sample shall be considered to have failed.

4.6.2.1.3 <u>Disposition of samples</u>. Samples which have been subjected to group C inspection shall not be delivered on the contract or purchase order.

4.6.2.2 <u>Group D inspection</u>. Group D inspection shall consist of the tests, as applicable, specified in table XII. Group D inspection shall be made on samples selected from inspection lots which have passed the groups A and B inspections.

TABLE XII.  Group D inspection

| Test | Requirement paragraph | Method paragraph |
|---|---|---|
| Resistivity of water extract 1/ | 3.2.2.2 | 4.7.3.2 |
| Solder powder mesh size 2/ | 3.2.3.1 | 4.7.4 |
| Viscosity 2/ | 3.2.3.2 | 4.7.5 |
| Fluxing action: | 3.4 | 4.7.7 |
| Spread factor 1/ 3/ | 3.4.2 | 4.7.7.2 |
| Dryness 1/ | 3.5 | 4.7.8 |
| Effect on copper mirror 4/ | 3.6 | 4.7.9 |

1/ Applicable only to flux type R, RMA, and RA.
2/ Applicable only to solder paste.
3/ Applicable only to composition Sn60.
4/ Applicable only to flux types R and RMA.

4.6.2.2.1 <u>Sampling plan</u>. One pound of flux-cored solder or solder paste shall be selected for each flux mixture from the first lot and thence from one lot in every 50 lots, or once each 24 months, whichever is more frequent.

4.6.2.2.2 <u>Failures</u>. If the sample fails to pass any of the group D inspections, as applicable, the sample shall be considered to have failed.

4.6.2.2.3 <u>Disposition of samples</u>. Samples which have been subjected to group D inspection shall not be delivered on the contract or purchase order.

4.6.2.3 <u>Noncompliance</u>. If a sample fails to pass groups C or D inspections, the supplier shall take corrective action on the materials or processes, or both, as warranted, and on all units of product which can be corrected and which were manufactured under essentially the same conditions, with essentially the same materials, processes, etc., and which are considered subject to the same failure. Acceptance of the product shall be discontinued until corrective action, acceptable to the Government, has been taken. After the corrective action has been taken, groups C and D inspections shall be repeated on additional samples (all inspection, or the inspection which the original sample failed, at the option of the Government). Groups A and B inspections may be reinstituted; however, final acceptance shall be withheld until the groups C and D reinspections have shown that the corrective action was successful. In the event of failure after reinspections, information concerning the failure and corrective action taken shall be furnished to the cognizant inspection activity and the qualifying activity.

4.6.3 <u>Inspection of preparation for delivery</u>. Sample packages and packs and the inspection of the preservation-packaging, packing, and marking for shipment and storage shall be in accordance with the requirements of section 5 and the documents specified therein.

4.7 <u>Methods of examination and test</u>.

13

QQ-S-571E

4.7.1 <u>Visual and dimensional examination</u>.

4.7.1.1 <u>Ribbon and wire solder (solid and flux-cored)</u>. Ribbon and wire solder shall be examined to verify that the dimensions, unit weight, and workmanship are in accordance with the applicable requirements (see 3.7, 3.8, and 3.10).

4.7.1.1.1 <u>Core (flux types R, RMA, RA, and AC) (see 3.2.2.4)</u>. The core shall be examined for end sealing. Two-inch pieces of flux-cored solder shall be cut at 2-foot intervals along the ribbon or wire until five such 2-inch pieces are obtained from each spool, coil, or cut length, as applicable. For core condition D, care shall be taken that the solder be cut over a contrasting color surface so that any spilled powder is visible. Both ends of each 2-inch piece shall be examined visually for continuity, homogeneity, dimensional uniformity, and core condition.

4.7.1.2 <u>Solder paste</u>. Solder paste shall be examined for smoothness of texture (no lumps), caking, separation of flux and solder, unit weight, and workmanship in accordance with the applicable requirements (see 3.2.3, 3.8, and 3.10).

4.7.1.3 <u>Bar and ingot solder</u>. Bar and ingot solder shall be examined to verify that the unit weight, marking, and workmanship are in accordance with the applicable requirements (see 3.8 to 3.10, inclusive).

4.7.2 <u>Alloy composition (see 3.2.1)</u>.

4.7.2.1 <u>Sample preparation</u>.

4.7.2.1.1 <u>Solid, ribbon and wire solder</u>. Each sample of solid, ribbon or wire solder shall be melted in a clean container, mixed thoroughly, and poured into a cold mold, forming a bar approximately 1/4-inch thick. The mixed sawings shall be prepared for analysis as specified in 4.7.2.1.3.

4.7.2.1.2 <u>Flux-cored, ribbon and wire solder and solder paste</u>. Each sample of flux-cored, ribbon or wire solder or solder paste shall be melted in a clean container and mixed thoroughly. After the flux has risen to the top, the alloy shall be poured carefully into a cool mold (care being taken to allow the flux and alloy to separate completely), forming a bar approximately 1/4-inch thick. The bar shall be cleaned of flux residue, and the mixed sawings shall be prepared for analysis as specified in 4.7.2.1.3.

4.7.2.1.3 <u>Bar and ingot solder</u>. Each sample piece shall be cut in half, and one half marked and held in reserve. The remaining half shall be melted in a clean container, mixed thoroughly, and poured into a cool mold, forming a bar approximately 1/4-inch thick. Saw cuts shall then be made across the 1/4-inch bar at equal intervals of not more than 1 inch throughout its length. If it is impracticable to melt the bar or ingot as specified above, saw cuts shall be made across each piece at equal intervals of not more than 1 inch throughout its length. No lubricants shall be used during sawing. The sawings shall be mixed thoroughly. The sample shall consist of not less than 5 ounces of mixed sawings.

4.7.2.2 <u>Method</u>. The alloy composition shall be determined by any suitable method, including wet-chemical or spectrochemical-analysis techniques, or both. When wet-chemical or spectrochemical-analysis techniques are used, they shall be in accordance with methods 111 and 112, respectively, of Fed. Test Method Std. No. 151.

4.7.3 <u>Flux (applicable only to flux types R, RMA, and RA)</u>.

4.7.3.1 <u>Flux extraction</u>.

14

4.7.3.1.1 <u>Flux-cored solder</u>. The flux core shall be extracted as follows: Cut a length of the flux-cored solder weighing approximately 150 grams and seal the ends. Wipe the surface clean with a cloth moistened with acetone. Place the sample in a beaker, add sufficient distilled water to cover the sample, and boil for 5 to 6 minutes. Rinse the sample with acetone and allow to dry. Protecting the solder surface from contamination, cut the sample into 3/8-inch (max) lengths without crimping the cut ends. Place the cut lengths in an extraction tube of a chemically clean soxhlet extraction apparatus and extract the flux with reagent grade, 99 percent isopropyl alcohol until the return condensate is clear. The resistivity of water extract and effect on copper mirror tests shall be performed using a test solution prepared by concentrating the solids content in the flux extract solution to approximately 35 percent by weight by evaporation of the excess solvent. The exact solids content of the test solution shall be determined on an aliquot, dried to constant weight in a circulating air oven maintained at 85° ±3°C. This test solution shall also be used for the chlorides and bromides test.

4.7.3.1.2 <u>Solder paste</u>. The flux shall be extracted as follows: Place 200 cubic centimeters of reagent grade, 99 percent isopropyl alcohol in a chemically clean erlenmeyer flask. Add 40 ±2 grams of solder paste to the flask, cover with a watch glass, and boil for 10 to 15 minutes using medium heat. Allow the powder to settle for 2 to 3 minutes and decant the hot solution into a funnel containing filter paper, collecting the flux extract in a chemically clean vessel. (Note: The solution in isopropyl alcohol does not necessarily have to be clear.) The solder powder shall be saved for the mesh size determination. The resistivity of water extract and effect on copper mirror tests shall be performed using a test solution prepared by concentrating the solids content in the flux extract solution to approximately 35 percent by weight by evaporation of the excess solvent. The exact solids content of the test solution shall be determined on an aliquot, dried to constant weight in a circulating air oven maintained at 85° ±3°C. This test solution shall also be used for the chlorides and bromides test.

4.7.3.2 <u>Resistivity of water extract (see 3.2.2.2)</u>. The resistivity of water extract shall be determined using the flux test solution (see 4.7.3.1.1 or 4.7.3.1.2, as applicable). Five watch glasses and five acid/alkali resistant, tall form graduated beakers shall be thoroughly cleaned by washing in hot water detergent solution, rinsing several times with tap water followed by at least five rinses with distilled water. CAUTION: All beakers shall be covered with the watch glasses to protect the contents from contaminants. The beakers' dimensions shall be such that when the conductivity cell is immersed in 50 milliliters (ml) of liquid contained therein, the electrodes are fully covered. Each cleaned beaker shall be filled to the 50 ml mark with distilled water. The beakers shall be immersed in a water bath maintained at 23° ±2°C. When thermal equilibrium is reached, the resistivity of the distilled water in each beaker shall be measured at this temperature with a conductivity bridge using a conductivity cell with a cell constant of approximately 0.1. The resistivity of the distilled water in each beaker shall not be less than 500,000 ohm-cm. If the resistivity of the water in any beaker is less than 500,000 ohm-cm, the complete process above shall be repeated. Two of these beakers shall be retained as controls. Add 0.100 ± 0.005 cubic-centimeters of the flux test solution to each of the other three beakers by means of a calibrated dropper or microliter syringe. The heating of all five beakers shall be started simultaneously. As the contents of each beaker comes to a boil, the boiling shall be timed for 1 minute followed by quick cooling of the beakers, to the touch, under running tap water or by immersing in ice water. The cooled, covered beakers shall be placed in a water bath maintained at 23° ±2°C. When the thermal equilibrium is reached, the resistivity in each of the five beakers shall be determined at this temperature as follows:

(a) Thoroughly wash the conductivity cell with distilled water and immerse it in the water extract of one sample. Make instrument reading.

(b) Thoroughly wash conductivity cell in distilled water and immerse in a water control. Make instrument reading.

(c) Thoroughly wash conductivity cell in distilled water and immerse in a water extract. Make instrument reading.

(d) Thoroughly wash conductivity cell in distilled water and continue measuring resistivities of the remaining control and water extract.

The resistivity of each of the controls shall not be less than 500,000 ohm-cm. If the control value is less than 500,000 ohm-cm, it indicates that the water was contaminated with water-soluble ionized materials and the entire test shall be repeated. The mean of the specific resistivities of the water extracts of the flux shall be calculated.

QQ-S-571E

4.7.3.3 Chlorides and bromides (see 3.2.2.3). One drop of flux test solution (see 4.7.3.1.1 or 4.7.3.1.2, as applicable) (approximately 0.05 ml/drop) shall be placed on a small dry piece of silver chromate test paper. The drop shall remain on the test paper for 15 seconds prior to immersing the test paper in clean isopropyl alcohol for 15 seconds to remove residual organic materials. The test paper shall dry for 10 minutes. The test paper shall be visually examined for color change.

4.7.4 Solder powder mesh size (applicable only to solder paste) (see 3.2.3.1). The solder powder obtained as specified in 4.7.3.1.2 shall be dried completely so that all particles are separated. A minimum of 80 percent of the powder shall pass through the appropriate size sieve (see 1.2.1.5) in order to be classified for that mesh size.

4.7.5 Viscosity (applicable only to solder paste) (see 3.2.3.2). A Brookfield Model RVT Viscosimeter equipped with a Brookfield Model C Helipath Stand using a TF Spindle operated at 4 revolutions per minute or other suitable viscosimeter shall be used. The solder paste temperature shall be 25° ±1°C. The solder paste shall be stirred with a spatula for 3 minutes. The bottom point of the spindle shall be set in the center of the sample and the speed shall be set at four revolutions per minute. The rotation and descent of the spindle shall be started. The dial reading shall be taken 2 minutes after the descent of the spindle into the paste. The dial reading shall be multiplied by 20,000 to obtain the viscosity in centipoises.

4.7.6 Flux percentage (applicable only to flux types R, RMA, RA, and AC) (see 3.3). A minimum of 20 grams of solder (weighed within an accuracy of 5 milligrams (mg)) shall be placed into a clean, preweighed, porcelain crucible. (For flux-cored solder, the sheared ends shall be sealed, and the solder shall be coiled into a small ball by winding it upon itself.) The weight of the solder alone shall be denoted as Wa. The crucible and solder shall be heated until the solder is completely molten. The molten solder shall be stirred several times to free any entrapped flux. (For solders containing a low percentage of flux, an oil having a high boiling point may be used to aid in the separation of the flux from the solder.) The solder shall be allowed to cool until it solidifies. The solder shall be cleaned thoroughly, using chemical solvents for the flux until the solder is free from any flux residues. The solder shall be dried and its weight determined in air in grams within an accuracy of 5 mg. This weight shall be denoted as Wb. Flux percentage shall be calculated as follows:

$$\text{Flux percentage, by weight} = \frac{Wa - Wb}{Wa} \times 100$$

4.7.7 Fluxing action.

4.7.7.1 Solder pool (applicable only to composition Sn60, flux types R, RMA, RA, and AC) (see 3.4.1). For each sample being tested, three coupons 1.5 inches square shall be cut from 0.063-inch-thick sheet copper in accordance with QQ-C-576. For flux type AC only, the coupons shall be cut from cold-rolled commercial sheet steel, approximately 0.063-inch-thick conforming to finish number 2 regular bright finish of QQ-S-698. The coupons shall be degreased by immersion in trichloroethylene or other suitable short-chain solvent. Both surfaces of each coupon shall be cleaned to a bright finish, using a 10 percent fluoroboric acid dip. The coupons shall be washed with tap water and dried thoroughly with a clean cloth. Approximately 0.2 grams of flux-cored solder or approximately 2 grams of solder paste shall be placed in the center of each coupon. (The area of the solder paste shall not exceed that of a 0.375-inch diameter circle.) The solder shall be melted in an oven maintained at 315° ±15°C. The solder pool shall be visually examined for thickness of edge. When the test is completed, each coupon shall be inspected for evidence of spattering of flux.

4.7.7.2 Spread factor (applicable only to composition Sn60, flux types R, RMA, and RA) (see 3.4.2).

16

QQ-S-571E

4.7.7.2.1 Preparation of coupon. Five coupons 2-inch square shall be cut from 0.005 inch-thick electrolytic copper sheets in accordance with MIL-F-55561. The coupons shall be cleaned in a 10 percent fluoroboric acid dip. One corner of each coupon shall be bent upwards to permit handling with tweezers. The coupons shall not be handled with bare hands. The coupons shall be vapor-degreased and then oxidized for 1 hour in an electric oven at 150° ±5°C for testing of flux types R, and RMA, and 205° ±5°C for testing of flux types RA. All coupons shall be at the same level in the oven. All coupons shall be removed from the oven and placed in tightly closed glass bottles until ready for use.

4.7.7.2.2 Procedure.

4.7.7.2.2.1 Flux-cored solder. Ten or more turns of 0.063-inch diameter flux-cored solder shall be tightly wrapped around a mandrel. The solder shall be cut through with a sharp blade along the longitudinal axis of the mandrel. The rings shall be slid off the mandrel and the helix removed by flattening each ring. The diameter of the mandrel shall be of such a size so as to produce a ring weighing 0.500 ± 0.025 gram. Ten rings shall be prepared. A solder ring shall be placed in the center of each one of the five coupons. The coupons shall be placed horizontally on a flat oxidized copper sheet in a circulating-air oven at 205° ±5°C for 6 minutes $^{+10}_{-0}$ seconds, with all coupons being at the same level. At the end of 6 minutes, the coupons shall be removed from the oven and allowed to cool. Excess flux residue shall be removed by washing with alcohol. The height, H, of the solder spot shall be measured to the nearest 0.001 cm, and the results averaged. Five additional solder-ring specimens shall be melted together in a small, porcelain combustion boat on a hot plate. The molten solder shall be stirred several times to free any entrapped flux. After cooling, the solder slab shall be removed from the boat, the excess flux removed by washing with alcohol, and the loss of weight in water determined to the nearest 0.001 gram.

4.7.7.2.2.2 Solder paste. The coupons shall be removed from the bottles and weighed to the nearest 0.001 gram. A metal washer with an internal diameter of 0.250 inch shall be placed in the center of each coupon and each opening shall be filled with solder paste. The excess solder paste shall be wiped off the washer using a spatula and then the washer shall be removed carefully. The coupons with solder paste shall be reweighed to the nearest 0.001 gram. (Note: The thickness of the washer shall be such that the solder weighs from 0.45 to 0.55 gram.) The coupons shall be placed horizontally on a flat oxidized copper sheet in a circulating-air oven at 205° ± 5°C for 6 minutes $^{+10}_{-0}$ seconds, with all coupons being at the same level. At the end of 6 minutes, the coupons shall be removed from the oven and allowed to cool. Excess flux residue shall be removed by washing with alcohol. The height, H, of the solder spot shall be measured to the nearest 0.001 cm, and the results averaged. An amount of solder paste equal to the total weight of solder paste on the five coupons shall be melted in a small, porcelain combustion boat on a hot plate. The molten solder shall be stirred several times to free any entrapped flux. After cooling, the solder slab shall be removed from the boat, the excess flux removed by washing with alcohol, and the loss of weight in water determined to the nearest 0.001 gram.

4.7.7.2.2.3 Calculation. The loss in weight of the solder slab in water shall be divided by five. This is the volume, V, of the solder to the nearest 0.001 cc. The diameter, D, of the equivalent sphere is $1.2407 \sqrt[3]{V}$. The spread factor shall be calculated in accordance with the following formula:

$$\text{Spread factor (percent)} = \frac{D - H}{D} \times 100$$

4.7.8 Dryness (applicable only to flux types R, RMA, and RA) (see 3.5). The dryness test shall be performed on samples prepared in accordance with 4.7.7.2.1 and 4.7.7.2.2 (as applicable), except that after heating the coupons in the oven, the flux residue shall not be removed. The coupons shall be allowed to cool for one-half hour. Powdered chalk shall be dusted onto the surface of the residual flux and the ability to remove the chalk from the surface of the flux by light brushing shall be observed.

QQ-S-571E

4.7.9 <u>Effect on copper mirror (applicable only to flux types R and RMA) (see 3.6)</u>.

4.7.9.1 <u>Preparation of the control-standard flux</u>. A control-standard flux shall be prepared by using 35 percent, by weight, of rosin conforming to class A, type II, grade WW, of LLL-R-626, and dissolving in reagent grade 99 percent isopropyl alcohol.

4.7.9.2 <u>Preparation of copper mirror</u>. A copper mirror shall consist of a vacuum-deposited film of pure copper metal on one surface of a flat sheet of clear, polished glass. The thickness of the copper film shall be uniform and shall permit 10 ±5 percent transmission of normal incident light of 5,000 angstrom units as determined with any suitable standard photoelectric spectrophoto- meter. To prevent oxidation of the copper mirror, it is recommended that the mirrors be stored in closed containers which have been flushed with nitrogen. Immediately prior to testing, the copper mirror shall be immersed in a 5 percent solution of ethylene diamine tetra acetic acid or similar chelating agent for copper oxide, rinsed thoroughly in running water, immersed in clean ethyl or methyl alcohol, and dried with clean, oil-free air. The copper film shall be examined in good light. The copper mirror is acceptable if no oxide film is visible and the copper film shall show no visible damage.

4.7.9.3 <u>Procedure</u>. Approximately 0.05 ml of the flux test solution (see 4.7.3.1.1 or 4.7.3.1.2, as applicable) and 0.05 ml of the control-standard flux (see 4.7.9.1) shall be placed adjacent to each other on the face of a flat, vacuum-deposited copper mirror (see 4.7.9.2). The dropper shall not be permitted to touch the copper surface, and the mirror shall be protected at all times from dirt, dust, and fingerprints. The mirror shall be placed in a horizontal position at $23° ±2°C$ and 50 ±5 percent relative humidity in a dust-free cabinet for $24 + 1/2$ hours. At the end of the 24-hour storage period, the test flux and the control standard flux shall be removed by immersing the copper mirror in clean isopropyl alcohol. The clean mirror shall be examined visually for compliance of the test flux and the control-standard flux with 3.6. If the control-standard flux does not comply with 3.6, the test shall be repeated using a new copper mirror.

5. PREPARATION FOR DELIVERY

5.1 <u>Preservation-packaging</u>. Preservation-packaging shall be level A or C, as specified (see 6.2).

5.1.1 <u>Level A</u>.

5.1.1.1 <u>Cleaning</u>. Solder shall be cleaned in accordance with MIL-P-116, process C-1.

5.1.1.2 <u>Drying</u>. Solder shall be dried in accordance with MIL-P-116.

5.1.1.3 <u>Preservative application</u>. Preservatives shall not be used.

5.1.1.4 <u>Unit packaging</u>. Solder shall be packaged by the procedures specified herein in accor- dance with MIL-P-116, method III, insuring compliance with the general requirements paragraph under methods of preservation (unit production) and the physical protection requirements paragraph therein.

5.1.1.4.1 <u>Ribbon and wire solder</u>. Unless otherwise specified (see 6.2), ribbon and wire solder shall be wound on metal spools. The quantity to be included on each spool shall be as specified (see 6.2). The outside layer of solder shall be located within the flanges of the spool and secured in a manner to prevent loosening.

5.1.1.4.2 <u>Pellet, powder, and paste solder</u>. This solder shall be packaged in type V, class 1 cans conforming to PPP-C-96 in quantities of one or five pounds net weight as specified (see 6.2). Plan B exterior coating shall be used.

5.1.1.4.3 <u>Bar and ingot solder</u>. Bar and ingot solder shall be bulk packaged in the quantities specified (see 6.2). The unit container shall also suffice as a shipping container and meet the re- quirements of 5.2 for the level of packing specified (see 6.2).

QQ-S-571E

5.1.1.4.4 <u>Special forms</u>. Preservation-packaging for special forms (other than pellets) shall be as specified in the contract or purchase order (see 6.2).

5.1.1.5 <u>Intermediate packaging</u>. Not required.

5.1.2 <u>Level C</u>. Solder shall be clean, dry and packaged in a manner that will afford adequate protection against corrosion, deterioration, and physical damage during shipment from supply source to the first receiving activity. For bar and ingot solder, the unit container shall also suffice as a shipping container and meet the requirements of 5.2 for the level of packing specified (see 6.2).

5.2 <u>Packing</u>. Packing shall be level A. B. or C. as specified (see 6.2).

5.2.1 <u>Level A</u>.

5.2.1.1 <u>Ribbon and wire solder</u>. The packaged ribbon and wire solder shall be packed in fiberboard containers conforming to PPP-B-636. class weather resistant, style optional, special requirements. In lieu of the closure and waterproofing requirements in the appendix of PPP-B-636, closure and waterproofing shall be accomplished by sealing all seams, corners, and manufacturer's joint with tape, two inches minimum width, conforming to PPP-T-60, class 1, or PPP-T-76. Banding (reinforcement requirements) shall be applied in accordance with the appendix to PPP-B-636, using nonmetallic or tape banding only.

5.2.1.2 <u>Pellet, powder, and paste solder</u>. Cans of solder shall be packed in accordance with the level A packing requirements for filled cans in the appendix of PPP-C-96. Solder shall be packed as specified in 5.2.1.1 when packaged in containers other than cans.

5.2.1.3 <u>Bar and ingot solder</u>. Bar and ingot solder shall be packed in snug fitting nailed wood boxes conforming to PPP-B-621, class 2. The boxes shall be modified to include full double end panels. Blocking shall be provided as necessary. The gross weight shall not exceed 200 pounds. Box closures and strapping shall be in accordance with the appendix of PPP-B-621.

5.2.1.4 <u>Special forms</u>. Packing for special forms (other than pellets) shall be as specified in the contract or purchase order (see 6.2).

5.2.2 <u>Level B</u>.

5.2.2.1 <u>Ribbon and wire solder</u>. The packaged ribbon and wire solder shall be packed in containers conforming to PPP-B-636, class domestic, style optional, special requirements. Closures shall be in accordance with the appendix thereto.

5.2.2.2 <u>Pellet, powder, and paste solder</u>. Cans of solder shall be packed in accordance with the level B packing requirements for filled cans in the appendix of PPP-C-96. Solder shall be packed as specified in 5.2.2.1 when packaged in containers other than cans.

5.2.2.3 <u>Bar and ingot solder</u>. Bar and ingot solder shall be packed as specified in 5.2.1.3 except that the nailed wood boxes shall conform to PPP-B-621. class 1.

5.2.2.4 <u>Special forms</u>. Packing for special forms (other than pellets) shall be as specified in the contract or purchase order (see 6.2).

5.2.3 <u>Level C</u>. Solder. packaged as specified in 5.1.1 or 5.1.2. shall be packed in shipping containers in a manner that will afford adequate protection against damage during direct shipment from the supply source to the first receiving activity. These packs shall conform to the applicable carrier rules and regulations.

5.2.4 <u>Unitized loads</u>. Unitized loads, commensurate with the level of packing specified in the contract or order, shall be used whenever total quantities for shipment to one destination equal 40 cubic feet or more. Quantities less than 40 cubic feet need not be unitized. Unitized loads shall be uniform in size and quantities to the greatest extent practicable.

QQ-S-571E

5.2.4.1 <u>Level A</u>. Solder, packed as specified in 5.2.1, shall be unitized on pallets in conformance with MIL-STD-147, load type I, with a fiberboard cap (storage aid 4) positioned over the
load.

5.2.4.2 <u>Level B</u>. Solder, packed as specified in 5.2.2, shall be unitized as specified in 5.2.4.1
except that the fiberboard caps shall be class domestic.

5.2.4.3 <u>Level C</u>. Solder, packed as specified in 5.2.3, shall be unitized with pallets and caps
of the type, size, and kind commonly used for the purpose and shall conform to the applicable
carrier rules and regulations.

5.3 <u>Marking</u>. In addition to any special marking required by the contract or purchase order
(see 6.2), spools, coils, cans, and unit containers shall be marked with the type designation, manufacturer's code, and net weight; also, spools and coils shall be marked with the dimensions of the
ribbon or the outside diameter of the wire, as applicable.

5.3.1 <u>Civil agencies</u>. For civil agencies marking for shipment shall be in accordance with
Fed. Std. No. 123.

5.3.2 <u>Military activities</u>. For military activities each unit package, exterior container, and
unitized load shall be marked in accordance with MIL-STD-129.

5.4 <u>General</u>.

5.4.1 <u>Exterior containers</u>. Exterior containers (see 5.2.1, 5.2.2, and 5.2.3) shall be of a
minimum tare and cube consistent with the protection required and shall contain equal quantities of
identical stock numbered items to the greatest extent practicable.

5.4.2 <u>Army procurements</u>.

5.4.2.1 <u>Level A and B packing</u>. For level A packing when quantities per destination are less than
a unitized load, the fiberboard containers shall not be banded but shall be placed in a close fitting
box conforming to PPP-B-601, overseas type; PPP-B-621, class 2, style 4 or PPP-B-585, class 3,
style 2 or 3. Closure and strapping shall be in accordance with applicable container specification
except that metal strapping shall conform to QQ-S-781, type I, class B. When the gross weight
exceeds 200 pounds or the container length and width is 18 x 24 inches or more and the weight exceeds 100 pounds, 3 x 4 inch skids (laid flat) shall be applied in accordance with the requirements
of the container specification. If not described in the container specification, the skids shall be
applied in a manner which will adequately support the item and facilitate the use of material handling
equipment. For level B packing, fiberboard boxes shall be weather resistant as specified in level A
and the containers shall be banded (see 5.2.1.1, 5.2.1.2, 5.2.2.1 and 5.2.2.2).

5.4.2.2 <u>Level A and B unitization</u>. For level A and B unitization, the fiberboard caps shall be
weather resistant and softwood pallets conforming to NN-P-71, type IV, size 2 shall be used (see
5.2.4.1 and 5.2.4.2).

5.4.3 <u>Navy procurements</u>. For Navy procurements the use of polystyrene loose fill material
(such as strips, strands and beads) is prohibited for packaging and packing applications.

6. NOTES

6.1 <u>Intended use</u>.

6.1.1 <u>Flux type</u>.

6.1.1.1 <u>Type R</u>. Type R is intended for use in the preparation of soldered joints for electrical
and electronic applications.

6.1.1.2 <u>Type RMA</u>. Type RMA provides a slightly more active fluxing action than type R.

6.1.1.3 <u>Type RA</u>. Type RA provides more active fluxing action than type RMA. It should be used only for soldering joints which are readily accessible so that the residues can be removed by cleaning agents and procedures specified in the governing document (drawing or specification) for the assembly in which the joints are used. Since the fumes and particulates given off during solder- ing may also be corrosive and contaminate the area surrounding the joint, this too must be suscep- tible to effective cleaning by the combination of materials and procedures to be used.

6.1.1.4 <u>Type AC</u>. Type AC is intended for use exclusive of that in electrical or electronic cir- cuits in the preparation of soldered connections for all common metals or alloys, other than alu- minum and magnesium and their alloys.

6.1.2 <u>Alloy compositions</u>.

6.1.2.1 <u>Sn96</u>. Composition Sn96 is a special-purpose solder with a higher joint strength than tin-lead solders. It is intended for use in the food-processing industry because of its nontoxic characteristic.

6.1.2.2 <u>Sn70</u>. Composition Sn70 is a special-purpose solder where high tin content is necessary. It is intended for soldering zinc and for coating metal, etc.

6.1.2.3 <u>Sn63</u>. Composition Sn63 is the tin-lead eutectic. It is used for soldering printed cir- cuits where temperature limitations are critical and in applications where an extremely short melting range is required.

6.1.2.4 <u>Sn62</u>. Composition Sn62 is a special-purpose solder widely used for soldering silver- coated ceramics.

6.1.2.5 <u>Sn60</u>. Composition Sn60 corresponds closely to the tin-lead eutectic (see 6.1.2.3) and has a short melting range. It is preferred for soldering electrical or electronic connections and for coating metals.

6.1.2.6 <u>Sn50</u>. Composition Sn50 is the customary "half-and-half" solder, intended for use in bit soldering and sweated joints in plain, tinned, or galvanized iron or steel, copper and copper alloys, etc. It is for use with soldered fittings in copper water tubing.

6.1.2.7 <u>Sn40</u>. Composition Sn40 can be used for the same purposes as composition Sn50, but is not as workable in bit soldering or sweating as is composition Sn50. Composition Sn40 is frequently used for dip soldering and as a wiping solder.

6.1.2.8 <u>Sn35 (see 6.1.2.16)</u>. Composition Sn35 is the customary wiping or plumber's solder. Higher antimony content in wiping solders promotes fine grain size and greater strength.

6.1.2.9 <u>Sn30 (see 6.1.2.15)</u>. Composition Sn30 is used as an automobile-body solder for filling dents and seams.

6.1.2.10 <u>Sn20 (see 6.1.2.14)</u>. Composition Sn20 is also widely used as an automobile-body solder for filling dents and seams, and for general purposes such as protective coatings on steel sheet where a high-tin-content alloy is not required.

6.1.2.11 <u>Sn10</u>. Composition Sn10 is a solder with a high melting point and is intended for use in making electrical or electronic connections in high-ambient-operating-temperature equipments. Its uses are similar to those of the silver solders (see 6.1.2.17 to 6.1.2.19, inclusive) but it provides better resistance to corrosion under humid conditions.

6.1.2.12 <u>Sn5</u>. Composition Sn5 is a solder used for applications similar to composition Sn10, but for applications having a slightly higher operating range.

QQ-S-571E

6.1.2.13 Sb5. Composition Sb5 is used for electrical or electronic connections subjected to peak temperatures of approximately 240°C, and for sweating copper-tube joints in refrigeration equipment.

6.1.2.14 Pb80. Composition Pb80 is an automobile-body solder similar to composition Sn20, but with a lower antimony content.

6.1.2.15 Pb70. Composition Pb70 is an automobile-body solder similar to composition Sn30, but with a lower antimony content.

6.1.2.16 Pb65. Composition Pb65 is a plumber's solder similar to composition Sn35, but with a lower antimony content.

6.1.2.17 Ag1.5. Composition Ag1.5 is used interchangeably with composition Ag2.5, but has a better shelf life and does not develop a black surface deposit when stored under humid environmental conditions.

6.1.2.18 Ag2.5. Composition Ag2.5 cannot be satisfactorily applied on black, uncoated, steel sheet by any of the current soldering techniques. This composition requires higher soldering temperatures and the use of a flux having a zinc-chloride base to produce a good joint on untinned surfaces. A rosin flux is unsatisfactory for soldering untinned copper, or brass, or steel with this solder. This composition is susceptible to corrosion under humid environmental conditions.

6.1.2.19 Ag5.5. Composition Ag5.5 will develop a shearing strength of 1,500 pounds per square inch at 177°C. When soldering hard-drawn brass or copper, the application temperature should not exceed 454°C. A typical application is on thermocouples for aircraft engines where relatively high operating temperatures will not affect strength of the solder. In other respects, precautions noted for composition Ag2.5 (see 6.1.2.18) also apply.

6.1.3 Soldering of zinc and cadmium. Inasmuch as zinc and cadmium appear to form inter-metallic alloys with the antimony in the solder, compositions Sn35, Sn30, and Sb5 should not be used for soldering zinc or cadmium, or zinc-coated or cadmium-coated iron or steel. These inter-metallic alloys have high melting points which inhibit the flow of the solder, resulting in brittle joints.

6.2 Ordering data. Purchasers should select the preferred options permitted herein and include the following information in procurement documents:

(a) Title, number, and date of this specification.
(b) Type designation (see 1.2.1).
(c) Detail requirements for special forms (see 1.2.1.2).
(d) Dimensions of ribbon and wire solder (see 3.7).
(e) Unit weight (see 3.8).
(f) Levels of preservation-packaging and packing required (see 5.1 and 5.2).
(g) Preservation-packaging and packing requirements for special forms (see 5.1.1.4.4, 5.2.1.4, and 5.2.2.4).
(h) Special marking, if required (see 5.3).

6.3 Commercially-available sizes of forms (see table I and 6.2). Approximate dimensions and weights are as indicated in 6.3.1 and 6.3.2. Other sizes may be available and can be procured under this specification.

6.3.1 Forms R (ribbon) and W (wire). Form R is available in widths up to 2 inches and in thickness from 0.003 inch to 0.092 inch; form W is available in the sizes indicated in table XIII. Forms R and W are furnished on spools or cards weighing 1, 5, 10, 25, 30, and 50 pounds.

QQ-S-571E

TABLE XIII.  Form W

| Wire diameter |
| --- |
| inches |
| 0.200 |
| 0.180 |
| 0.160 |
| 0.140 |
| 0.125 |
| 0.112 |
| 0.100 |
| 0.090 |
| 0.080 |
| 0.071 |
| 0.063 |
| 0.056 |
| 0.050 |
| 0.045 |
| 0.040 |
| 0.036 |
| 0.032 |

6.3.2  Forms B (bar) and I (ingot).  Forms B and I are available in the dimensions and weights indicated in table XIV.

TABLE XIV.  Forms B and I

| Form | Length | Width | Thickness | Weight |
| --- | --- | --- | --- | --- |
| | inches | inches | inches | pounds |
| Bar: | | | | |
| Top - - - - - - - | 13-1/2 | 3/4 | 3/8 | 1 |
| Bottom - - - - - | 13-1/2 | 5/8 | 3/8 | 1 |
| Ingot: | | | | |
| Top - - - - - - - | 5-1/2 | 2-1/2 | 1-1/2 | 5 |
| Bottom - - - - - | 4-1/2 | 1-1/2 | 1-1/2 | 5 |

6.4  Qualification.  With respect to products requiring qualification, awards will be made only for such products as have, prior to the time set for opening bids, been tested and approved for inclusion in the applicable Qualified Products List whether or not such products have actually been so listed by that date.  The attention of the suppliers is called to this requirement, and manufacturers are urged to arrange to have the products that they propose to offer to the Federal Government, tested for qualification, in order that they may be eligible to be awarded contracts or orders for the products covered by this specification.  The activity responsible for the Qualified Products List is the U.S. Army Electronics Command, Fort Monmouth, New Jersey 07703; however, information pertaining to qualification of products may be obtained from the Defense Electronics Supply Center (DESC-E), 1507 Wilmington Pike, Dayton, Ohio 45401.

6.4.1  Copies of the SD-6, "Provisions Governing Qualification", may be obtained upon application to Commanding Officer, Naval Publications and Forms Center, 5801 Tabor Avenue, Philadelphia, Pennsylvania  19120.

QQ-S-571E

6.5 **Transportation description.** Transportation description and minimum weights applicable to this commodity are:

Rail:

Solder, not otherwise indexed by name.
(Specify tin content.)
Carload minimum weight 40,000 pounds.

Motor:

Solder, not otherwise indexed.
(Specify tin content.)

Truckload minimum weight 36,000 pounds,
subject to Rule 115, National Motor Freight
Classification.

MILITARY CUSTODIANS:

Army - EL
Navy - EC
Air Force - 11

Review activities:

Army -
Navy - AS, EC, YD
Air Force -
DSA - IP

User activities:

Army -
Navy - MC, OS
Air Force - 80

Preparing activity:

Army - EL

Agent:

DSA - ES

(Project 3439-0195)

☆U.S. GOVERNMENT PRINTING OFFICE: 1981/707-968

QQ-S-571E
AMENDMENT 4
16 August 1986
SUPERSEDING
INTERIM AMENDMENT 3 (ER)
15 November 1985
SUPERSEDING
AMENDMENT 2
16 July 1975

FEDERAL SPECIFICATION

SOLDER; TIN ALLOY, TIN-LEAD ALLOY, AND LEAD ALLOY

This amendment, which forms a part of Federal Specification QQ-S-571E, dated May 5, 1972, is approved by the Commissioner, Federal Supply Service, General Services Administration, for the use of all Federal Agencies.

PAGE 4

• Paragraph 2.2: Add new paragraph as indicated:

•    2.2 Other publications. The following documents form a part of this specification to the extent specified herein. Unless a specific issue is identified, the issue in effect on the date of invitation for bids or request for proposal shall apply.

SD-6  Provisions Governing Qualification  (see 6.4.1).

• Paragraph 3.1: Delete text in its entirety and substitute:

•    3.1 Qualification. Solder furnished under this specification shall be products which are authorized by the qualifying activity for listing in the applicable qualified products list (QPL) at the time set for opening of bids (See 4.5 and 6.4).

Paragraph 3.2.2.2, line 4:  Delete "45,000" and substitute "50,000".

PAGE 6 AND 7

TABLE V:  In "Silver" column, replace all "---" with "0.015 max"; in "Cadmium, max" column, replace all "---" with "0.001".

PAGE 9

Paragraph 4.5.1, line 1:  Delete "inch diameter" and substitute "nominal thickness".

PAGE 11

• Paragraph 4.5.5:  Delete text in its entirety and substitute:

•    4.5.5 Extent of qualification. Qualification by testing of Sn60 flux-cored solder of a specific flux type and percentage shall be accepted as the basis for qualification by similarity of all other flux-cored solder alloys of the same flux type and for qualification of solid-core solder. Qualification by testing of one solid-core solder alloy shall also be accepted as the basis for qualification by similarity of all other solid-core solders. Qualification by testing of Sn60 solder paste of a specific flux type and percentage shall be accepted as the basis for qualification by similarity for all other solder paste alloys of the same flux type.

AMSC N/A                                                                    FSC 3439

QQ-S-571E
AMENDMENT 4

PAGE 11 (Cont'd)

* Paragraph 4.5.5.1: Add new paragraph as indicated:

*   4.5.5.1 Qualification_by_similarity. Requests for qualification of solder products by similarity shall be submitted in accordance with procedures for submittal of requests for qualification of products in SD-6. In addition to the information and certifications required by SD-6, applications for qualification by similarity shall identify the type designation for each solder for which qualification by similarity is requested and shall identify the notification letter for the product qualified by testing which is to serve as the basis for the qualification by similarity being requested.

PAGE 18

Paragraph 5.1.1.4.1: Delete first sentence and substitute: "Unless otherwise specified (See 6.2), ribbon and wire solder shall be wound on spools having sufficient strength and durability to maintain proper support."

PAGE 19

Paragraph 5.2.2.3: Add to end of paragraph: "Alternately, the box shall conform to PPP-B-636, class domestic. The box shall be closed in accordance with the appendix to PPP-B-636."

PAGE 21

Paragraph 6.1.1.3: Delete text in its entirety and substitute:

    6.1.1.3 Type_RA. (CAUTION) Type RA flux may contain materials which will promote corrosion and adversely affect electrical properties. It is mandatory that type RA flux residues be completely removed after soldering and that an appropriate test be used to determine the post cleaning absence of residual substances.

PAGE 23

* Paragraph 6.3.2: Delete text in its entirety and substitute:

*   6.3.2 Forms_B_(Bar)_and_I_(Ingot). Forms B and I solder products are available in but not limited to the nominal dimensions and weights indicated in Table XIV. Solder bars and ingots are formed in a variety of sizes and shapes. Bars are generally long and slender while ingots tend to be short and thick. The weight of bars and ingots will vary significantly from the nominal values listed in Table XIV. This results primarily from manufacturers using one set of molds for all solder alloys, from depth tolerances in pouring molten solder into the molds, and from differencies in the densities of solder alloys (due to the relative amount of lead in the various alloys).

* TABLE XIV: Add the following note at the bottom of the table:

*   Tolerances: Weight: +0.5/-0 lbs.;   Length: ± 20%:
              Width and Thickness: ± 33%.

The margins of this amendment are marked with an asterisk (*) to indicate where changes from the previous amendment were made. This was done as a convenience only and the government assumes no liability whatsoever for any inaccuracies in the notation. Bidders and contractors are cautioned to evaluate the requirements of this document based on the entire contents irrespective of the marginal notation and relationship to previous amendments.

2

QQ-S-571E
AMENDMENT 4

Military Custodians:

    Navy - EC
    Air Force - 20
    Army - ER
Review Activities:

    Air Force - 84, 99

User Activities:

Preparing Activity:

    Army - ER

Agent:

    DLA - GS

Civil Agency Coordinating
  Activity:

    GSA-FSS

(Project No. 3439-A602)

# Ternary Phase Diagram: Pb-Ag-Sn, Sn-Pb-Bi

# Phase Diagram
## Pb-Ag-Sn Lead-Silver-Tin*

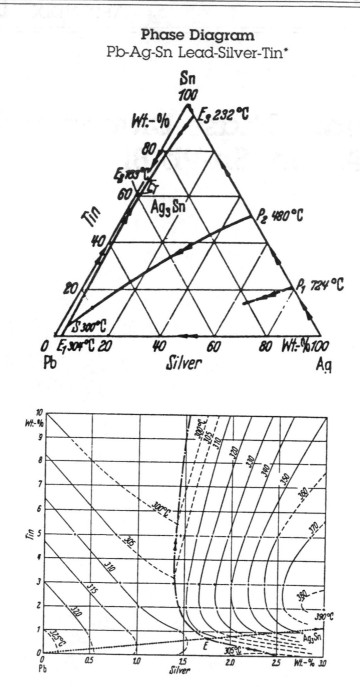

*After W. Hoffman, "Lead and Lead Alloys," Springer-Verlag, New York, 2nd Edition, 1970.

## Phase Diagram
### Sn-Pb-Bi Tin-Lead-Bismuth*

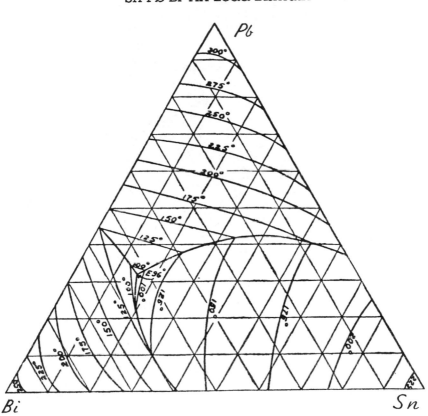

*After W. R. Lewis, Tin Research Institute, 4th Edition, 1961, England.

# Military Specification MIL-P-28809A: Printed Wiring Assemblies

MIL-P-28809A
5 October 1981
SUPERSEDING
MIL-P-28809
21 March 1975

MILITARY SPECIFICATION

PRINTED WIRING ASSEMBLIES

This specification is approved for use by all Depart-
ments and Agencies of the Department of Defense.

1. SCOPE

1.1 Scope. This specification covers conformally coated printed wiring assemblies
(circuit-card assemblies) consisting of rigid printed wiring boards on which
separately manufactured parts have been added (see 6.2).

1.2 Classification.

1.2.1 Types. Printed wiring assemblies shall be of the types shown in table I, as
specified (see 6.2).

TABLE I. Types.

| Type designator | Board type |
|---|---|
| 1 | Single-sided board |
| 2 | Double-sided board |
| 3 | Multilayer board |

2. APPLICABLE DOCUMENTS

2.1 Government documents. Unless otherwise specified, the following specifications and standards,
of the issue listed in that issue of the Department of Defense Index of Specifications and
Standards specified in the solicitation, form a part of this specification to the extent
specified herein.

SPECIFICATIONS

FEDERAL

| | | |
|---|---|---|
| QQ-S-571 | - | Solder; Tin Alloy; Lead-Tin Alloy; and Lead Alloy. |
| QQ-S-781 | - | Strapping, Steel, Flat and Seals. |
| QQ-W-343 | - | Wire, Electrical (Uninsulated). |
| PPP-B-566 | - | Boxes, Folding, Paperboard. |
| PPP-B-585 | - | Boxes, Wood, Wirebound. |
| PPP-B-601 | - | Boxes, Wood, Cleated-Plywood. |
| PPP-B-621 | - | Boxes, Wood, Nailed and Lock-Corner. |
| PPP-B-636 | - | Boxes, Shipping, Fiberboard. |
| PPP-B-676 | - | Boxes, Setup. |
| PPP-C-1842 | - | Cushioning Material, Plastic, Open Cell (For Packaging Applications). |
| PPP-T-60 | - | Tape, Packaging, Waterproof. |
| PPP-T-76 | - | Tape, Pressure-Sensitive Adhesive Paper, (For Carton Sealing). |

Beneficial comments (recommendations, additions, deletions) and any pertinent data
which may be of use in improving this document should be addressed to: Naval
Electronic Systems Command, ATTN: ELEX 8111, Washington, DC 20360, by using the
self-addressed Standardization Document.Improvement Proposal (DD Form 1426)
appearing at the end of this document or by letter.

MIL-P-28809A

MILITARY

| | | |
|---|---|---|
| MIL-P-116 | - | Preservation, Methods of. |
| MIL-P-13949 | - | Plastic Sheet, Laminated, Metal Clad (For Printed Wiring Boards), General Specification for. |
| MIL-F-14256 | - | Flux, Soldering, Liquid (Rosin Base). |
| MIL-I-46058 | - | Insulating Compound, Electrical (For Coating Printed Circuit Assemblies). |
| MIL-P-55110 | - | Printed Wiring Boards. |
| MIL-B-81705 | - | Barrier Materials, Flexible, Electrostatic-Free, Heat Sealable. |

STANDARDS

MILITARY

| | | |
|---|---|---|
| MIL-STD-105 | - | Sampling Procedures and Tables for Inspection by Attributes. |
| MIL-STD-129 | - | Marking for Shipment and Storage. |
| MIL-STD-202 | - | Test Methods for Electronic and Electrical Component Parts. |
| MIL-STD-275 | - | Printed Wiring for Electronic Equipment. |
| MIL-STD-794 | - | Parts and Equipment, Procedures for Packaging and Packing of. |
| MIL-STD-810 | - | Environmental Test Methods. |
| MIL-STD-1188 | - | Commercial Packaging of Supplies and Equipment. |
| MIL-STD-45662 | - | Calibration Systems Requirements. |

(Copies of specifications, standards, handbooks, drawings, and publications required by manufacturers in connection with specific acquisition functions should be obtained from the contracting activity or as directed by the contracting officer.)

2.2  Other Government documents, drawings, and publications. The following other Government documents, drawings, and publications form a part of this specification to the extent specified herein.

INSTITUTE FOR INTERCONNECTING AND PACKAGING ELECTRONIC CIRCUITS

| | | |
|---|---|---|
| ANSI/IPC-T-50 | - | Terms and Definitions. |
| IPC-S-815 | - | General Requirements for Soldering of Electrical Connections and Printed Wiring Assemblies. |

(Application for copies should be addressed to the Institute for Interconnecting and Packaging Electronic Circuits, 3451 Church Rd., Evanston, Illinois  60203.)

BUREAU OF MEDICINE AND SURGERY (BUMED)

BUMED INST 6270.3 - Personnel Exposure Limit Values for Health Hazardous Air Contaminants.

(Application for copies should be addressed to the Chief, Bureau of Medicine and Surgery, Department of the Navy, Washington, DC  20372.)

3.  REQUIREMENTS

3.1  General requirements. Printed wiring assemblies furnished under this specification shall be a product which meets the requirements of this specification and the applicable assembly drawing (see 6.1 and 6.2). The design features of the printed wiring assemblies shall be in accordance with MIL-STD-275 and the approved assembly drawing.

3.1.1  Conflict. In the event of any conflict between the approved assembly drawing and the requirements of this specification, the provisions of the assembly drawing(s) shall govern. Changes to the approved assembly drawing(s) shall be processed in accordance with the requirements of MIL-STD-275.

MIL-P-28809A

3.2 <u>First article</u>. Printed wiring assemblies furnished under this specification shall be products which have passed the first article inspection specified in 4.4. Alternatives provided in this specification do not constitute authority to produce production units using different materials or processes other than those used on the first article sample.

3.3 <u>Terms and definitions</u>. Terms and definitions shall be in accordance with th· Appendix, paragraph 30. In the event of conflict, ANSI/IPC-T-50 shall govern.

3.4 <u>Materials</u>. Materials furnished as part of the printed wiring assembly shall be as specified herein and in the applicable assembly drawing. Such materials shal be non-toxic (threshold limit values for toxicity per BUMED Instruction 6270.3) anc shall meet the certification requirements of 4.3. When data is not available to support such certification, it will be necessary to generate data showing complianc with these requirements. For materials not listed in BUMED Instruction 6270.3, it may be necessary to establish acceptable limits of toxicity. The responsibility fc inspection of material quality shall be as specified in 4.1.

3.4.1 <u>Materials and processes compatibility</u>. It shall be the responsibility of the manufacturer to select those processes and materials which are compatible with one another and which best suit the end product desired by the contract.

3.4.2 <u>Printed wiring boards</u>. Printed wiring assemblies shall use rigid printed wiring boards in accordance with MIL-P-55110 and MIL-STD-275.

3.4.3 <u>Component leads and wires</u>. Incoming inspection should be performed on al component leads and wires per the appropriate component/wire specification(s). Pr to component assembly all component leads and wires shall show evidence of good solderability in accordance with 4.3.1 or shall be tinned. Component leads and wi shall be appropriately stored prior to assembly. Leads and wires that were tested for solderability shall be assembled within six months of testing; leads and wires that have been tinned shall be assembled within one year of tinning. Component le and wires exceeding the time requirements shown above shall be retested or retinne prior to assembly.

3.4.3.1 <u>Component parts and wires</u>. Component parts and wires shall be as specified on the approved assembly drawing(s) and associated parts list(s).

3.4.3.2 <u>Gold-plated leads and wires</u>. All gold-plated leads and wires that are hand soldered or planar mounted shall be tinned.

3.4.4 <u>Solder</u>. The solder used shall be in accordance with composition Sn 60, S 62, or Sn 63 of QQ-S-571. For wave or dip soldering, use bar solder, form B. For hand soldering, use solder wire form W, either solid metal, type S, or with a cor‹ flux of either type R, RMA, or RA of QQ-S-571, with the exception that RA flux sha not be used on stranded wire. (With special contract approval type, RA flux may be used on Army ERADCOM, CORADCOM, and AVRADCOM contracts).

3.4.5 <u>Soldering flux</u>. Soldering flux shall be a liquid flux conforming to MIL-F-14256, type R, RMA, or RA, with the exception that RA flux shall not be use stranded wire. (With special contract approval, type RA flux may be used on Army ERADCOM, CORADCOM, and AVRADCOM contracts.)

3.4.6 <u>Conformal coating</u>. Conformal coating material shall be as specified on approved assembly drawing and shall be in accordance with MIL-I-46058.

3.4.7 <u>Buffer material (see 3.6.8)</u>. The buffer material shall be a thin, pliar material such as polyvinylidene fluoride, polyethlene terephthalate, or silicone rubber, and be nonreactive with the conformal coating material and all parts to w it comes in contact. The buffer material shall be fungus and flame resistant, ar clear or transparent so markings on the components are visible (see 3.8.9 and 3.c

MIL-P-28809A

3.5  Design principles and production criteria.  Part mounting and attachment shall
_e in accordance with this specification, and the approved assembly drawing (see
:.1), or other documents of the assembly manufacturer referenced on the approved
:ssembly drawing.  The requirements of 3.5.1 through 3.5.2.4 shall apply to the
:ounting of parts on the printed wiring assembly.

3.5.1  Part mounting.  Each part shall be mounted in the location specified on the
:pproved assembly drawing.  All parts shall be correctly located, oriented, mounted,
:nd attached.  All parts shall be correctly soldered (see 3.5.2 and 3.5.3).

3.5.1.1  Location.  Parts shall be mounted so as to avoid the occurrence of
moisture traps.

3.5.1.2  Conductive areas.  The part shall be mounted so that subsequent conformal
_oating will cover the conductive area under the part except where thermal
dissipation or electrical conduction is required.  When conformal coating will not
:over conductive areas under the part, the conductive areas shall be insulated and
protected against moisture entrapment by applying, and curing a resin coating,
:aminating low flow prepreg material in accordance with MIL-P-13949, or by a solder
mask coating over the area prior to mounting of the parts.

3.5.1.3  Spacing.  Parts shall be mounted and spaced so that any part can be
~emoved from the board without having to remove any other part, unless otherwise
specified on the approved assembly drawing.

3.5.1.3.1  Electrical spacing.  The minimum spacing between component leads, wires,
:onductor patterns, and other conductive material (such as conductive markings or
nounted hardware) shall be in accordance with table II.

TABLE II.  Electrical spacing.

| Voltage between conductors (dc or ac peak) | Minimum spacing |
|---|---|
| Volts | |
| 0-15 | 0.005 inch(0.13 mm) |
| 16-30 | 0.010 inch(0.25 mm) |
| 31-50 | 0.015 inch(0.38 mm) |
| 51-100 | 0.020 inch(0.51 mm) |
| 101-300 | 0.030 inch(0.76 mm) |
| 301-500 | 0.060 inch(1.52 mm) |
| Greater than 500 | 0.00012 inch(.00305 mm) (per volt) |

3.5.1.4  Alinement of part leads.  Axial part leads shall coincide to the
centerline through their respective land areas whenever possible.  The part lead may
overhang the land area only if the resultant electrical spacing between adjacent
conductors meets the requirements specified in table II.

3.5.1.5  Stress relief bends.  Parts shall be mounted or provided with stress
relief bends in such a manner that the leads cannot overstress the part-lead
interface when subjected to the conditions of 3.8.3 through 3.8.6.  The straight lead
length adjacent to the component body shall be in accordance with figure 1.  NOTE:
For solder in the stress relief band of axial-leaded parts see 3.5.1.7.

3.5.1.6  Lead bend radius.  Minimum bend radius for leads shall be in accordance
with figure 1.

4

MIL-P-28809A

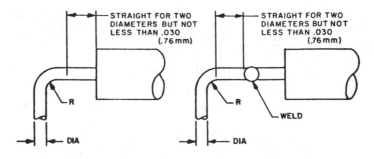

**A. STANDARD BEND**                                    **B. WELDED BEND**

| Lead diameter in inches | Minimum radius (R) inch |
|---|---|
| ≤.027(.69 mm) | 1 diameter |
| ≥.028(.71 mm) ≤ .047(1.19 mm) | 1.5 diameter |
| ≥.048(1.22 mm) | 2 diameters |

FIGURE 1.  Lead bend.

3.5.1.7  Axial-leaded parts.  Axial-leaded parts shall be mounted as specified on the approved assembly drawing or mounted so that a portion of the body is as close to the printed wiring board as possible.  The leads shall be shaped in accordance with 3.5.1.5 and 3.5.1.6.  This does not apply to parts mounted on standoff terminals (see 3.5.2.1).  Solder which extends into the lead bend radius of axial-leaded parts is acceptable only into the lead bend radius on one lead of the part (the lead which is closest to the board).  Solder shall not extend beyond the land (see figure 2).

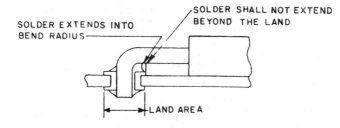

FIGURE 2.  Solder in bend radius.

3.5.1.7.1  Perpendicular mounting.  When specified on the approved assembly drawing, axial-leaded parts weighing less than 0.50 ounce shall be mounted perpendicular to the board.  The part shall be mounted to provide a minimum of 0.015 inch between the end of the part body and the mounting surface.  The end of the part is defined to include any coating meniscus, solder seal, solder or weld bead, or any other extension.  The maximum vertical misalinement of the part's vertical axis shall be 15 degress in any direction from a line perpendicular to the mounting surface. The maximum allowed vertical height of the part from the board mounting surface shall be 0.55 inch (see figure 3).

5

MIL-P-28809A

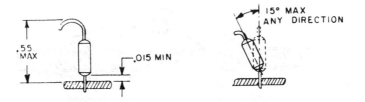

FIGURE 3.  Perpendicular part mounting.

3.5.1.8  Nonaxial-leaded parts.  Non-axial leaded parts shall be mounted with the surface from which the lead projects (end of the part) a minimum of 0.010 inch above the board mounting surface.

3.5.1.8.1  Multiple-leaded components.  Multiple-leaded components (components with three or more leads), except multiple-leaded components mounted to thermal planes or heat sinks, shall be mounted in such a manner that spacing is provided under the body of the part to facilitate cleaning.

3.5.1.9  Heat dissipating parts.  All parts dissipating 1 watt or more shall be mounted as specified on the approved assembly drawing or in such a manner that the body of the part is not in direct contact with the printed wiring board unless either a clamp or thermal groundplane, or both, is used which will dissipate sufficient heat so that the maximum allowable operating temperature of the printed wiring board is not exceeded.

3.5.1.10  Jumper wires.  Jumper wires shall be short as practical and shall not be applied over or under other parts.  Jumper wires less than 0.5 inch in length whose path does not pass over conductive areas and does not violate the spacing requirements of 3.5.1.3.1 may be uninsulated.

3.5.1.11  Interfacial connections.

3.5.1.11.1  Clinched wires.  Interfacial connections on type 2 boards may be made by the use of uninsulated solid wire in accordance with QQ-W-343, type S, coated, extending through a hole and clinched.  The wire shall make contact with the conductor pattern on each side of the printed wiring assembly before soldering (see figure 4) and the end shall not extend beyond the edge of its land area or its electrically connected conductor pattern in violation of the minimum spacing requirements.  Part lead wires do not qualify for interfacial connections.  The top and bottom portions of the wire need not be alined in the same vertical plane.

3.5.1.11.2  Plated-through holes.  Plated-through holes used for interfacial connections, internal or interlayer connections shall not be used for mounting of eyelets, standoff terminals, rivets, or other devices which put the plated-through hole in compression.  Interfacial connections on type 2 boards may also be made by the use of plated-through holes.  Interfacial connections on type 3 boards shall be made only by the use of plated-through holes.

3.5.2  Part attachment.  Part attachment shall be in accordance with the requirements of IPC-S-815.

MIL-P-28809A

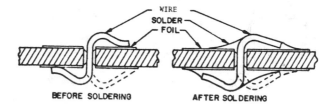

FIGURE 4. Clinched wire (interfacial connection) (direction of clinch optional) (Type 2 boards only).


3.5.2.1 Standoff terminals. Part attachment to standoff terminals shall be in accordance with the requirements of IPC-S-815.

3.5.2.2 Clinched leads. Clinched leads shall be in accordance with the requirements of IPC-S-815.

3.5.2.3 Straight-through, partially clinched or swaged leads. Straight-through, partially clinched or swaged leads shall be in accordance with the requirements of IPC-S-815.

3.5.2.4 Ribbon leads. Flat-wire ribbon leads shall be attached in accordance with IPC-S-815.

3.5.3 Soldering. Soldering shall be in accordance with IPC-S-815, class III. Solder shall not be used on surfaces specified to be free of solder. Solder and flux shall be in accordance with 3.4.4 and 3.4.5.

3.5.3.1 Metal. All metal surfaces shall be free of corrosion and contamination. All printed conductors shall be firmly bonded to the printed wiring board.

3.5.3.2 Solder plugs applicability.

3.5.3.2.1 When applicable. Solder plugs are required in:

        a.  All electrically functional plated-through holes with a lead.
        b.  98 percent of all plated-through holes (electrically functional or not, without a lead) that are subjected to wave or dip soldering.

3.5.3.2.2 When not applicable. Solder plugs are not required in:

        a.  Unsupported holes with a lead.
        b.  Non-functional plated-through holes (hand soldered).
        c.  Any electrically functional plated-through hole (without a lead) not subjected to wave or dip soldering.

3.5.3.3 Post soldering cleaning. Printed wiring assemblies shall be cleaned within 1 hour after completion of soldering using solvents or combinations of solvents or other solutions which will remove polar and nonpolar contaminants (see appendix). After cleaning, printed wiring assemblies shall not be contaminated by handling or environment prior to conformal coating. After cleaning there shall be no visual evidence of flux residue or other contamination. "Other contamination" includes particles of foreign matter which may result in insulation breakdown or change in electrical characteristics or degradation of mechanical integrity (e.g., improper bonding of conformal coating). This solder cleaning requirement shall also apply after rework.

MIL-P-28809A

3.5.3.4 <u>Measling and crazing</u>. After soldering and cleaning processes have been completed, measling or crazing, or both, on the printed wiring assembly shall not bridge more than 50 percent of the distance between electrical conductors and/or not exceed 3 percent of the total printed wiring board surface area on one side. A separate measurement and determination shall be made for each side of the printed wiring assembly (see 4.8.1). If measling or crazing, or both, of any extent occurs on more than 5 percent of a printed wiring assembly lot sample examined (see 4.6.1.2.2), the lot shall be rejected and corrective action is required.

3.5.3.5 <u>Delamination, blistering or softening</u>. There shall be no delamination, blistering, or softening of the plastic materials.

3.6 <u>Conformal coating</u>. Conformally coated printed wiring assemblies furnished under this specification shall meet the requirements of this specification before and after repair (see 3.7). The coated assemblies shall have no blisters, cracking, crazing, peeling, wrinkles, mealing, or evidence of reversion or corrosion. A pin-hole or bubble and/or a combination of pinhole(s) and bubble(s) may bridge up to 50 percent of the distance between conductors, provided that the minimum dielectric spacing requirement is not violated.

3.6.1 <u>Coating area</u>. Printed wiring assemblies, except those using a board material of polytetrafluoroethylene shall be conformally coated with a coating material that conforms to 3.4.6. The coating shall be applied to both sides of the cleaned printed wiring assembly.

3.6.2 <u>Adjustable components</u>. Printed wiring assemblies having adjustable components shall not have the adjustable portion covered with coating, unless otherwise specified on the approved assembly drawing.

3.6.3 <u>Mating surfaces</u>. Electrical and mechanical mating surfaces, such as probe points, screw threads, bearing surfaces, etc., shall not be coated.

3.6.4 <u>Masking</u>. The masking material used to prevent coating in unwanted areas shall have no deleterious effects on the printed wiring assembly.

3.6.5 <u>Compatibility</u>. The conformal coating shall be compatible with all parts and material of the printed wiring assembly.

3.6.6 <u>Thickness</u>. The thickness of the conformal coating shall be as follows for the type specified, when measured on a flat unencumbered surface (see MIL-I-46058).

   a.  Types ER, UR, and AR:  0.003 ±0.002 inch.
   b.  Type SR:  0.005 +0.003 inch.
   c.  Type XY:  0.0005 to 0.002 inch.

3.6.7 <u>Electrical performance</u>. Printed wiring assemblies shall be constructed, adequately masked, or otherwise protected in such a manner that application of conformal coating does not degrade the electrical performance of the assembly. Electrical testing shall be accomplished (see 6.1).

3.6.8 <u>Buffer material (see 3.4.7)</u>. Buffer material shall be as required on the approved assembly drawing.

3.6.9 <u>Surfaces to be free of coating</u>. Surfaces specified on the approved assembly drawing to be free of conformal coating shall be suitably masked and protected from coating, coating residues, and masking residues.

3.6.10 <u>Cleanliness (see 4.8.2)</u>. When tested in accordance with 4.8.2 and sampled in accordance with 4.5.2, printed wiring assemblies shall have no evidence of flux residues, ionic and other contaminants before applying the coating. Cleaning compounds shall have no deleterious effects on any part of the printed wiring assembly. In the case of printed wiring boards requiring permanent solder-mask, conformal or other coating, the uncoated boards shall be free of ionic contaminants or flux residue (see 6.4) prior to the application of conformal coating. Prior to the application of any coating, the Government reserves the right to require confirmation that the uncoated printed wiring assemblies were inspected for cleanliness. The cleanliness test shall be accomplished (see 6.1).

8

MIL-P-28809A

3.6.10.1 <u>Resistivity of solvent extract (see 6.6)</u>. The test solution used to wash uncoated printed wiring assemblies shall have a resistivity not less than 2,000,000 ohm - centimeter (or equivalent) when tested in accordance with 4.8.2 as required by in-process inspection of 4.5  The equivalent test methods and factors specified in 6.6.1 may be used in lieu of the method specified in 4.8.2  Other equivalent test methods not specified in 6.6.1 may be used in lieu of 4.8.2 only when specifically approved by the government procuring activity.  Such approval will be determined on the basis that the alternate method is demonstrated to have equal or better sensitivity, and employs solvents with the ability to dissolve flux residue as does the alcohol-water solution specified in 4.8.2.

3.7 <u>Rework, Repair and Modification</u>.  Rework, repair and modification shall be accomplished as described in the following paragraphs.  Assemblies reworked, repaired or modified by these methods shall be processed as normal material.

3.7.1 <u>Rework</u>.  The rework of defective solder connections and replacement of defective components is permissible (see Appendix 30.1.11).  Defective solder plugs in plated through holes which do not contain a component lead, and within the 2 percent limit of 3.5.3.2.1.b, need not be reworked.  After rework, the assembly shall meet the requirements of 3.8.

3.7.2 <u>Repairs</u>.  Standard repairs (see 30.1.12) described by appendix paragraph 50.2 and authorized by paragraph 10.2.1 may be performed within the limitations set forth.  Such repairs shall be documented as specified in the appendix.  Proposed methods of repair other than the standard repairs shall be submitted to the Government procuring activity for approval.  Approval for such repairs are applicable only to the contract under which the approval was granted.

3.7.3 <u>Modifications</u>.  Modifications (see 30.1.9) may be made to prototype or production assemblies in accordance with the appendix.  Modifications require written authorization (see 10.2.2) of the Government procuring activity and shall be documented as specified in the appendix (see 40.4).

3.8 <u>Printed wiring assembly performance requirements</u>.  Upon completion of final assembly, the printed wiring assemblies shall meet all visual, electrical, and all other operational requirements in accordance with the requirements of the approved assembly drawing(s) and test specifications.

3.8.1 <u>Bow and twist</u>.  When tested as specified in 4.8.3, the maximum allowable bow and twist shall be 1.5 percent, unless otherwise specified on the approved assembly drawing.

3.8.2 <u>Electrical parameters</u>.  When tested as specified in 4.8.4, printed wiring assemblies shall function as specified on the approved assembly drawing.

3.8.3 <u>Vibration (when specified)</u>.  When tested as specified in 4.8.5, printed wiring assemblies shall be capable of continuous operation as specified (see 3.8), and there shall be no evidence of physical damage.

3.8.4 <u>Shock (when specified)</u>.  When tested as specified in 4.8.6, printed wiring assemblies shall be capable of continuous operation as specified (see 3.8), and there shall be no evidence of physical damage.

3.8.5 <u>Thermal shock</u>.  After testing as specified in 4.8.7, printed wiring assemblies shall be capable of operation as specified (see 3.8), and there shall be no evidence of physical damage.

3.8.6 <u>Temperature-altitude (when specified)</u>.  When tested as specified in 4.8.8, printed wiring assemblies shall be capable of continuous operation as specified (see 3.8), and there shall be no evidence of physical damage.

3.8.7 <u>Humidity</u>.  When tested as specified in 4.8.9, there shall be  no evidence of corrosion on any part of the printed wiring assembly.  No crazing or measling (in excess of that allowed in 3.5.3.4), blistering, cracking, delamination, embrittlement, mealing, or softening shall become evident in the conformal coating or other constituent parts and materials used in the assembly.  The electrical performance of the assembly shall not be degraded (see 3.8).

9

MIL-P-28809A

3.8.8 Salt fog (when specified). When tested as specified in 4.8.10, there shall be no evidence of corrosion of the printed wiring assemblies.

3.8.9 Fungus (when specified). When tested as specified in 4.8.11, the materials or combination of materials used in the production of printed wiring assemblies shall not serve as nutrients to fungi.

3.9 Marking. Printed wiring assemblies shall be marked as specified on the approved assembly drawing(s). All assemblies shall be serialized for traceability. The board may be marked at the discretion of the contractor for use as manufacturing aids. covering such marking by the components at assembly shall not be cause for rejection.

3.10 Workmanship. Printed wiring assemblies shall be clean and show no evidence of dirt, foreign matter, oil, fingerprints, corrosion, salts, flux residues, and contaminants. The completed printed wiring assembly shall also be free of defects as defined in 4.8.1.1.

4. QUALITY ASSURANCE PROVISIONS

4.1 Responsibility for inspection. Unless otherwise specified in the contract or purchase order, the contractor is responsible for the performance of all inspection requirements specified herein. Except as otherwise specified in the contract or purchase order, the contractor may use his own or any other facilities acceptable to the Government procuring activity, which are suitable for the performance of the inspection requirements specified herein. The Government procuring activity reserves the right to perform any of the inspections set forth in the specification, where such inspections are deemed necessary to assure conformance to prescribed requirements.

4.1.1 Test equipment and inspection facilities. Test and measuring equipment and inspection facilities of sufficient accuracy, quality and quantity to permit performance of the required inspection shall be established and maintained by the contractor. The establishment and maintenance of a calibration system to control the accuracy of the measuring and test equipment shall be in accordance with MIL-STD-45662.

4.1.2 Inspection conditions. Unless otherwise specified herein, all inspections shall be performed in accordance with the test conditions specified in the "General Requirements" of MIL-STD-202 and MIL-STD-810, and acceptability inspection of IPC-S-815, as applicable.

4.2 Classification of inspections. The inspections specified herein are classified as follows:

    a. Incoming component and materials inspection (see 4.3).
    b. First article inspection (see 4.4).
    c. In-process inspection (see 4.5).
    d. Quality conformance inspection (see 4.6).

4.3 Incoming component and materials inspection. Inspection shall, as a minimum, consist of certification supported by verifying data that the components and materials listed in table III, used in fabricating the printed wiring assemblies, are in accordance with the applicable referenced documents, specifications, and requirements prior to such fabrication. Additional materials inspection shall be as specified in applicable engineering drawing(s), as specified in 3.4.3, 3.4.3.1, and 4.3.1.

4.3.1 Solderability of component leads and wires. All component leads and wires of all electronic and electrical components shall be tinned or shall be inspected in accordance with 4.3.1.1 and shall meet the solderability requirements of MIL-STD-202, test method 208 (see table III).

MIL-P-28809A

4.3.1.1 Sampling plan for solderability test (see 4.3.1). Lot sampling shall be
in accordance with MIL-STD-105, special inspection level S-4, with an AQL of 2.5.
Samples for solderability test may be selected from components that failed to meet
incoming electrical testing (see 3.8).

TABLE III. Component and materials inspection.

| Component and materials | Requirement paragraph | Applicable specification |
|---|---|---|
| Printed wiring board | 3.4.2 | Master drawing |
| Component lead and wire solderability | 3.4.3 | MIL-STD-202, Test method 208 |
| Components parts and wire | 3.4.3.1 | As specified on assembly drawings |
| Solder | 3.4.4 | QQ-S-571 |
| Soldering flux | 3.4.5 | MIL-F-14256 |
| Conformal coating | 3.4.6 | Assembly drawing MIL-I-46058 |
| Buffer material | 3.4.7 | Assembly drawing |

4.4 First article inspection. First article inspection shall be performed by the
contractor, at a location acceptable to the government procuring activity. First
article inspection shall be performed on sample printed wiring assemblies which have
been produced with material, equipment, processes, and procedures which shall be used
in production. Production is defined as one or more printed wiring assemblies
delivered on a contract. First article inspection is divided into two categories,
Design and Type, as defined in paragraph 4.4.1.1 and 4.4.1.2. Minor modifications to
design, production processes or techniques do not necessarily require complete
repetition of all first article design inspections. The extent of retest
necessitated by such changes shall be determined by the contractor and approved by
the government procuring activity. Where two or more printed wiring assemblies are
electrically direct wired (joined without the use of connectors) and mechanically
packaged together to form a replaceable functional assembly, the electrical and
environmental tests shall be performed at the assembly level. This level is normally
identified as the lowest replaceable unit from the standpoint of equipment
maintenance.

First article approval is valid only on those contracts or purchase orders so
designated by the Government procuring activity concerned. Failure to pass the tests
defined in table IV per the sampling procedure shown in 4.4.1.1 and 4.4.1.2 shall
necessitate corrective action to all assemblies that are to be delivered under the
contract. Any corrective action requiring changes to the assembly procedure, design,
process, etc. shall be reflected in the printed wiring assembly drawing and must be
approved by the Government procuring activity.

4.4.1 Inspection routine. The first article inspection shall consist of the tests
defined in table IV and on the approved assembly drawing, contract, or purchase order.

MIL-P-22809A

TABLE IV.  First article inspection.

| Test or Inspection | Requirement paragraph | Method paragraph |
|---|---|---|
| Components and materials | 3.4 and see table I | 4.3 |
| Visual and mechanical examination  Bow and twist | 3.5 to 3.6.10 (inclusive) and 3.8.1 | 4.8.1 to 4.8.3 |
| Electrical parameters | 3.8.2 | 4.8.4 |
| Vibration 1/ | 3.8.3 | 4.8.5 |
| Shock 1/ | 3.8.4 | 4.8.6 |
| Thermal shock | 3.8.5 | 4.8.7 |
| Temperature-altitude 1/ | 3.8.6 | 4.8.8 |
| Humidity | 3.8.7 | 4.8.9 |
| Salt fog 1/ | 3.8.8 | 4.8.10 |
| Fungus (separate sample) 1/ | 3.8.9 | 4.8.11 |

1/  When specified, the tests in table IV must be conducted on the completed printed wiring assemblies. Additionally, the contractor may wish to conduct certain of these tests at earlier stages of fabrication to ensure progress toward an acceptable assembly.

4.4.1.1  First article design inspection.  One printed wiring assembly of each new assembly design shall be subjected to the first article inspection in accordance with table IV.  Significant changes to existing designs may require additional design testing to assure design integrity is maintained.  The contractor may wish to conduct certain of these tests at earlier stages of fabrication to ensure progress toward an acceptable assembly.

4.4.1.2  First article type inspection.  Subsequent procurements of printed wiring assemblies which have previously passed first article design inspection shall be submitted to first article type inspection.  Printed wiring assembly types are defined in 1.2.1.  When type testing is to be performed, one printed wiring assembly of each type, a representative of the most complex assembly of each type in a single contract shall be selected by the contractor, and agreed to by the Government representative.  The assembly selected for type testing shall be the first part produced by the production process.  This type representative shall be subjected to the first article inspection of table IV.

4.4.2  Failures.  When one or more samples fail to pass the first article inspection (see table IV) this shall be cause for refusal to grant first article approval.

4.4.3  Disposition of samples.  Unless otherwise specified by the Government procuring activity, sample assemblies, which have been subjected to and have passed first article inspection, shall be disposed of in accordance with the manufacturer's disposal procedures.

4.4.4  First article approval.  Approval of the first article in no way relieves the contractor of responsibility for complying with all requirements of the specifications, applicable assembly drawings, and all other terms and conditions of the contract, nor shall an approved first article be construed as altering or taking precedence over any of these requirements.

12

MIL-P-28809A

4.5 **In-process inspection.** In-process inspection shall consist of the examinations of 4.5.1 and the test of 4.5.2 as shown in table V.

4.5.1 **Soldered connections inspections.** Each soldered connection on each printed wiring assembly shall be visually inspected to the requirements of 3.5.3, using an optical apparatus or aid which provides a minimum magnification equivalent to a lens of 3 diopters (approximately 3/4X). Referee inspections may be accomplished at a magnification of 10x. Defective solder connections shall be reworked per 3.7.1 prior to further processing. Assemblies which do not meet the requirements of 3.5.3.2.1b shall be rejected.

4.5.2 **Inspection lot for cleanliness verification.** An inspection lot for cleanliness verification shall consist of all printed wiring assemblies processed through the cleaning process during a single shift. Immediately after cleaning and prior to conformal coating, five printed wiring assemblies per production shift shall be selected and subjected to the test of table V.

4.5.3 **Failures.** If one or more of the five printed wiring assemblies fail to meet the cleanliness and resistivity of solvent extract test of table V, the lot shall be rejected (see 4.5.4).

4.5.4 **Rejected lots.** When a lot is rejected as a result of a failure to pass the test specified in table V, the manufacturer shall withdraw the lot, take corrective action in connection with the cleaning materials and procedures, reclean the lot, and resubmit the lot to the test of table V. Such lots shall be separated from new lots, and shall be clearly identified as reinspected lots.

TABLE V.    In-process inspection.

| Test or inspection | Requirement paragraph | Method paragraph | Sampling plan |
|---|---|---|---|
| Cleanliness and resistivity of solvent extract | 3.6.10 | 4.8.2 | 4.5.2 |
| Visual of soldered connections | 3.5.3 | 4.5.1 | 100% |

4.6 **Quality conformance inspection.**

4.6.1 **Inspection of printed wiring assemblies for delivery.** Inspection of printed wiring assemblies for delivery shall consist of group A. Except as specified in 4.6.1.3.5, delivery of printed wiring assemblies which have passed the group A inspection shall not be delayed pending the results of the group B inspection.

4.6.1.2 **Group A inspection.** Group A inspection shall consist of the inspections specified in table VI.

4.6.1.2.1 **Group A inspection lot.** An inspection lot for group A inspection shall consist of all printed wiring assemblies on a specific contract fabricated from the same design, using the same processing procedures, and produced under essentially the same conditions within a maximum period of 90 days and offered for inspection at one time.

4.6.1.2.2 **Sampling plan.** Statistical sampling and inspection shall be in accordance with MIL-STD-105 for general inspection level II. The acceptable quality level (AQL) shall be as specified in table VI. Major and minor defects shall be as defined in MIL-STD-105 and 4.8.1.1.

TABLE VI.    Group A inspection.

| Test or inspection | Requirement paragraph | Method paragraph | AQL (percent defective) | |
|---|---|---|---|---|
| | | | Major | Minor |
| Visual and mechanical | 3.5 to 3.6.10 incl, 3.9, and 3.10 | 4.8.1 4.8.1.1 | 1% | 4% |
| Electrical parameters | 3.8.2 | 4.8.4 | 100% | test |

13

MIL-P-28809A

4.6.1.2.3 <u>Rejected lots</u>. If an inspection lot (see 4.6.1.2.1) is rejected, the manufacturer shall withdraw the lot and take corrective action, or screen out the defective units and reinspect. Such lots shall be separate from new lots, and shall be clearly identified as reinspected lots.

4.6.1.2.4 <u>Disposition of sample units</u>. Samples subjected to group A inspection may be delivered with the order, if the inspection lot passes.

4.6.1.3 <u>Group B inspection</u>. Group B inspection shall consist of the tests specified in table VII, in the order shown and other such tests specified on the approved assembly drawing.

TABLE VII.  <u>Group B inspection</u>.

| Test or inspection | Requirement paragraph | Method paragraph |
|---|---|---|
| Thermal shock | 3.8.5 | 4.8.7 |
| Humidity | 3.8.7 | 4.8.9 |
| Electrical parameters | 3.8.2 | 4.8.4 |

The tests in table VII must be conducted on the completed printed wiring assemblies. Additionally, the contractor may wish to conduct certain of these tests at earlier stages of fabrication to ensure progress toward an acceptable assembly.

4.6.1.3.1 <u>Group B inspection lot</u>. An inspection lot for group B inspection shall consist of all printed wiring assemblies of the same type (see 1.2.1) which have passed group A inspection. Board assemblies of the same type and fabricated to this specification may be grouped from different contracts to form a group B inspection lot.

4.6.1.3.2 <u>Sampling plan</u>. Once every 60 days, one printed wiring assembly of each assembly type (see 1.2.1) representative of the most complex assembly of each type shall be selected by the contractor and agreed upon by the Government representative and subjected to tests in table VII. The assembly selected shall have passed group A inspection and may be from one or more contracts unless otherwise specified on the approved assembly drawing.

4.6.1.3.3 <u>Failures</u>. If one or more samples fail to pass group B inspection, the lot shall be considered to have failed.

4.6.1.3.4 <u>Disposition of samples</u>. Unless otherwise specified by the Government procuring activity, sample assemblies which have been subjected to and have passed group B inspection shall be disposed of in accordance with the manufacturer's disposal procedures.

4.6.1.3.5 <u>Noncompliance</u>. If a sample fails to pass group B inspection, the manufacturer shall take corrective action on the materials or processes, or both, as warranted, and on all printed wiring assemblies which can be corrected and which were manufactured under esentially the same conditions, with essentially the same materials, processes, etc., and which are considered subject to the same failure. Acceptance of the printed wiring assemblies shall be discontinued until corrective action acceptable to the Government has been taken. After the corrective action has been taken, the group B inspection shall be repeated on additional sample assemblies. Group A inspection may be reinstituted; however, final acceptance shall be withheld until the group B reinspection has shown that the corrective action was successful. In the event of failure, after reinspection, information concerning the failure and corrective action taken shall be furnished to the Government procuring activity for those printed wiring assemblies warranting corrective action.

4.7 <u>Inspection of packaging</u>. Except when industrial packaging is specified, the sampling and inspection of the preservation and interior package marking shall be in accordance with the group A and B quality conformance inspection requirements of MIL-P-116. The sampling and inspection of the packing and marking for shipment and storage shall be in accordance with the quality assurance provisions of the applicable container specification and the marking requirements of MIL-STD-129. The inspection of industrial packaging shall be as specified in the contract (see 6.2).

14

MIL-P-28809A

## 4.8 Methods of examination and test.

4.8.1 <u>Visual and dimensional examination</u>. Completed printed wiring assemblies shall be examined to verify that the materials, construction, marking, and workmanship are in accordance with the applicable requirements (see 3.1, 3.4, 3.5, 3.9, and 3.10). Examination shall be accomplished utilizing an optical apparatus or aid which provides a minimum magnification equivalent to a lens of 3 diopters (approximately 3/4X). Referee inspections may be accomplished at a magnification of 10X.

4.8.1.1 <u>Classificaton of defects</u>. Unless otherwise specified on the assembly drawing, the classification of defects for visual and dimensional examination shall be as specified herein. A suggested coding system is indicated to allow the use of an automatic data processing system so that a particular coded number will be applicable only to a specific kind of defect. The letter "A" is for major defects and "B" for minor defects.

| Defect code number | Major defects |
|---|---|
| A1 | Wrong parts used. |
| A2 | Wrong printed wiring board used. |
| A3 | Solder bridging between adjacent circuits, protrusions, or peaks that reduce the distance between an element of one circuit and an adjacent circuit or conducting material below the minimum specified on the printed wiring board assembly drawing. |
| A4 | Solder on component side of single-sided boards. |
| A5 | Solder on surfaces designated to be free of solder. |
| A6 | Features, conductor patterns, interfacial connections, jumpers, and components not in accordance with the assembly drawing or approved drawing referenced therein. |
| A7 | Holes in the board which are not specified on the assembly drawing or master drawing referenced therein. |
| A8 | Corrosion on the metal surfaces. |
| A9 | Printed wiring conductors loose or missing. |
| A10 | Components mounted in the wrong locations on the board. |
| A11 | Wrong orientation of polarized components. |
| A12 | Misalinement of lead wires with respect to land areas (see 3.5.1.4). |
| A13 | Jumper wires not as specified on the assembly drawing. |
| A14 | Jumper wires that terminate at locations other than terminal areas. |
| A15 | Component leads used as jumper wires. |
| A16 | Jumper wires routed over or under components. |
| A17 | Inadequate spacing between uninsulated jumper wires or lead wires and adjacent conductors. |
| A18 | Absence of insulation sleeving on jumper wires, when specified, (see 3.5.1.10). |
| A19 | Poor wetting of solder to the basis metal, as evidenced by convex fillets, nonwetting and dewetting (not in excess of that allowed in IPC-S-815), cold joints, rosin joints, etc. |
| A20 | Cracked solder joints, as evidenced by cracks or other discontinuities. |
| A21 | Excess solder on joints. |
| A22 | Insufficient solder on joints. |
| A23 | Welds not in compliance with the assembly drawing. |
| A24 | Bow and twist in excess of that permitted by the assembly drawing. |
| A25 | Inadequate cleanliness of the printed wiring assembly as evidenced by the presence of dirt, foreign matter, oil, fingerprints, corrosion, salts, flux residues, and contaminants. |
| A26 | Measling in excess of that allowed in 3.5.3.4. |
| A27 | Conformal coating containing bubbles or pinholes in excess of that allowed in 3.6. |
| A28 | Conformal coating exhibiting blisters, cracking, crazing, mealing, peeling, wrinkles, or reversion. |
| A29 | Unauthorized repair. |

MIL-P-28809A

| Defect code number | Major defects |
|---|---|
| A30 | Printed wiring assemblies that are charred, burned, blistered, chipped, gouged, delaminated, or otherwise damaged. |
| A31 | Identification markings illegible or missing from the printed wiring assembly. |
| A32 | Incorrect identification marking on the printed wiring assembly. |
| A33 | Leakage from oil-impregnated or electrolytic components. |
| A34 | Physical damage to parts resulting from the straightening, cutting, bending, inserting, or clinching of wire leads. |
| A35 | Chipped, cracked, or broken parts. |
| A36 | Wire leads which have been broken or nicked exposing basis metal. |
| A37 | Loose parts not securely attached or supported on the board. |
| A38 | Deformation of lead diameter greater than 10 percent. |
| A39 | Elecrically functional plated-through hole with a lead, without solder plug. |
| A40 | Greater than 2 percent of all plated-through holes electrically functional or not, without a lead) that are subjected to wave or dip soldering without a solder plug. |

| | Minor defects |
|---|---|
| B1 | Part polarity markings illegible (except as permitted for automatic insertion and lead forming equipment). |
| B2 | Part identification markings illegible (except as permitted for automatic insertion and lead forming equipment). |
| B3 | Thickness of conformal coating not within specified limits. |
| B4 | Inadequate coverage of conformal coating on the printed wiring assembly. |

4.8.2  Cleanliness and resistivity of solvent extract (see 3.6.10).

4.8.2.1  Preparation of solvent extract test solution.  Prepare a test solution of 75 percent by volume ACS reagent grade isopropyl alcohol and 25 percent by volume distilled/deionized water.  Pass this solution through a mixed bed deionizer cartridge (Barnstead D8902, Ultra-Purse, Hose-Nipple Cartridge, or equal).  After passage through the cartridge, typical resistivity of the solution will be 25 x $10^6$ ohm-cm (conductivity - 0.04 micromho/cm).  Replacement of the deionizer cartridge shall be required when the resistivity of the solution is of value less than 6x$10^6$ ohm-cm (conductivity - greater than 0.166 micromho/cm) (see 6.6).  Replacement of the solvent extract solution shall be required when the resistivity of the solution is of a value less than 2x$10^6$ ohm-cm (conductivity - greater than 0.50 micromho/cm) (see 3.6.10.1).

4.8.2.2  Preparation for test.  Position a convenient sized polyethylene funnel over a suitable polyethylene container.  Premark the container for the volume of test solution required for the test.  Suspend the printed wiring assembly within the funnel.

4.8.2.3  Test procedure.  Direct the test solution, in a fine stream, onto both sides of the assembly until 10 ml of test solution is collected for each square inch of assembly area.  Assembly area includes the area of both sides of the board plus an estimate of the area of the components mounted thereon.  Wash the assembly for a minimum of 1 minute.  It is imperative that the initial washings be included in the test sample.  Measure the resistivity/conductivity of the collected test solution with a conductivity bridge or other instrument of equivalent range and accuracy.  NOTE:  All laboratory ware must be scrupulously clean.  Preferably, laboratory ware used for this test should be reserved for this test and not used elsewhere (see 6.6).  Alternate test methods specified in 3.6.10.1 and 6.6.1 may be used.

MIL-P-28809A

4.8.3 <u>Bow and twist (see 3.8.1)</u>. Bow, twist, or any combination thereof, shall be determined by physical measurement and percentage calculation. The calculation shall be based on the formula:

$$\% = \frac{D}{L} \times 100, \text{ where:}$$

D = the measured deviation, worst case, of the assembly mounting surface from a true plane best approximating the mounting surface. Use the cumulative plus and minus planer deviations.
L = the measured breadth dimension of the board along the direction of the board's greatest degree of curvature.

4.8.3.1 <u>Practical measurement of "D"</u>. Stand the assembly on two extremeties of the board, at a zero planar reference. Rock the board so the two other points of greatest deviation from zero are equally displaced from zero ("twist" predominant), or rock the board so the two other diagonal extremes closest to zero are equally displaced from zero ("bow" predominant). Measure the displacement from zero at the point on the mounting surface farthest from zero.

4.8.3.2 <u>Practical measurement of "L"</u>.
    a.  Twist predominant; measure the diagonal of the board along the direction of greatest apparent curvature.
    b.  Bow predominant; measure the width or length of the board along the side most greatly curved.

4.8.4 <u>Electrical parameters (see 3.8.2)</u>. Completed printed wiring assemblies shall be tested as specified on the assembly drawing.

4.8.5 <u>Vibration (see 3.8.3)</u>. Completed printed wiring assemblies shall be tested in accordance with method 514, procedure I, of MIL-STD-810 with the following details:.

    a.  Curve - E, unless otherwise specified (see 6.2).
    b.  Fixture - Hard mount.
    c.  Test procedures - The accelerometer shall be mounted at the center of the unit. The "g" input to the board shall be reduced so that the maximum unit output, at the center of the board, does not exceed 100 g's.
    d.  Electrical tests - Unless otherwise specified (see 6.2), the electrical tests shall be conducted as specified in 4.8.4 after the vibration test.

4.8.6 <u>Shock (see 3.8.4)</u>. Completed printed wiring assemblies shall be tested in accordance with method 516, procedure I, of MIL-STD-810 with the following details:

    a.  Shock pulse - Half sine, 6.5 ±0.1 ms; 100 g's, unless otherwise specified (see 6.2).
    b.  Fixture - Hard mount.
    c.  Electrical tests - Unless otherwise specified (see 6.2), the electrical tests shall be conducted as specified in 4.8.4 after the shock test.

4.8.7 <u>Thermal shock (see 3.8.5)</u>. Completed printed wiring assemblies shall be tested in accordance with method 107, Test condition A-3, of MIL-STD-202.

4.8.8 <u>Temperature-altitude (see 3.8.6)</u>. Completed printed wiring assemblies shall be tested in accordance with method 504 of MIL-STD-810. Temperature and altitude shall be as specified in the engineering drawing (see 6.2).

4.8.9 <u>Humidity (see 3.8.7)</u>. Completed printed wiring assemblies shall be tested in accordance with method 507, procedure I, of MIL-STD-810.

4.8.10 <u>Salt fog (see 3.8.8)</u>. Completed printed wiring assemblies shall be tested in accordance with method 509, procedure I, of MIL-STD-810.

4.8.11 <u>Fungus (see 3.8.9)</u>. Completed printed wiring assemblies shall be tested in accordance with method 508, procedure I, of MIL-STD-810. Data available on materials as to conformance with this test will suffice.

17

MIL-P-28809A

## 5. PACKAGING

5.1 <u>Preservation</u>.  Preservation shall be level A, C, or industrial, or as specified (see 6.2).

5.1.1 <u>Level A</u>.

5.1.1.1 <u>Cleaning</u>.  Printed wiring assemblies shall be cleaned in accordance with MIL-P-116, process C-1.

5.1.1.2 <u>Drying</u>.  Printed wiring assemblies shall be dried in accordance with MIL-P-116.

5.1.1.3 <u>Preservative application</u>.  Preservatives shall not be used.

5.1.1.4 <u>Unit packs</u>.  Each printed wiring assembly shall be individually unit packed in accordance with MIL-P-116, submethod IA-8 insuring compliance with the applicable requirements of that specification.  Cushioning shall conform to PPP-C-1842, type III.  When electrostatic and/or electromagnetic protection is required, the unit container shall be in accordance with the requirements of MIL-B-81705, type I.  Each unit pack shall be placed in a supplementary container conforming to PPP-B-566 or PPP-B-676.

5.1.1.5 <u>Intermediate packs</u>.  Intermediate packs are not required.

5.1.2 <u>Level C</u>.  The level C preservation for printed wiring assemblies shall meet the requirements specified for level A except that nonspecification versions of the electrostatic protective cushioning and supplementary containers may be used.

5.1.3 <u>Industrial</u>.  The industrial preservation of printed wiring assemblies shall be in accordance with the requirements of MIL-STD-1188.

5.2 <u>Packing</u>.  Packing shall be level A, B, C, or industrial, or as specified (see 6.2).

5.2.1 <u>Level A</u>.  The packaged printed wiring assemblies shall be packed in fiberboard containers conforming to PPP-B-636, class weather resistant, style optional, special requirements.  In lieu of the closure and waterproofing requirement in the appendix of PPP-B-636, closure and waterproofing shall be accomplished by sealing all seams, corners and manufacturer's joints with tape, two inches minimum width, conforming to PPP-T-60, class 1 or PPP-T-76.  Banding (reinforcement requirements) shall be applied in accordance with the appendix to PPP-B-636 using nonmetallic or tape banding only.

5.2.2 <u>Level B</u>.  The packaged printed wiring assemblies shall be packed in fiberboard containers conforming to PPP-B-636, class domestic, style optional, special requirements.  Closures shall be in accordance with the appendix thereto.

5.2.3 <u>Level C</u>.  The level C packing for printed wiring assemblies shall conform to the MIL-STD-794 requirements for this level.

5.2.4 <u>Industrial</u>.  The preserved printed wiring assemblies shall be packed in accordance with the requirements of MIL-STD-1188.

5.3 <u>Marking</u>.  In addition to any special or other identification marking required by the contract (see 6.2), each unit, supplementary and exterior container shall be marked in accordance with MIL-STD-129.  Industrial marking shall be in accordance with the requirements of MIL-STD-1188.  Regardless of type of packaging specified, the sensitive electronic device symbol and associated caution label shall be marked as specified in MIL-STD-129 on all units, supplementary and exterior containers.

5.4 <u>General</u>.

5.4.1 <u>Exterior containers</u>.  Exterior containers (see 5.2.1, 5.2.2 and 5.2.3) shall be of a minimum tare and cube consistent with the protection required and shall contain equal quantities of identical stock numbered items to the greatest extent practicable.

MIL-P-28809A

5.4.2 <u>Packaging inspection</u>. The inspection of these packaging requirements shall be in accordance with 4.7.

5.4.3 <u>Army procurements</u>.

5.4.3.1 <u>Level A unit packs</u>. All supplementary containers shall be either weather (or water) resistant or overwrapped with waterproof barrier materials (see 5.1.1.4).

5.4.3.2 <u>Level A and level B packing</u>. For level A packing the fiberboard containers shall not be banded but shall be placed in a close fitting box conforming to PPP-B-601, overseas type; PPP-B-621, class 2, style 4 or PPP-B-585, class 3, style 2 or 3. Closure and strapping shall be in accordance with applicable container specification except that metal strapping shall conform to QQ-S-781, type I, finish A. When the gross weight exceeds 200 pounds or the container length and width is 48 x 24 inches or more and the weight exceeds 100 pounds, 3 x 4 inch skids (laid flat) shall be applied in accordance with the requirements of the container specification. If not described in the container specification, the skids shall be applied in a manner which will adequately support the item and facilitate the use of material handling equipment. For level B packing, fiberboard boxes shall be weather resistant as specified in level A and the containers shall be banded (see 5.2.1 and 5.2.2).

6. NOTES

6.1 <u>Waiver of testing</u>. The Government procuring activity may wish to waive certain environmental tests based on testing the next higher assembly; however, the cleanliness test and electrical tests should not be waived. Relying solely on tests at the next higher assembly may result in inadequate or unproven documentation for procurement of replacement items. This consideration should be made prior to request for quote.

6.2 <u>Ordering data</u>. The acquisition document should specify the following:

    a.  Title, number, and date of this specification.
    b.  Type of printed wiring assembly required (see 1.2.1).
    c.  Title, number, and date of applicable assembly drawing (see 3.1).
    d.  Environmental tests, if any, which may be deferred to testing of the next
        higher assembly (see 3.1 and 6.1).
    e.  Salt fog, if required (see 3.8.8).
    f.  Thermal shock test condition required (see 3.8.5).
    g.  Temperature-altitude 1/, if required (see 3.8.6.).
    h.  Vibration frequency, if other than specified; maximum size of accelerometer,
        where critical; normal mounting means where used in lieu of the hard
        mounting specified (see 4.8.5).
    i.  Continuous electrical operation during vibration, if required (see 3.8.3).
    j.  Shock force, if other than specified (see 3.8.4).
    k.  Continuous electrical operation during shock, if required (see 3.8.4).
    l.  Delivery of first article samples (see 4.4).
    m.  Disposition of group B samples (see 4.6.1.3.4).
    n.  Inspection of industrial packing (see 4.7).
    o.  Levels of preservation and packing required (see 5.1 and 5.2).
    p.  If special or other identification marking is required (see 5.3).

1/  Normally for airborne applications only.

6.3 <u>First article inspection</u>. Information pertaining to first article inspection of products covered by this specification should be obtained from the procuring activity for the specific contracts involved.

6.4 <u>Flux removal</u>. Selection of procedures for flux removal is at the contractor's discretion. A procedure must be chosen which will enable the printed wiring assembly fabricator to produce results enabling compliance with 3.5.3.3. Both polar and nonpolar solvents may be required to effect adequate flux removal.

MIL-P-28809A

6.5 **Plastic bags.** Where plastic bags are used for packaging printed wiring assemblies, they shall be clean and free from ionic contaminants. (The resistivity of solvent test in 4.8.2 can be used to determine the degree of ionic contamination.) NOTE: Printed wiring assemblies which contain parts sensitive to electrostatic damage should not be packaged in plastic bags, unless the bags have been treated to permanently prevent build-up of electrostatic charges (see 5.1).

6.6 **Resistivity of solvent extract (see 4.8.2).** This test procedure, including solution preparation and a laboratory ware cleaning procedure, is documented in Materials Research Report No. 3-72, "Printed-wiring assemblies; detection of ionic contaminants on". Application for copies of this report should be addressed to the Commander, Naval Avionics Facility, Indianapolis, IN 46218.

6.6.1 **Alternate methods.** The following methods of determining the cleanliness of printed wiring assemblies have been shown to be equivalent to the resistivity of the solvent extract method in 6.6:

     a. The Kenco Alloy and Chemical Company, Incorporated, "Omega Meter $^{TM}$, Model 200".
     b. Alpha Metals Incorporated, "Ionograph $^{TM}$".
     c. E. I. Dupont Company, Incorporated, "Ion Chaser $^{TM}$."

Test procedures and calibration techniques for these methods are documented in Materials Research Report 3-78. "Review of Data Generated With Instruments Used to Detect and Measure Ionic Contaminants on Printed-Wiring Assemblies". Application for copies of this report should be addressed to the Commander, Naval Avionics Center, Indianapolis, IN 46218. Table VIII lists the equivalence factors for these methods in terms of microgram equivalents of sodium chloride per unit area:

TABLE VIII. Equivalence factors.

| Method | $\overline{X}$ $\mu gNaCl/in^2$ | | Equivalence factor | Instrument "Acceptance limit" | |
|---|---|---|---|---|---|
| | | | | $\mu gNaCl/Cm^2$ | $\mu gNaCl/in^2$ |
| MIL-P-28809-<br>Beckman | 7.47 | 7.545 | $\frac{7.545}{7.545} = 1$ | 1.56 | 10.06 |
| MIL-P-28809-<br>Markson | 7.62 | | $\frac{7.545}{7.545} = 1$ | 1.56 | 10.06 |
| Omega Meter | 10.51 | | $\frac{10.51}{7.545} = 1.39$ | 2.2 | 14 |
| Ionograph | 15.20 | | $\frac{15.20}{7.545} = 2.01$ | 3.1 | 20 |
| Ion Chaser | 24.50 | | $\frac{24.50}{7.545} = 3.25$ | 5.1 | 32 |

6.7 **Ultrasonic cleaning.** Ultrasonic cleaning may damage certain component parts, particularly integrated circuits and semiconductors, and should not be used.

6.8 **Sensitive-component handling.** To prevent damage by static electricity, persons coming into contact with or handling of electro-sensitive components, such as some semiconductors should be grounded prior to touching or installing the component. Certain electro-sensitive devices may require additional precautions and should be handled in accordance with manufacturer's recommendations.

6.9 **Changes from previous issue.** Asterisks are not used in this revision to denote changes with respect to the previous issue, due to the extensiveness of the changes.

20

MIL-P-28809A

Custodians:
  Army - ER
  Navy - EC
  Air Force - 17

Review activities:
  Army - MI, AR, ME, AT
  Navy - SH, OS
  Air Force - 85, 99
  DLA - ES
  NS - S23

User activity:
  Navy - AS, MC

Agent:
  DLA - ES

Preparing activity:
  Navy - EC

(Project 5999-0103)

MIL-P-28809A


APPENDIX

REPAIR AND MODIFICATION OF PRINTED WIRING ASSEMBLIES

10.  SCOPE

10.1  Purpose.  This appendix establishes requirements for the repair and
modification of printed wiring assemblies produced in accordance with this
specification.

10.2  Authorization.

10.2.1  Repair authorization.  Standard repairs (see 30.1.12 and 50.2) covered by
this appendix may be used when the contractor has Material Review Board (MRB)
delegation on the contract.  Authorization for these standard repairs shall be
individually authorized by the contractors material review procedure, with
concurrence by the cognizant government representative.

Alternately, the contractor may request authorization to perform standard repairs
using in-house control systems other than MRB action.  These alternate methods
require approval of the Government procuring activity and must demonstrate that
standard repairs are documented and that there is traceability to the assemblies to
which the repairs are applied.

Repairs other than the standard repairs contained herein require written
authorization of the Government procuring activity.  Authorization of other repairs
is applicable only to the contract under which the repair authorization was granted.

10.2.2  Modification authorization.

10.2.2.1  Prototype modification authorization.  Modification of prototype
assemblies may be performed by the contractor when the details of the modification
are determined and defined in sufficient detail, acceptable to the Government
procuring activity, to assure adequate documentation of the modification(s).  Details
of the modifications shall be added to the design package and shall be submitted to
the Government procuring activity for post incorporation review.  Prototype
modifications which are to be incorporated in the production articles shall be
incorporated into the design prior to the start of production.

10.2.2.2  Production modification.  Modification of production printed wiring
assemblies requires written authorization of the Government procuring activity.
Documentation of modifications shall be as specified herein (see 40.4).  Modification
authorizations are limited to the contract under which the authorization was granted,
until the printed wiring master drawing and/or the printed wiring assembly drawing
has been changed and approved to reflect the modification.

10.3  Unassembled boards.  The repair or modification of unassembled (bare) printed
wiring boards shall not be permitted.

20.  APPLICABLE DOCUMENTS

20.1  Government documents.  Unless otherwise specified, the following specifications and standards,
of the issue listed in that issue of the Department of Defense Index of Specifications and Standards
specified in the solicitation, form a part of this appendix to the extent specified herein.


DRAWINGS

MILITARY

NAVAIR 200AS107-1 - Adhesive, Epoxy, Flexible.

MIL-P-28809A

REPORTS

MILITARY

| | |
|---|---|
| NAFI Materials Research Report No. 3-70 | - Welded Repair of Severed Conductors on Printed Wiring Assemblies. |
| NAFI Materials Research Report No. 3-72 | - Printed Wiring Assemblies; Detection of Ionic Contaminants On. |
| Naval Air Center Materials Research Report No. 3-78 | - Review of Data Generated with Instruments used to Detect and Measure Ionic Contaminants on Printed Wiring Assemblies. |

(Copies of specifications, standards, handbooks, drawings, and publications required by manufacturers in connection with specific acquisition functions should be obtained from the contracting activity or as directed by the contracting officer.)

30.  DEFINITIONS

30.1  Terms and definitions.  The definitions of all terms used herein shall be as specified in ANSI/IPC-T-50 and the following:

30.1.1  Automated component insertion.  Automated component insertion is the act or operation of assembling discrete components to printed boards by means of computer-controlled component-insertion equipment.

30.1.2  Buffer material.  A resilient material which is used to protect crack-sensitive components from excessive stresses generated by the conformal coating.

30.1.3  Circuit card assembly.  A circuit card assembly is a grouping of two or more physically connected or related electrical and/or electronic parts capable of disassembly.  Each component of the assembly must be capable of functioning in accordance with its own item name.  Consists of a printed-wiring board upon which are mounted separately manufactured electronic componenets, such as capacitors, inductors, resistors, and the like.  It may also include printed electronic components.  (Source:  H6).

30.1.4  Component.  A component is a separate part of a printed-circuit assembly or printed-wiring assembly which performs a circuit function (e.g. resistor, capacitor, transistor, transformer, etc.).

30.1.5  Component mounting.  Component mounting is the act  of mechanically attaching the component to the printed board, or the manner in which they are attached, or both.

30.1.6  Component orientation.  Component orientation is the direction in which the components, on a printed board or other assembly, are lined-up electrically with respect to the polarity of polarized components and also with respect to one another and to the board.

30.1.7  Cracking.  Cracking is that phenomena manifest in coatings by a break extending through to the base surface.

30.1.8  Hard wiring.  Hard wiring is electrical wiring that interconnects two or more parts or assemblies into an assembly which is inseparable without the use of special tools and techniques.

30.1.9  Modification.  Modification is defined as a revision to the interconnect features on a printed wiring assembly accomplished by interrupting conductors, adding/deleting components and/or adding wires.  Modification of a printed wiring assembly is done in lieu of using a new-design assembly with the changes incorporated in the conductor pattern, drilled features, or other characteristics changed by the revision.

MIL-P-28809A

30.1.10 <u>Repair</u>. A repair is any correctional process not sanctioned by the assembly's design documentation. The purpose of a repair is to restore the usability of a damaged or defective assembly without having to totally replace the discrepant member of the assembly.

30.1.11 <u>Rework</u>. Reworking is the act of reestablishing the functional and physical characteristics of an assembly without deviating from the original design drawing, specifications, or contract requirements. The reworking of defective solder connections or replacement of defective component is not considered to be a modification or repair and is permissible.

30.1.12 <u>Standard repairs</u>. Standard repairs are those repair techniques described by this specification, not exceeding the numerical limits set.

30.1.13 <u>Standoff terminal</u>. A standoff terminal is a terminal generally postlike, having an axial portion of its body designed for projecting through or into a board for mounting. (See figure 1).

FIGURE 1.  <u>Standoff terminal</u>.

30.1.14 <u>Straight-through lead</u>. A straight-through lead is a component lead wire that is not clinched or swaged after insertion in a hole of a printed-wiring board. (See figure 2).

FIGURE 2.  <u>Straight through lead</u>.

40. GENERAL

40.1 <u>Quality</u>. Limited repair or modification of printed wiring assemblies may be necessary in the interest of economy or delivery. It is essential that such repairs and modifications be accomplished in a manner which will not degrade the quality of the products.

40.2 <u>Performance</u>. Repaired or modified printed wiring assemblies shall meet the performance requirements and quality assurance provisions of this specification.

MIL-P-28809A

40.3 Spacings. Repair shall not reduce circuitry spacing below that provided by design. The cross sectional area of wires, leads, or copper strips shall be equal to or larger than the replaced conductor.

40.4 Documentation of modification. Documentation of modification on prototypes and printed wiring assemblies under production shall be as specified herein.

40.4.1 Documentation of prototypes. Details of the modification to a prototype of a printed wiring assembly shall be prepared and added to the data design package.

40.4.2 Documentation of production. Upon approval from the Government procuring activity of a modification change, the board assembly fabricator shall take the appropriate action to reflect the approved change and production will continue on the next assemblies. At all times the data package shall be kept updated with documentation to reflect modified or unmodified assemblies and their approved changes. Part numbers shall be changed if the modified or unmodified assemblies are not physically or electrically interchangeable (i.e., form, fit or function).

40.5 Cleanliness. Coated and uncoated, modified or repaired printed wiring assemblies shall be free of flux, flux residues, and other contaminants prior to the application of conformal coating (see 40.6.2.1). The Government reserves the right to require confirmation that all coated and uncoated, modified or repaired printed wiring assemblies were inspected for flux residues after modification or repair.

40.5.1 Resistivity of solvent extract. When required by the Government procuring activity, coated or uncoated, modified or repaired printed wiring assemblies shall be tested as specified in 50.3. The resistivity shall be not less than 2,000,000 ohm-cm (or equivalent) (see 50.3.4.1). The equivalent test methods and factors specified in 50.3.4.1 may be used in lieu of the method specified in 50.3. Other equivalent test methods not specified in 50.3.4.1 may be used in lieu of 50.3 only when specifically approved by the Government procuring activity. Such approval will be determined on the basis that the alternate method is demonstrated to have equal or better sensitivity and employs solvents with the ability to dissolve flux residue as does the alcohol water solution specified in 50.3.

40.6 Materials. Materials used in modification and repair shall be as specified herein or on the assembly drawing.

40.6.1 Solder. The solder used shall be in accordance with composition Sn 60, Sn 62 or Sn 63 of QQ-S-571. For wave or dip soldering, use bar solder, form B. For hand soldering, use solder wire form W, either solid metal, type S, or with a core of either type R, RMA, or RA of QQ-S-571, with the exception that RA flux shall not be used on stranded wire. (With special approval type RA flux may be used on Army ERADCOM, CORADCOM and AVRADCOM contracts).

40.6.2 Soldering flux. Soldering flux shall be a liquid flux conforming to MIL-F-14256, type R, RMA, or RA, with the exception that RA flux shall not be used on stranded wire. (With special contract approval, type RA flux may be used on Army ERADCOM, CORADCOM, and AVRADCOM contracts). If the resistivity of solvent extract test or equivalent cannot be performed, type RA flux shall not be used (see 40.5.1).

40.6.2.1 Flux removal. Selection of procedures for flux removal is at the contractor's discretion. A procedure shall be chosen which will enable the printed wiring assembly to be in compliance with 40.5. Both polar and nonpolar solvents may be required to effect adequate flux removal. The procedure chosen should not degrade markings, components, or board materials.

40.6.3 Adhesive. Epoxy structural adhesive shall be in accordance with NAVAIR Drawing No. 200AS107-1 or as specified on the approved assembly drawing or approved standard repair procedure. Conformal coating shall not be relied upon in place of adhesive.

MIL-P-28809A

40.6.4  Conformal coating. Conformal coating shall be type UR, ER, AR, SR, or XY per MIL-I-46058. The thickness of the conformal coating shall be as follows for the type specified. Measurement shall be on a flat unencumbered surface (see MIL-I-46058). Coating thickness in any areas of overlap may be within twice the specified coating thickness.

 a. Types ER, UR, and AR:  0.003 ±0.002 inch.
 b. Type SR:  0.005 ±0.003 inch.
 c. Type XY:  0.0005 to 0.002 inch.

40.6.5  Hook-up wire. Hook-up wire for added wires shall be:

 a.  Solid copper conductor, tin-plated, with a compatible jacket.
 b.  Tin-coated copper wire, per QQ-W-343, type S, soft or drawn and annealed.

40.6.6  Insulation tubing. Polytetrafluoroethylene tubing used for insulating hook-up wires shall be in accordance with MIL-I-22129, etched for bonding.

40.7  Coating area. Repaired circuitry shall be covered with epoxy adhesive or conformal coating for a minimum of 0.030 inch beyond the end of the repair area except in special areas (e.g., printed edgeboard contacts, etc.,) where the assembly drawing specifies that these areas be exposed. The conformal coating shall be a material that conforms to 40.6.4.

40.8  Component replacement. Repairs shall be made so that components may be replaced without damaging the repaired area

40.9  Workmanship. Modified or repaired printed wiring assemblies shall be clean and free of dirt, foreign matter, oil, fingerprints, corrosion, salts, flux residues, and other contaminants, and meet the requirements of this specification.

50.  DETAIL REQUIREMENTS

50.1  Standard modifications.

50.1.1  Conductor removal.

50.1.1.1  Minimum removal. Unless otherwise limited by design constraints, a minimum of 0.030 inch of conductor shall be removed where the circuit is to be interrupted (see figure 3).

50.1.1.2  Circuit junctions. Unless otherwise limited by design constraints, conductors shall not be cut or removed within 0.010 inch of land areas or circuit junctions (see figure 3).

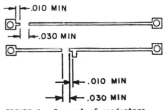

FIGURE 3.  Removal of conductors.

50.1.1.3  After removal. The area under the circuits removed shall be inspected to assure that all traces of conductor are removed. The glass cloth shall not be scraped into and the coupling agent shall not be scraped away from the glass filaments. This area shall then be covered with epoxy adhesive or approved conformal coating (see 40.6.4).

50.1.2  Added wires.

50.1.2.1  Insulation and sleeving. All added wires greater than 0.50 inch in length shall be insulated or sleeved.

MIL-P-28809A

50.1.2.2 <u>Number of attachments</u>.  A maximum of two wires or leads may be attached to any termination except:

  a.  Wires or leads shall be attached to any flat pack lead in accordance with figures 4 and 4A.
  b.  When wires or leads are to be attached to DIP type components, terminations shall be made in accordance with figure 5.
  c.  No more than one wire shall be added to each connector tang and shall be attached in accordance with figure 6.
  d.  When large standoff terminals are used which have provision for additional attachments.
  e.  No wires shall be attached to the mating contact surfaces of connectors.

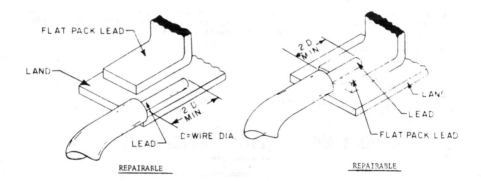

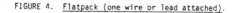

FIGURE 4.  <u>Flatpack (one wire or lead attached)</u>.

MIL-P-28809A

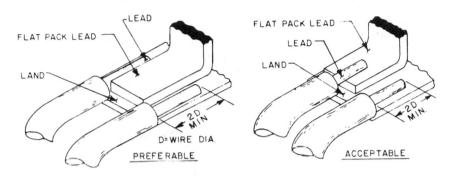

FIGURE 4A.  Flatpack (2 wires or leads attached).

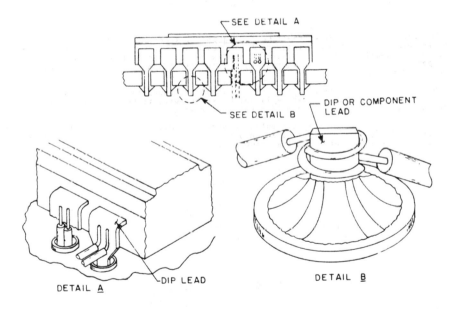

FIGURE 5.  Dip type component.

28

MIL-P-28809A

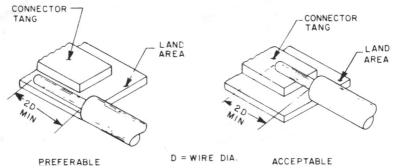

FIGURE 6.  Connector tang.

50.1.2.3  Routing.  Wires shall be routed in the X and Y directions (no diagonal lines permitted) by the shortest practical route and minimizing wire crossings unless otherwise specified by the revision.  Added wires shall not cover plated-through component mounting holes.  Wire routings on boards having the same part number shall be routed the same.

50.1.2.4  Preferred termination.  Wires shall be connected at a point where heating to remove an adjacent component will not cause the wire to become unsoldered.  Unused through holes are the preferred termination for wires.

50.1.2.5  Land area distance.  Unless otherwise limited by design constraints, when wires are soldered directly to the conductor there shall be a minimum distance of 0.050 inch from the land area (see figure 7).

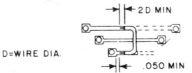

FIGURE 7.  Positioning wiring on circuits.

50.1.2.6  Wire diameter.  The diameter of the wire soldered directly to a conductor path shall not exceed the width of the conductor path and shall be positioned, where possible, as shown in figure 7 and in such a manner that the minimum electrical spacing is maintained.  The wire shall contact the conductor path a minimum of two wire diameters at each end.

MIL-P-28809A

50.1.2.7 **Securing of wires.** Wires shall be secured to the board except where they pass through pin fields or where a 1 inch or shorter wire has both ends terminated in plated-through holes.

50.1.2.8 **Existing terminals.** When wires or leads are added to existing terminals, they shall be attached above the conformally coated area or the conformal coating shall be removed prior to attachment.

50.1.3 **Added components.**

50.1.3.1 **Land area distance.** When component leads are soldered directly to the conductor, there shall be a minimum clearance of 0.050 inch from a land area (see figure 7).

50.2 **Standard repairs.**

50.2.1 **Land area.**

50.2.1.1 **Maximum permitted.** The maximum number of land area repairs permitted per board shall be in accordance with table I.

TABLE I. **Maximum number of land repairs.**

| Board size (x) | Maximum number allowed |
|---|---|
| Square inches | |
| $x < 20$ | 3 |
| $20 < x < 50$ | 6 |
| $50 \leq x < 100$ | 9 |
| $100 \leq x$ | 12 |

50.2.1.2 **Connecting land areas.** Repair of both land areas on both sides of a hole which connects internal circuits is prohibited (see figure 8).

DAMAGED LAND AREA

GOOD LAND AREA

REPAIRABLE

DAMAGED LAND AREA

DAMAGED LAND AREA

NONREPAIRABLE

FIGURE 8. **Repair limitations.**

50.2.1.3 **Lifted land areas.** Any land which has been separated, loosened, lifted or which has otherwise become unbonded from the base material may be repaired by rebonding with an approved epoxy adhesive (see 40.6.3), provided that the maximum feeler gauge penetration is equal to 1/2x (see figure 9).

MIL-P-28809A

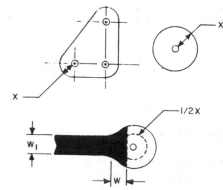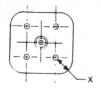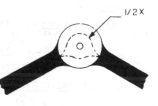

W Minimum width of connected circuitry
within 1/4" of hole center,but not over
one-half land area diameter.

$W_1 = W$

FIGURE 9. Land areas.

50.2.1.4 Repair visibility. All repairs shall be visible after soldering.
Repairs to land areas that will subsequently be covered by flush mounted components
(transistors, transformers, etc.) shall be inspected prior to installation of said
components.

50.2.2 Conductor.

50.2.2.1 Maximum permitted. The maximum number of conductor repairs permitted per
board shall be in accordance with table II.

TABLE II. Maximum number of conductor repairs.

| Board size (x) | Maximum number allowed |
|---|---|
| Square inches | |
| x < 20 | 3 |
| 20 < x < 50 | 6 |
| 50 ≤ x < 100 | 9 |
| 100 ≤ x | 12 |

50.2.2.2 Unbonded conductors. Unless otherwise limited by design constraints,
unbonded conductors no more than 0.500 inch in length may be rebonded to the base
laminate with approved epoxy adhesive material (see 40.6.3) extending a minimum of
0.030 inch beyond the lifted area in all directions.

MIL-P-28809A

50.2.2.3 <u>Conductor breaks and defects</u>. Unless otherwise prohibited by design constraints, conductor breaks, scratches, or similar defects no more than 0.500 inch in length may be repaired by use of a repair conductor, which shall overlap the original conductor by 0.125 to 0.250 inch at each end. If the conductor is lifted in the defective area, it shall be trimmed back to where a good bond exists. The repair conductor shall be centered over the original conductor. The repair conductor shall be formed to the board unless the conductor break is small (less than 0.100 inch) (see figure 10). Welded repairs shall be made as specified in NAFI Materials Research Report No. 3-70.

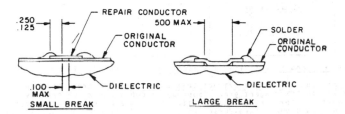

FIGURE 10.  <u>Repair of conductor breaks</u>.

50.2.2.4 <u>Sleeved conductors</u>. Conductor defects of any length may be repaired by routing a sleeved conductor between the breaks or between terminations. When attached to a conductor line only, the repair conductor shall contact the original conductor a minimum of 0.125 inch from each end, unless otherwise limited by design constraints. The sleeved conductor shall be firmly secured to the board by an approved epoxy adhesive (see 40.6.3). Jumper wires shall be on the component side of the board and shall be routed in the X and Y directions by the shortest practical route minimizing wire crossings.

50.2.3 <u>Plated through holes with no internal connections</u>.

50.2.3.1 <u>Maximum permitted</u>. The maximum number of open plated-through hole repairs permitted per board shall be in accordance with table III.

TABLE III.  <u>Maximum number of open plated-through hole repairs</u>.

| Board size (x) | Maximum number allowed |
|---|---|
| Square inches | |
| x < 20 | 3 |
| 20 < x < 50 | 6 |
| 50 < x < 100 | 9 |
| 100 < x | 12 |

50.2.3.2 <u>Shorted plated-through holes</u>. Shorted plated-through holes in multilayer boards shall not be repaired.

50.2.3.3 <u>Open plated-through holes</u>.

50.2.3.3.1 <u>Double-sided boards</u>. Open plated-through holes in double-sided boards may be repaired by insertion of a wire or flat ribbon through the hole, clinching it on both sides, or a funnel flanged eyelet. With soldered holes, solder must completely fill the hole and must form a fillet around the wire or eyelet on both sides of the board (see 3.5.3.2).

MIL-P-28809A

50.2.4 <u>Measling and crazing</u>. Repair is not permitted.

50.2.5 <u>Holes and slots in boards</u>. Repair is not permitted.

50.2.6 <u>Internal circuits</u>. Repair is not permitted.

50.2.7 <u>Total maximum repairs permitted</u>. The total maximum number of repairs of all types permitted per board shall be in accordance with table IV.

TABLE IV. ·Total number of repairs of all types permitted.

| Board size (x) | Maximum number allowed |
|---|---|
| Square inches | |
| x < 20 | 6 |
| $20 \leq x < 50$ | 12 |
| $50 \leq x < 100$ | 18 |
| $100 \leq x$ | 24 |

50.3 <u>Cleanliness and resistivity of solvent extract (see 4.8.2)</u>.

50.3.1 <u>Preparation of solvent extract test solution</u>. Prepare a test solution of 75 percent by volume ACS reagent grade isopropyl alcohol and 25 percent by volume distilled/deionized water. Pass this solution through a mixed bed deionizer cartridge (Barnstead D8902, Ultra-Purse, Hose-Nipple Cartridge, or equal). After passage through the cartridge, typical resistivity of the solution will be $25 \times 10^6$ ohm-cm (conductivity - 0.04 micromho/cm). Replacement of the deionizer cartridge shall be required when the resistivity of the solution is of a value less than $6 \times 10^6$ ohm-cm (conductivity - greater than 0.166 micromho/cm) (see 50.3.4). Replacement of the solvent extract solution shall be required when the resistivity of the solution is of a value less than $2 \times 10^6$ ohm-cm (conductivity - greater than 0.50 micromho/cm) (see 40.5.1).

50.3.2 <u>Preparation for test</u>. Position a convenient sized polyethylene funnel over a suitable polyethylene container. Premark the container for the volume of test solution required for the test. Suspend the printed wiring assembly within the funnel.

50.3.3 <u>Test procedure</u>. Direct the test solution, in a fine stream, onto both sides of the assembly until 10 ml of test solution is collected for each square inch of assembly area. Assembly area includes the area of both sides of the board plus an estimate of the area of the components mounted thereon. Wash the assembly for a minimum of 1 minute. It is imperative that the initial washings be included in the test sample. Measure the resistivity/conductivity of the collected test solution with a conductivity bridge or other instrument of equivalent range and accuracy. NOTE: All laboratory ware must be scrupulously clean. Preferably, laboratory ware used for this test should be reserved for this test and not used elsewhere (see 50.3.4). Alternate test methods specified in 40.5.1 and 50.3.4.1 may be used.

50.3.4 <u>Resistivity of solvent extract (see 4.8.2)</u>. This test procedure, including solution preparation and a laboratory ware cleaning procedure, is documented in Materials Research Report No. 3-72, "Printed wiring assemblies; detection of ionic contaminants on". Application for copies of this report should be addressed to the Commander, Naval Avionics Facility, Indianapolis, IN 46218.

50.3.4.1 <u>Alternate methods</u>. The following methods of determining the cleanliness of printed wiring assemblies have been shown to be equivalent to the resistivity of the solvent extract method in 50.3.4.

   a. The Kenco Alloy and Chemical Company, Incorporated, "Omega Meter TM, Model 200".

MIL-P-28809A

Test procedures and calibration techniques for these methods are documented in
Materials Research Report 3-78, "Review of Data Generated with Instruments Used to
Detect and Measure Ionic Contaminants on Printed-Wiring Assemblies". Application for
copies of this report should be addressed to the Commander, Naval Avionics Center,
Indianapolis, IN 46218. Table V lists the equivalence factors for these methods in
terms of microgram equivalents of sodium chloride per unit area:

TABLE V. Equivalence factors.

| Method | $\bar{X}$ µgNaCl/in² | Equivalence factor | Instrument "Acceptance limit" | |
| | | | µgNaCl/Cm² | µgNaCl/in² |
|---|---|---|---|---|
| MIL-P-28809-Beckman | 7.47 7.545 | $\frac{7.545}{7.545} = 1$ | 1.56 | 10.06 |
| MIL-P-28809-Markson | 7.62 | $\frac{7.545}{7.545} = 1$ | 1.56 | 10.06 |
| Omega Meter | 10.51 | $\frac{10.51}{7.545} = 1.39$ | 2.2 | 14 |
| Ionograph | 15.20 | $\frac{15.20}{7.545} = 2.01$ | 3.1 | 20 |
| Ion Chaser | 24.50 | $\frac{24.50}{7.545} = 3.25$ | 5.1 | 32 |

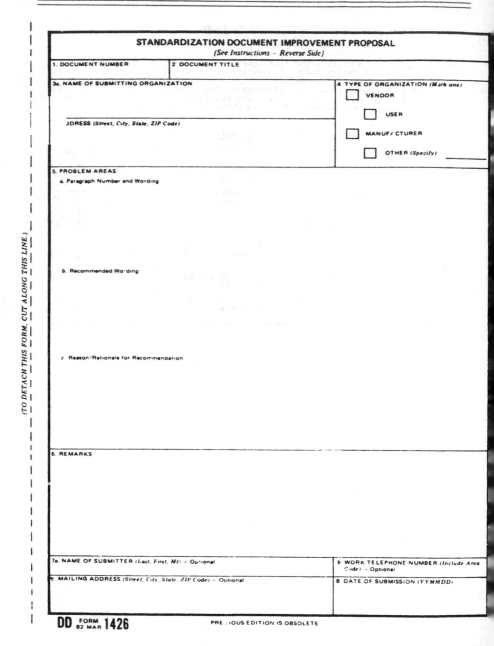

**STANDARDIZATION DOCUMENT IMPROVEMENT PROPOSAL**
*(See Instructions – Reverse Side)*

| 1. DOCUMENT NUMBER | 2 DOCUMENT TITLE |
|---|---|

3a. NAME OF SUBMITTING ORGANIZATION

4 TYPE OF ORGANIZATION *(Mark one)*
- [ ] VENDOR
- [ ] USER
- [ ] MANUFACTURER
- [ ] OTHER *(Specify)* _____

ODDRESS *(Street, City, State, ZIP Code)*

5. PROBLEM AREAS

a. Paragraph Number and Wording

b. Recommended Wording

c. Reason/Rationale for Recommendation

6. REMARKS

| 7a. NAME OF SUBMITTER *(Last, First, MI) – Optional* | b WORK TELEPHONE NUMBER *(Include Area Code) – Optional* |
|---|---|
| c MAILING ADDRESS *(Street, City, State, ZIP Code) – Optional* | 8 DATE OF SUBMISSION *(YYMMDD)* |

**DD** FORM 82 MAR **1426**          PREVIOUS EDITION IS OBSOLETE

**INSTRUCTIONS:** In a continuing effort to make our standardization documents better, the DoD provides this form for use in submitting comments and suggestions for improvements. All users of military standardization documents are invited to provide suggestions. This form may be detached, folded along the lines indicated, taped along the loose edge *(DO NOT STAPLE)*, and mailed. In block 5, be as specific as possible about particular problem areas such as wording which required interpretation, was too rigid, restrictive, loose, ambiguous, or was incompatible, and give proposed wording changes which would alleviate the problems. Enter in block 6 any remarks not related to a specific paragraph of the document. If block 7 is filled out, an acknowledgement will be mailed to you within 30 days to let you know that your comments were received and are being considered.

*NOTE:* This form may not be used to request copies of documents, nor to request waivers, deviations, or clarification of specification requirements on current contracts. Comments submitted on this form do not constitute or imply authorization to waive any portion of the referenced document(s) or to amend contractual requirements.

*(Fold along this line)*

*(Fold along this line)*

DEPARTMENT OF THE NAVY

OFFICIAL BUSINESS
PENALTY FOR PRIVATE USE $300

## BUSINESS REPLY MAIL
FIRST CLASS    PERMIT NO 12503    WASHINGTON D C

POSTAGE WILL BE PAID BY THE DEPARTMENT OF THE NAVY

NO POSTAGE
NECESSARY
IF MAILED
IN THE
UNITED STATES

Commanding Officer
Naval Air Engineering Center
Engineering Specifications & Standards
Department (ESSD) Code 93

# Quantitative Determination of Rosin Residues on Cleaned Electronics Assemblies

# Quantitative Determination of Rosin Residues on Cleaned Electronics Assemblies

EXPERIMENTAL

TURBIDIMETRIC OR NEPHELOMETRIC ANALYSIS

## Determination of Rosin Residue on a PWA

Step 1: Measure the PWA in inches to find the total surface area. Minimize contamination due to handling.

Step 2: Place the circuit board in a good quality anti-static bag. If a high "blank" (see Step 5) is obtained, it may be necessary to pre-clean the bags in order to determine very low levels of surface contamination. Add a volume of isopropanol (in mls) equal to the total area determined in Step 1.

Step 3: Fold over the top of the bag 2-3 times, then shake the bag vigorously for 2-3 minutes. Ensure that the entire circuit board is washed well with the isopropanol.

Step 4: Transfer 1 ml of isopropanol extract from the anti-static bag (using a syringe or pipet) to a small flask or bottle. Then add 8 mls of 0.5 N HCl, shake well and allow to stand for 10 minutes. Add 1 ml of a stabilizer solution, 0.1% poly(vinyl alcohol) in water. Mix well. Place the suspension in the nephelometer/turbidity meter and record the value. It will take 2 or 3 minutes for the reading to stabilize. Note: If the instrument has large curvets, it may be necessary to double each of the volumes given above.

Step 5: Perform a blank determination by using exactly the same procedure, with the exception that a circuit board is not placed in the bag.

Step 6: Subtract the value for the blank from the value obtained in Step 4. From the previously prepared calibration curve, read the concentration (in ug rosin/ml) of the wash solution. Since the volume of isopropanol was the same as the total area of the circuit board, this value is also the residual rosin level in ug rosin/in.$^2$.

## Preparation of a Calibration Curve

Step 1: Determine the percent solids in the solder flux being used (supplied by the flux manufacturer). Weigh out into a flask or beaker an amount of flux in grams given by the formula (0.1 / % solids) x 100).

After Wesley L. Archer, Tim D. Cabelka, and Jeffrey J. Nalazek, Dow Chemical, U.S.A.

Example:  For a flux that is 35% solids, weigh out
(0.1 / 35%) x 100 = 0.286 g.

Alternatively, evaporate off the flux solvent by
placing it in a well-ventilated space for a few hours.
The remaining flux solid may be weighed out directly.

Step 2:    Preheat the flux weighed out in Step 1 to simulate the
actual solder time and temperature (e.g. $250^{\circ}C$ for
15 seconds).  There must be no splatter loss.  For
example, this can be accomplished by use of a suitable
oven or a solder pot.  Allow the flux to cool.
CAUTION:  Avoid using an open flame because of the
flammability of the flux solvent.

Step 3:    Add about 25 mls of isopropanol to the flask or beaker
to re-dissolve the flux solids.  Transfer the flux
solids/isopropanol to a 100 ml volumetric flask.
Continue to slowly add additional isopropanol and to
swirl until all the flux solids are in solution and a
final volume of precisely 100 ml is obtained.  This
solution, the "stock solution," contains 1000 $\mu$g
rosin/ml.  Keep the stock solution in a stoppered flask
or bottle to minimize evaporation losses.

Step 4:    Prepare a 120 $\mu$g/ml "standard solution" by diluting
6 ml of the stock solution with additional pure
isopropanol to give a final volume of 50 mls (the most
convenient way to do this is to utilize a volumetric
flask).  In a similar manner, prepare 100, 80, 60, 40,
20 and 10 $\mu$g/ml standard solutions by performing 5:50,
4:50, 3:50, 2:50, 1:50 and 1:100 dilutions of stock
solution, respectively.  Note:  The more concentrated
standard solutions need not be prepared if it is known
that contamination level on the circuit boards is low.

Step 5:    For each of the standard solutions prepared in Step 4,
combine 1 ml of the standard solution with 8 mls of
0.5 N HCl in a suitable flask or bottle, stopper, mix
well and allow to stand for 10 minutes.  Then add 1 ml
of a stabilizer solution, 0.1% poly(vinyl alcohol) in
water.  Mix well.  Place the suspension in the
nephelometer/turbidity meter and record the value.  It
will take 2 or 3 minutes for the reading to stabilize.
Note:  If the instrument has large curvets, it may be
necessary to double each of the volumes given above.
Plot a calibration curve of instrument reading (e.g.
NTU) versus ug rosin/ml in the final solution.  (The
concentration in the final solution is 1/10 that of the
standard solution used to prepare it.)  A linear
calibration curve should be obtained.

ULTRAVIOLET SPECTROPHOTOMETRIC ANALYSIS

## Determination of Rosin Residue on a PWA

### Extraction Step:

Step 1:   Place an accurately measured volume (using a pipet, repipet dispenser or syringe of high performance liquid chromatographic (HPLC) grade or spectral grade isopropanol into each of two anti-static plastic bags. Pure isopropanol or aqueous isopropanol with up to 25 volume percent deionized water may be used.  Choose an initial volume in milliliters equal to the surface area in inches of the circuit assembly to be tested.

Step 2:   Place the defluxed circuit assembly into one of the two bags containing isopropanol.

Step 3:   Seal the zipper closures on the bags and fold the bags as needed to maximize contact between solvent and circuit board surfaces.

Step 4:   Shake both bags by hand, or by mechanical shaker if the bags offer EMI/RFI protection, for 10 minutes.

Step 5:   Remove circuit assembly from its bag.

Step 6:   Pour the contents of each bag into labelled glass sample bottles, with tightly closed screw caps, for storage until ready to make the UV absorbance measurements.

### Absorbance Measurement (241 nm)

Step 1:   Follow the spectrophotometer manufacturer's instructions for operating the instrument.  Calibrate it with the quartz sample cell in place to allow the instrument to compensate for absorbance by the cell.

Step 2:   Set the wavelength to 241 nm (absorbance maximum may vary from 240.5-242 nm).

Step 3:   Set the absorbance range to the maximum for the instrument.

Step 4:   While holding the quartz sample cell by the frosted
          surfaces, use a disposable glass pipet to carefully
          rinse the cell with the solution as follows:  rinse the
          inside walls of the cell with solution stream from
          pipet, filling cell to about 1/3 level, then remove all
          of the rinse solution from the cell with the pipet and
          discard solution.  Repeat rinse procedure three times,
          being careful to remove all of the rinse solution each
          time.

Step 5:   After sample cell has been thoroughly rinse as
          described in Step 4 above, fill the cell with the
          solution to be measured.  Place filled cell into sample
          compartment of the instrument, close sample door and
          read absorbance.

Step 6:   The solution from the bag in which the circuit assembly
          was shaken is the sample; the other solution is the
          blank.

Step 7:   If the absorbance for the sample is the same as that of
          the blank, repeat the entire procedure, but use another
          circuit assembly and less volume of isopropanol.

          Absorbance is linear with concentration of absorbing
          species, so one may estimate the volume required to
          obtain an absorbance reading in the range 0.4 to 0.6,
          where the instrument is the most responsive.

Step 8:   If the absorbance for the sample is greater than 1.4,
          repeat the procedure, but use another circuit assembly
          and a greater volume of isopropanol (see Step 7 above).

## Calculation of Residual Rosin ($\mu g/in.^2$)

Step 1:   Subtract the absorbance reading obtained from the blank
          from that of the sample.

Step 2:   Using the calibration curve, find the concentration of
          residual rosin (ppm rosin) correlating to the
          absorbance difference obtained in Step 1.

Step 3:    Substitute the appropriate numerical values in the
           equation:

$$\mu g\ rosin/in.^2 = \frac{ppm\ rosin\ x\ density\ (g/ml)\ x\ volume\ (ml)}{board\ surface\ area\ (in^2)}$$

           Density of 100% isopropanol = 0.785 g/ml
           Density of 75% isopropanol/water = 0.85 g/ml

## Calibration Procedure

### Stock Solution Preparation

Step 1:    Weigh exactly 0.0785 g gum rosin or flux solids:
           (1) directly into a one-liter volumetric flask, or
           (2) into a smaller container, such as a beaker, then
           quantitatively transfer the rosin using high
           performance liquid chromatographic (HPLC) or spectral
           grade isopropanol into a one-liter volumetric flask.
           Rinse the beaker with the isopropanol at least
           10 times, pouring all the rinse solution into the
           volumetric flask.

Step 2:    Fill the flask within 1 inch of the calibration line on
           its neck.

Step 3:    Shake the stoppered flask vigorously, turning it upside
           down several times to dissolve the rosin and to
           thoroughly mix the solution.

Step 4:    Let the flask sit for about 30 minutes to allow liquid
           to drain from the neck. Bring the volume to the
           one-liter mark by adding isopropanol dropwise. This is
           the stock 100 ppm rosin solution from which dilutions
           are prepared for constructing the calibration curve.
           Keep solutions tightly stoppered to prevent
           evaporation.

### Dilutions

Step 1:    Using an appropriately sized volumetric pipet, transfer
           exactly the volume of the stock solution indicated in
           the table below into a 100 ml flask for each dilute
           solution desired.

| ppm rosin in final (dilute) solution | 1 | 2 | 5 | 10 | 20 | 50 |
|---|---|---|---|---|---|---|
| ml stock solution | 1 | 2 | 5 | 10 | 20 | 50 |

Step 2:  Add HPLC isopropanol to each flask to a level within
0.5 inches of the calibration mark. Shake thoroughly
before filling to the mark following the same technique
outlined above in Steps 3 and 4.

## Calibration Curve Construction

Step 1:  Follow the procedure for absorbance measurements
(241 nm). The blank is HPLC isopropanol from the
reagent bottle.

Step 2:  Plot the absorbance difference (sample-blank) versus
ppm rosin. The points should fall on a straight line
which passes through zero.

Step 3:  Points for concentrations greater than 50 ppm rosin may
lie slightly below the calibration line, but these
concentrations are unlikely to be encountered.

CAUTION: Use proper handling and safety precautions
while conducting this analysis.

NOTE: The information herein is presented in good
faith, but no warranty expressed or implied, is given
nor is freedom from any patent owned by The Dow
Chemical Company or by others to be inferred.

# Index